Shaping Technology, Guiding Policy: Concepts, Spaces and Tools

Shaping Technology, Guiding Policy

Concepts, Spaces and Tools

Edited by

Knut H. Sørensen

Professor of Sociology, Norwegian University of Science and Technology, Norway

Robin Williams

Professor of Socio-Economic Research, The University of Edinburgh, UK

Edward Elgar
Cheltenham, UK • Northampton, MA, USA

Published by
Edward Elgar Publishing Limited
Glensanda House
Montpellier Parade
Cheltenham
Glos GL50 1UA
UK

Edward Elgar Publishing, Inc.
136 West Street
Suite 202
Northampton
Massachusetts 01060
USA

A catalogue record for this book
is available from the British Library

Library of Congress Cataloguing in Publication Data

Shaping technology, guiding policy : concepts, spaces, and tools / edited by Knut H. Sørensen, Robin Williams
p. cm.
Includes bibliographical references and index.
1. Technology--Social aspects. I. Sørensen, Knut H. II. Williams, Robin, 1952 Nov. 13-

T14.5 .S493 2002
303.48'3--dc21

2002023624

ISBN 1 84064 649 7

Printed and bound in Great Britain by MPG Books Ltd, Bodmin, Cornwall

Contents

List of Figures

Contributors

Christian Clausen is Associate Professor at the Department of Manufacturing Engineering and Management/Technology and Environment Studies Unit at the Technical University of Denmark. His research include studies in action-oriented technology assessment concerned with working life and the social shaping of computer aided production managment and management concepts. A main interest is with the studies of political processes in the social shaping of technology and organisations.

Christian Koch, is Associate Professor at the Technical University of Denmark, Department of Civil Engineering (the group for construction management). Has for a number of years written and researched on the management of technology, information technology, management concepts, organisation and human resources. He has published articles in various journals including, amongst others, the *Journal of Organisational Change Management, Technology Analysis and Strategic Management* and *International Journal of Innovation Management*. He is author of *ERP-systemer, erfaringer, ressourcer, forandringer* (ERP-systems, experiences, resources, change) (2001)

Frank Geels is a PhD student at the Centre for Studies of Science, Technology and Society at the University of Twente, The Netherlands. His research is about long-term and large-scale technological transitions and prospective methodologies to explore these.

Ulrik Jørgensen is Associate Professor at IPL, the unit of Technology and Environment Studies at the Technical University of Denmark. His research include studies in electronics and media industry and technology, environmental technology and regulation, transport technology and policy, wind turbine technology, and engineering competence and knowledge. Publications include the books *Generic Technology: IT* (1995, in Danish), *The International Position of Danish Wind Turbine Industry and its Potential* (1995, in Danish), and *The Danish Support for Cleaner Technology: Results, Diffusion and Future Activities* (1997).

Pål Næsje is a Researcher at SINTEF Industrial Management, an independent contract research group in Trondheim, Norway. He holds a Ph.D. in Sociology. His research addresses aspects of energy technologies, especially the interface between policy, environmental issues, use and implementation of alternative technologies. Currently, he is involved in technology assessment for the Norwegian Board of Technology, research on the Hydrogen Economy, and on distributed energy production. Previously, he has been Research Fellow and Postdoctoral Fellow at the Norwegian University of Science and Technology (NTNU), and Visiting Scholar at Cornell University, USA.

Werner Rammert is Director of the Center of Technology and Society and Professor of Sociology and Technology Studies at the Technical University of Berlin. He is engaged in the 'Socionics' research programme and in European and national research projects about the assessment of CCTV, the inquiry and design of hybrid Multi-agent Systems and the risks of socio-technical systems in hospitals. Books include: *Sociology and Artificial Intelligence* (1996); *Knowledge Machines* (1998), *Technology and Social Theory* (1998); *Innovation - Virtuality - Culture: Technology from a Sociological Perspective* (2000);

Arie Rip is Professor and Head of Department, Philosophy of Science and Technology, at the University of Twente. His research and publications range across the sociology of scientific knowledge, science policy studies (including cross-national comparisons), technology dynamics, management of technology and technology assessment. With Johan Schot, he has put Constructive Technology Asessment on the intellectual and policy map; see Rip, Misa & Schot, *Managing Technology in Society. The Approach of Constructive Technology Assessment* (Pinter Publishers, 1995).

Stewart Russell is Senior Lecturer in Science, Technology and Society at the University of Wollongong, Australia, and a member of the Centre for Research Policy and Innovation Studies in the University's International Business Research Institute. His research interests and published work range across social shaping theories, technology development and policies in the energy sector, hazards and risk, and technology, employment and skills.

Johan Schot is Professor in Social History of Technology at Eindhoven University of Technology and the University of Twente. He is scientific director of the Foundation for the History of Technology, and leads the National Research Program on the *History of Technology in the Netherlands in the 20th century* which includes the publication of a series of seven volumes exhibitions and diverse other outputs. He is co-founder (together with Kurt Fischer) of the *Greening of Industry Network* and leads several EU funded international projects. His research work and publications range from history of technology, science and technology studies, innovation and diffusion theory, constructive technology assessment, environmental management, and policy studies.

Ole Henning Sørensen has been working as an IT consultant, managing customer support, after finishing his engineering education in software engineering and technology management. He finalised a PhD study at the Technical University of Denmark with a dissertation entitled *Strategic Management of Technology and the Structuring of Industrial Networks* (2000).

Professor Knut H. Sørensen is Head of the Department of Interdisciplinary Studies of Culture and Director of the Centre for Technology and Society at the Norwegian University of Science and Technology (NTNU). Books published include *The Spectre of Participation. Technology and Work in a Welfare State* (Scandinavian University Press, 1998), *Making Technology Our Own? Domesticating Technology into Everyday Life* (edited with Merete Lie, Scandinavian University Press, 1996) and *Frankensteins dilemma. En bok om teknologi, miljø og verdier* (with Håkon With Andersen, Ad Notam Gyldendal, 1992),

Hendrik Storstein Spilker is currently a Ph. D. student at the Centre for Technology and Society/ Institute for Crossdisciplinary Cultural Studies at the Norwegian University for Science and Technology in Trondheim. As well as his dissertation entitled *From cult to culture: The future internet-user in the making*, he has been Research Fellow on several EU-funded projects, including Social Learning in Multimedia and TeleCityVision.

Lars Strunge is working as an Advisory Officer at the Danish Society of Engineers (IDA), where he is providing policy analysis cover issues of innovation, technology policy, and energy policy. Publications include the books *The liberalisation of the electricity sector - environmental and industrial consequences* (1999, in Danish) and *Can biotechnology further a sustainable environment*? (2001, in Danish).

Dr. Karl Matthias Weber is Head of the Technology Policy Department at Austrian Research Centers (ARC) in Seibersdorf. Before joining ARC at the end of 2000, he has been working for several years at the European Commission's Institute for Prospective Technological Studies (IPTS) in

Seville. Apart from several scientific and policy-oriented articles in the field of innovation and technology policy, he has published two books: *Innovation Diffusion and Political Control of Energy Technologies*, Springer/Physica, and *Experimenting with Sustainable Transport Innovations. A Workbook for Strategic Niche Management*, European Commission/University of Twente (with Hoogma, Lane and Schot).

Professor Robin Williams is Director of the Research Centre for Social Sciences/Technology Studies Unit at the University of Edinburgh, where he convenes a programme of interdisciplinary research into 'the social shaping of technology'. Books published include *Policies for Cleaner Technology: A New Agenda for Government and Industry?* (with Clayton, and Spinardi, Earthscan, 1999), *Exploring Expertise*, (co-edited with Faulkner and Fleck, Macmillan, 1998), *The Social Shaping of Information Superhighways: European and American Roads to the Information Society* (co-edited with Kubicek and Dutton, Campus Verlag, 1997); and *Expertise and Innovation: Information Technology Strategies in the Financial Services Sector*, (with Fincham *et al.*) Oxford University Press, 1995.

Preface

This book reviews the intellectual achievements of the 'Social Shaping of Technology' (SST) perspective which has allowed important recent advances in understanding the relationship between technology and society. In particular, the book explores the various ways in which such advances may be applied to the analysis and to the practice of technology policy. In this respect, it is breaking new ground by providing a stronger analytical foundation for policy intervention in the area of technology. As well as highlighting shortcomings of existing policy instruments, the contributions in this book point to new opportunities and methods of intervention, the challenges involved and some of the benefits that they may bring.

The book is indebted to the European COST A4 Action on the Social Shaping of Technology 1991-9, in bringing together leading scholars from Europe and beyond, as well as engaging with practitioners and policymakers who shared these interests. COST – European Co-operation in the Field of Scientific and Technical Research – is a framework for scientific and technical co-operation, allowing the co-ordination of national research on a European level, which from 1990 was extended into the social sciences. COST A4 on the Social Shaping of Technology was one of the first social science actions. A series of COST A4 workshops analysed social shaping in a wide range of social and technical contexts, resulting in a series of specific volumes. The very success of SST, both empirical and conceptual, has in turn posed problems in drawing together insights gained from addressing particular technical fields, contexts and analytic concerns. Important parallels and complementarities in approach between different scholars and centres suggested the need to bring these diverse insights together.

At the conclusion of the COST A4 programme, support was obtained for a final short study (Focused Study on The Social Shaping of Technology COST – STY – 98-4018) to address these questions, and in particular:

1. To review the array of concepts advanced to explain 'the social shaping of technology', examine the relationship between them and explore their relevance in explaining technological change in different settings and fields.
2. To examine the policy relevance of the SST perspective.
3. To present these to external, especially policy communities.

The study was undertaken by a small group from four centres most closely involved: Ulrik Jørgensen, Technical University of Denmark, Prof. Arie Rip and Johan Schot, University of Twente, Prof Knut H. Sørensen, Norwegian University of Science and Technology, and Prof. Robin Williams, University of Edinburgh, who led the project and provided day to day co-ordination. A set of papers was selected to review empirical and conceptual advances, and discussed at a series of workshops. This resulted in a report to the European Commission: *Concepts, Spaces and Tools: Recent developments in social shaping research* (University of Edinburgh: Edinburgh). Knut H. Sørensen and Robin Williams transformed and edited these into the current volume. Dr. Stewart Russell, on secondment from the University of Wollongong provided valuable inputs throughout, and we benefited from feedback from the other contributors and other workshop participants, especially: Prof. Ian Miles University of Manchester, Dr Mikel Olazaran, University of the Basque Country, Dr Fred Steward, University of Aston, Dr Sally Wyatt, University of Amsterdam and Prof. Judy Wajcman, Australian National University.

We thank Dr Gudrun Maass, (European Commission DGXII) Scientific Officer for COST Social Science Actions for her encouragement and advice, Barbara Silander (University of Edinburgh) for administering the project and with Gerdien M. Linde-de Ruiter (University of Twente) for organising the various workshops and meetings.

PART ONE

Overview of developments in the field and their implications for policy

1. Introduction[1]

Robin Williams

This book explores the implications of recent advances in analysing the complex relationship between technology and society. Theories and concepts developed by various European researchers in the field of 'the social shaping of technology' extend our understanding of the range of actors and social spaces in which technology can be influenced. For example technological innovation is not restricted to the laboratory and the world of technical specialists; non-specialists engaging with new technologies at work and in everyday life also play a role in redefining artefacts, their uses and social significance. Some exciting new concepts have emerged which highlight both the scope for local discretion in the way technologies are designed and used as well as the broader structures and systems that may serve to restrict choice.

By applying these concepts to case-studies of various social and technical settings, the book explores their utility for understanding how contemporary technologies are developed and applied and how they influence society. It further examines the implications of these insights for those involved in and concerned by technology including public policy-makers, managers, technologists and citizens. In this way this book seeks to broaden our view of the avenues and processes of technology policy and explore new tools for analysing and intervening. By way of an introduction this opening chapter briefly reviews these developments – in terms of the emergence of new concepts and their policy contribution – and outlines the structure and contents of the book.

BACKGROUND TO THIS WORK

The rapid development and proliferating social implications of technologies are attracting increasing attention. The pace of technological change is increasing. New technologies are no longer remote and confined to laboratories and high-technology industries but are becoming an increasingly important part of everyday life across an ever growing range of social activities. For example, Information and Communication Technologies

(ICTs), with their unprecedented rate of innovation and uptake, have been portrayed as a defining feature of our society – heralding a shift towards an 'Information Society'. ICT applications such as e-commerce are seen as crucial for economic competitiveness, while buzzwords such as cyber-space highlight the possibility of profound changes in how, when and where we carry out many kinds of social interaction. Similarly, developments in biotechnology hold out on the one hand great expectations of health benefits, more plentiful food and wealth creation and, on the other hand, have prompted widespread concerns about ethical implications and potential risks to health and ecosystems.

The stakes seem to be high. Technology invokes high levels of both anxiety and promise! Technologies are seen as a key to wealth creation and improving the quality of life. At the same time they may bring new risks, raising questions about how to identify and *control*, or at least ameliorate, unwanted consequences. Moreover there are also anticipated risks for those who *fail* to develop new technologies of missing out on what they could deliver. And since the efforts of governments and firms to develop and commercialise new technologies are frequently subject to setbacks and failures, particularly with radical technological innovation, the *promotion* of technology is also a concern. Public policy-makers and managers in industry thus increasingly need to engage with the issues and choices surrounding technological change. However, these issues are not just of concern to the narrow band directly involved in technology policy. As technology moves from the restricted domain of technological institutions and becomes more pervasive in economic and social life, it impinges upon the activities of a wide range of decision-makers with more general remits. Thus today, ICT has become an issue across virtually all policy domains and sectors. More significantly, an ever wider range of actors may be concerned – perhaps directly, in their role as consumers of technology, or indirectly as commentators and citizens – to assess emerging technological potentialities and their societal implications (Williams, Faulkner & Fleck 1998). These concerns underpin the increasing attention to the relationship between technology and society.

Technical domains present continued difficulties for social and policy analyses. Many traditional approaches remained 'outside' technological practice and simply addressed its societal context and/or its imputed socio-economic 'impacts' and outcomes (for example neo-classical economics which treats technical change as a residual factor in correlations between inputs and outputs). Such approaches are poorly equipped to capture the dynamism and uncertainties that surround technological innovation – since technological change may fail to deliver expected outcomes and may call into question established paradigms and socio-economic criteria (the 'circle of uncertainties' which Rammert analyses (Chapter 6, in this volume) in his the critique of traditional techno-economic approaches).

An important exception here is the vibrant strand in socio-economic research on technology that has sought to address the *process* of technological change in detail, and engage with the *content* of technological artefacts and practices. Early studies highlighted the choices available at every stage of innovation – choices which are influenced by 'social', economic, political and cultural factors, as well as narrowly 'technical' considerations – and the way in which these choices about new technologies and their outcomes may be 'socially shaped' and patterned by the contexts in which they are developed and deployed. We refer to this as the social shaping of technology (SST) perspective (MacKenzie & Wajcman 1985, Cronberg & Sørensen 1995, Williams & Edge 1996).[2]

CHANGING UNDERSTANDINGS OF THE SOCIAL SHAPING OF TECHNOLOGY

'The social shaping of technology' has proved to be a fruitful and dynamic area of study and debate. Our understanding of the character and social significance of technologies is developing rapidly, with increasing experience and a growing empirical research base, coupled with the development of new tools and analytical concepts.

Within the 'broad church' (Williams & Edge 1996) of SST, a wide range of perspectives and concepts have been advanced to explicate the relationship between technology and society. This has enabled deeper understanding of the process of technological change and its social and technical outcomes. For example early studies demonstrated the close interplay, and lack of clear boundaries, between 'social' and 'technical' elements and led to an understanding of the 'socio-technical' (Hughes 1983) character of these phenomena: both regarding the socio-economic influences upon technical choices and the material basis of social action. Thus Hughes (1983) showed that technical specialists needed to grapple with obdurate socio-economic and technical considerations (the price and the conductivity of copper wiring) in a process which Law (1988) described as *heterogeneous engineering*. However, beyond these limited areas of widely shared terminology, there is, as yet, no single unified explanatory schema. We can identify a number of reasons for this.

First, is the rate of change in our understanding of the relationship between technology and society. The social shaping analyses which emerged in the 1980s, through a critique of the prevalent technological determinist models of the technology-society relationship, have themselves been subject to re-assessment. For example early SST work was criticised, amongst other things, for giving undue attention to technology developers and prior technological design (Russell 1986, Russell & Williams 1987), thereby downplaying both the influence of the broader *social settings* in which these technologies were inserted and the *process of consumption*. Research into the latter has highlighted the scope for actors concerned with the implementation and use of artefacts to appropriate supplier offerings selectively and use them in different ways to those anticipated in design (e.g. Pinch & Bijker 1984, Cowan 1987, MacKay & Gillespie 1992).

Second, the various concepts have each been advanced to capture particular aspects and settings of the innovation process, which may differ significantly between technologies and contexts. For example, models developed to explain what happens in the laboratory will often differ from those to explain changes in industrial processes, or for that matter, the consumption and use of domestic technologies – even though there may be important parallels between these different phases and settings of innovation. There are also significant differences between technologies in terms of both their technical content and their social and institutional insertion, and, likewise, differences between emerging and established technical fields. Analysts have been drawn to account for particular features of innovation processes in different settings. Thus different sets of concepts have been articulated to explain the stability of already established large-scale socio-technical systems (see for example Hughes 1983), from those used to explain the early development of novel technical fields which emphasise the negotiability and interpretative flexibility (Pinch & Bijker 1984) surrounding artefacts and their uses.

Third, and in parallel with the above is the search for more comprehensive accounts of the settings of innovation – which is no longer restricted to the activities of engineers in the R&D laboratory but includes a broader range of actors, including 'intermediate' and 'final' consumers of artefacts. In addition to those actors directly involved, work has flagged the importance of the broader institutional context, in particular, the 'meso level' of interactions between organisations (Sørensen & Levold 1992), and the broader regulatory etc. setting (e.g. of regimes and broader techno-economic paradigms [Rip & Schot, Chapter 5 in this volume]). Conceptual developments have sought to grapple with the complexity of interactions amongst a wide range of players in innovation, and the consequent need to find modes of analysis that could cater for the interaction between the different phases of innovation and address the influence of both the local actors directly involved, and the broader institutional setting.

Finally a wide range of authors have been drawn to contribute to the field internationally. A key feature of this intellectual development has been the establishment of a dialogue between a range of scholars and research centres particularly within Europe and North America (Williams & Edge 1996). The analytical frameworks that have emerged have drawn on different disciplinary roots (including evolutionary economics, industrial sociology, history and philosophy of science, sociology of scientific knowledge, gender studies) whose traditions and relative strengths vary between countries and centres of research (Bijker & Law 1992, Cronberg & Sørensen 1995, Sørensen 1999).

This conceptual blossoming has by no means been random. As the preceding paragraphs make clear one largely shared concern has been to produce a richer, more integrated and comprehensive account of empirical developments. At the same time, these developments have also been patterned by certain underpinning theoretical and epistemological concerns. In 1996, Williams and Edge pointed to a number of lines of tension and debate in SST (see also Russell 1986, Russell & Williams 1987). One was between accounts which stressed the role and autonomy of local actors, versus broader social and institutional structures (and linked to this the emphasis on local negotiability versus broader patterning and stability in social and technological change). Another concerned the relationship between the material and symbolic dimensions of technologies. Although we did not see any evidence for convergence and closure, we did note some developments which sought to address and move beyond these dichotomised understandings.

When we look at the field today – as, for example, reviewed and presented in this volume – one thing that stands out is that these analytical dilemmas can indeed be seen to have underpinned many of the areas of conceptual development. Thus Rip & Schot (Chapter 5 in this volume) and Jørgensen & Sørensen (Chapter 7 in this volume) both seek to address the interplay between local actors and broader structural conditions in technological innovation, but in distinctive ways given their intellectual traditions. Rip & Schot draw from the rather generalised evolutionary economic accounts of 'selection environment' to focus 'down' from recognition of broad technology regimes to see how the circumstances for regime shifts can be created; Jørgensen & Sørensen drawing on the very different intellectual traditions of Actor Network Theory seek to retain its emphasis on local negotiability while extending it to understand technology development in an increasingly globalised setting in which a number of actor-worlds and networks interact. Likewise Weber (Chapter 12 in this volume) and Geels (Chapter 13 in this volume) share a concern to examine the detailed interplay between micro-level of innovation actors, the meso-level of interactions between organisations and the broader macro-level context. They develop policy tools that address these – and in particular the micro-meso dynamics which are often overlooked, resulting in analyses which fail to capture the complexity and uncertainties surrounding innovation processes and their outcomes.

This is not a simple process of closure or alignment of approaches: different analytical strategies and concepts are being advanced building from within distinct analytical traditions, but which broadly address the same analytical dilemmas and lacunae. The conceptual gap (e.g. between actor-centred and structural accounts) is being bridged by the articulation of new concepts.[3] And from these vantage points more nuanced debates and empirical analyses are made possible. Points of disputation would remain between the various authors with their different perspectives. And these disputes touch upon fundamental social scientific and epistemological debates which are perhaps unresolvable. However they have been sites for parallel processes of conceptual innovation in SST.

We thus argue that the overall proliferation of accounts in SST is a sign of intellectual dynamism and academic vigour. Diversity can be seen as a source of strength – in offering a rich resource for understanding the relationship between technology and society. These various frameworks have stimulated a vigorous programme of empirical inquiry. However, at the same time the field can be seen to have suffered from a lack of integration in its overall conceptual development. There has been little attempt to reconcile apparently analogous (or perhaps homologous) concepts articulated in relation to different fields and contexts of technological change. Addressing this creative diversity is one of the central goals of this book. The other concerns understanding its implications for policy and intervention.

THE POLICY APPLICATION OF SOCIAL SHAPING CONCEPTS

Scholars in the field are, of course, aware of each others' work, and there is substantial cross-fertilisation of ideas (though the synergies are not always properly flagged). However, the proliferation of concepts can be a barrier to the wider access and use of these ideas. It presents special problems for the external, non-academic audiences of SST – policymakers, practitioners and citizens – wanting to grasp the insights available from this body of work and deploy them to resolve particular concerns. There are few, if any, authoritative synthetic accounts.[4] The lack of such accounts from within the SST field in effect leaves it to the external audiences to try and make sense of a multiplicity of, apparently closely-related, concepts and tools.

A more fundamental problem surrounding the application of these concepts by practitioners, policymakers and publics derives from the kinds of understanding about the process of technological change that are emerging over two decades of socio-economic research on technology. While early SST accounts offered a rather straightforward metaphor of the politics of technology that emphasised the values that underpinned technological design and how these could become embedded in artefacts and reproduced when those artefacts were subsequently used. However Russell and Williams (Chapter 3 in this volume) point out the extension in the scope of recent SST accounts to include both a wider range of settings for innovation and a wider array of players involved and able to shape innovation outcomes. Policy intervention thus needs to address all the phases in the cycle of innovation from Research & Development to the implementation, consumption and use of artefacts, and likewise include a broader range of actors, including non-specialist users as well as specialists in technology supply. This development is part of a rejection of simplistic models of innovation – such as the traditional 'linear' model, driven by technology supply – towards recognising a complex system of interactions amongst a wide range of heterogeneous players. In such a context it is easier to see why technologies may fail, or may have unanticipated affects. Recent accounts emphasise the uncertainties that surround technology development and implementation processes and their outcomes. They likewise point to the contingencies surrounding technology developments in particular locales and historical settings (Sørensen, Chapter 2 in this volume). The consequence is a much more elaborate and sophisticated conception of the context and challenge for policymaking.

The concept of social shaping of technology has proved attractive to policymakers, notably in the UK and Europe (Williams & Edge 1996).[5] However, the challenge surrounding the wider policy application of the social shaping of technology perspective, arguably, concerns its ability to create generalisations about the process of technological change – and in particular to create concepts that will be sufficiently robust to guide, or at least inform, public policy (or commercial strategies or public responses) on technology.[6] There have been a number of attempts to explore these policy implications in relation to particular approaches within the social shaping approach, but none which have attempted this across the board.

STRUCTURE OF THE BOOK

In this brief introductory chapter we provide a brief overview of the structure and contents of this book.

This book is in two parts.

The first part reviews the development of SST as a research area and the main concepts and approaches that have emerged within this area. It pays special attention to the policy implications, and notes the difficulties surrounding attempts to make policy pronouncements from an approach that emphasises contingencies and the complexity and uncertainty of innovation processes and their social and technical outcomes. The review highlights in particular attempts to reconceptualise technology policy and the broader problems of intervention in technological innovation.

The second part of the book presents some of the new explanatory frameworks, concepts and tools that have recently emerged. These chapters each carry out three tasks. First they explicate the particular concept(s). Second they explore the utility of this framework in analysing specific developments in particular empirical settings. Finally the authors were encouraged to make explicit how their findings contribute to policy and public and commercial intervention around technological innovation.

Part I: Overview of developments in the field and their implications for policy

Part I of the book represents an attempt to review and analyse the main intellectual developments that have taken place and gained some wider currency in the field. It further explores implications for policy and for intervention.

In Chapter 2 (*Social Shaping on the Move? On the Policy Relevance of the Social Shaping of Technology Perspective),* Knut Sørensen lays out the issues at the heart of this book. The critique by SST of determinist accounts of technology as a driver of social change yields a view that change emerges through interaction and the weaving together of material and non-material elements. Outcomes arise from the ways in which artefacts become *domesticated,* and may become *entrenched,* within practice and culture. This, Sørensen argues, implies a redefinition of problems and concerns, away from individual artefacts towards clusters of socio-technical practices and a broader view of the policy tools potentially available, with increased emphasis on the need to combine policy instruments more creatively.

In Chapter 3 (*Social Shaping of Technology: Frameworks, Findings and Implications for Policy*) Stewart Russell and Robin Williams present a detailed literature review, highlighting a number of common aspects of recent developments in SST and related analyses (notably evolutionary economics). They stress: the sociotechnical, situated and systemic character of technologies; the complexity, contingency and dynamism of innovation processes; stability, continuity and patterning in innovation processes; the extension of the scope of innovation to include appropriation/use; a wider conception of relevant actors and the terrain of transformation; increasing attention to culture and knowledge in technology.
Noting that the proliferation of concepts and schemata may present a barrier to the wider uptake and application of these ideas, they present a selective Glossary of Social Shaping Concepts.

In Chapter 4 (*Concepts, Spaces and Tools for Action: Exploring the Policy Potential of the Social Shaping Perspective*) Russell and Williams explore the implications of this changing understanding of the process of socio-technical change for our view of the tasks, spaces and challenges of technology policy. The rejection of simplistic 'linear' models of technological innovation (e.g. technology-push or its converse, market-pull) and the articulation of complex, interactive innovation models coincides with a period in which the goals for policy intervention are becoming more complex, including the amelioration of undesirable effects as well as the promotion of innovation. New tools and strategies for intervention may be called for, e.g. ideas such as 'technology steering' which recognise the difficulties of achieving direct control in such complex sociotechnical settings. Intervention needs to address the application and consumption of technology as well as its supply. More integrated approaches to policy are needed; these may need to be centred around socio-economic goals rather than around technologies.

In Chapter 5 (*Identifying Loci for Influencing the Dynamics of Technological Development)* Arie Rip and Johan Schot present a broad framework, analysing how our understanding of technology dynamics could be used to create tools for intervention in innovation. This draws upon their efforts with co-workers at the University of Twente to combine evolutionary economic and constructivist perspectives to create tools for analysis and policy intervention in a context where entrenchment (e.g. through economies of scale, shared visions etc.) tends to create technology regimes that may suppress or exclude promising and beneficial technical options (Rip, Misa & Schot 1995). This attention to stabilisation in technological innovation also highlights the scope for dynamism. One example is the idea of Strategic Niche Management, geared towards building-up a protected space for 'hopeful monstrosities'. In this way, three loci for intervention are identified: building agendas and mobilising resources for a potential new artefact; technology demonstrators to learn about a technology and its impacts; intervening in the cumulative processes which may lead to regime changes.

Finally, in Chapter 6 (*The Cultural Shaping of Technologies and the Politics of Technodiversity),* Werner Rammert highlights cultural shaping – the role of different symbolic orientations, patterns of practices and institutional regimes – in the process of technological development. Generic *traditions and regimes*, *concepts and styles of engineering* and particular *visions of technology function and use*, open up options and paths of technological development and at the same time limit the range of other possibilities. Whilst traditional techno-economic approaches to technology all too often stifle technological diversity and choice, addressing cultural shaping opens up a space and offers tools for a new kind of technology policy that Rammert terms the *politics of technodiversity*. Rammert sees these interactive network relations – linking diverse, dispersed actors in technology supply, promotion and use – as constituting a new kind of technology policy and innovation regime. Technodiversity involves creating a space for multiple options and a variety of ideas and views and connecting them loosely – for example through joint innovative actions and recursive learning between networks of all the actors engaged (including even dissenting scientists and critical social movements).

Part II: New tools for analysing and intervening

In Chapter 7 (*Arenas of Development: A Space Populated by Actor-worlds, Artefacts, and Surprises*), Ulrik Jørgensen and Ole Sørensen address the complex geographical and social spaces in which innovation occurs. They advance the concept of a *development arena* – a space (constituted in geographical, social and cognitive terms) which brings together a number of particular actors and locales for innovation – to analyse the relationships of mutual influence and transformation amongst different local actor worlds. They apply this to the case of High Definition TeleVision (HDTV) to show how the introduction and presentation of a working prototype of the technology by a number of relatively small scale, economically marginal actors completely reshaped the space for the development of new television technology – in a way that was unanticipated and rather surprising given the competing established HDTV programmes in Japan, Europe and the USA. The arena concept is a sensitising device for the entry of new kinds of actors, their realignment and the fundamental reshaping of the arena which is hard to capture with existing theory (which exhibits a dichotomy in starting points for analyses either from particular locales of innovation or from generalised system level dynamics).

A better understanding of the relationship between local and broader settings is also pursued by Christian Clausen and Christian Koch in Chapter 8 (*Spaces and Occasions in the Social Shaping of Information Technologies: The Transformation of IT systems for Manufacturing in a Danish Context*). They present an historical account of the global shaping process of information technology (IT) for manufacturing – in which developments had been strongly influenced by supplier driven technological push and models of best practice strongly shaped by US corporate visions and practices. Within this they identify three occasions and spaces in which opportunities are opened up for the social choice around IT, namely:

Segmentation of IT-supply: new players emerge and establish links with Danish organisational users creating a diversity of systems;

Company internal dynamics: the local politics which surround and may reshape IT choice and implementation;

New technology supply/design strategies, with the emergence of mass produced systems designed for customisation (e.g. Enterprise Resource Planning): different levels of supplier-user co-operation offer more or less scope for customisation.
They argue that technology policy and actor strategies to influence development need to be alert to these varying opportunities for influence.

This work exemplifies a view that the improved understanding of the dynamics of technological change, emerging from SST, can highlight new opportunities for intervention, indicating areas where initiatives by policymakers or private actors may 'go with the grain' and other areas where the prospects of reshaping existing patterns are lower. Very similar conclusions emerge from a study of a strongly contrasting domain in Chapter 9 (*Feminism For Profit? Public and Private Gender Politics in Multimedia*). Hendrik Spilker and Knut Sørensen point to a number of private sector initiatives that have recently emerged in Norway to open up the new electronic media to women. They contrast this with the rather limited impacts of public policies to promote greater involvement of women in ICTs – which have often been based on rather formal models of women's exclusion (mainly providing training in ICT skills). The private sector initiatives have been geared to mobilise the interest of different groups of women – projecting a view that technology can be attractive and fun. Public policy has not yet woken up to the 'turn to entertainment' in ICT. Policies for greater inclusion around ICTs need to be more open to these opportunities, perhaps linking in to the efforts of private sector players. Closer engagement between policy and dynamics of technology development and use can make for more effective policy and broaden our view – of the avenues and processes of technology policy.

The last block of chapters (10 – 13) are all concerned with the energy and/or transport sectors and are underpinned by a shared concern to promote environmentally sustainable technologies and industrial systems.

The state is an extremely central player in the energy sector through detailed regulatory control of energy production and distribution (and also formerly through state ownership and provision of financial and technical support). This contrasts for example with the situation with computer technologies in which regulation is more fragmented (with many activities seemingly unregulated) and the activities of commercial players are more salient. The energy sector has also been the classic example of the entrenchment of technologies and particular paradigms of operation. Thus it is in relation to electricity supply that Hughes (1983) articulated his theory of the stabilisation of socio-technical systems. Callon (1980) presented the classic study of the difficulty of achieving radical technological changes in technology (shifting from the internal combustion engine to the electric powered vehicle) in the face of closely aligned and strongly entrenched interests. Attempts to intervene and transform such large scale socio-technical systems must address these issues of stabilisation and entrenchment.

On the other hand, today the energy sector is undergoing rapid and profound change – providing exciting opportunities to reassess our existing understandings. In particular we see a shift across the European Union (and in the global trade regime) towards the liberalisation and privatisation of energy. The early liberalisation in the UK promoted a rapid and unanticipated change in technology choice with important environmental etc. consequences – the 'dash to gas' with combined cycle generation technology (Winskel 1998). At the same time environmental concerns, and particularly fears about global warming, add new social demands about the performance of our energy producing and using industries.

This is an area in which technology studies has made a more sustained engagement with policy. In the three chapters presented here authors have sought, in different ways, to draw upon policy studies and to integrate it more effectively with SST.

In Chapter 10 (*Governing Measures: User-stories and Heat Pump Subsidies*), Pål Næsje explores the co-production of technological knowledge and political order which are simultaneously recreated and sustaining each other. He suggests that problem definitions and the data collected to support policy are, rightfully or not, skewed in specific directions. His study (of a Norwegian initiative to support heat pump installation) highlights the role of 'stories' in structuring policy discourses: serving to exclude some perspectives, creating meaning and rationalising policy outcomes, particularly in areas where little 'factual' information exists.

In Chapter 11 (*Restructuring the Power Arena in Denmark: Shaping Markets, Technology and Environment Priorities*) Ulrik Jørgensen and Lars Strunge examine the liberalisation of the Danish power sector in accord with EU Directives and coupled with strong commitments obliging Denmark to reduce CO2 emissions following the Kyoto agreement. Noting that many studies from a policy perspective have operated with a rather restricted understanding of technology, and *vice versa*, their account seeks to bring together an understanding of policy processes and technology innovation processes. They explore how the institutional structure of the power sector constrains innovation and technology choices and examine how these are conditioned by the interaction between regulation and liberalisation. In a period of transformation of the power sector, they identify elements of continuity in the Danish power arena (regarding environmental priorities and the local ownership structure) and restructuring (of companies and rules for accounting and regulation). The latter which were defined in terms of efficiency, are now to be defined in terms of market competition and price relations.

The final two chapters are concerned to understand the dynamics of socio-technical systems and to propose tools for intervention. In Chapter 12 (*The Political Control of Large Socio-technical Systems: New Concepts and Empirical Applications from a Multi-disciplinary Perspective*) Matthias Weber proposes an integrated approach to analysing large socio-technical systems (LSTS). His Politics-Economics-Technology systems model emphasises the interdependence of actors (involved in technology innovation, supply, implementation and use), technological elements and broader (economic, political, technological and informational) structures. He explores the interactions between elements and possible mechanisms that reinforce or delay change.

Examining the case of Combined Heat and Power (CHP) in Britain, the Netherlands and Germany, he argues that liberalisation has been a 'two-edged sword'. Private companies can far more easily set up local generation and supply. However CHP, though economically and environmentally beneficial in the long run, requires comparatively high initial investments and can be threatened by a highly competitive economic environment which forces the actors to address short-term returns. He concludes that policy strategies need to address the interdependence and co-evolution of the different structural elements of an LSTS. Well-timed and multi-level policy strategies can trigger quite fast processes of change though precise outcomes cannot be fully anticipated given the complexity of LSTS.

Finally, in Chapter 13 (*Towards Sociotechnical Scenarios and Reflexive Anticipation: Using Patterns and Regularities in Technology Dynamics*), Frank Geels seeks to overcome the weaknesses of traditional forecasting tools and methods which are based on rather simplistic assumptions about technology dynamics. For technology policies and strategies to take into account the complex social processes involved in technological development, better 'anticipation' tools are needed. Building on the Twente model of innovation (see Rip and Schot, Chapter 5 in this volume) he proposes the concept of 'socio-technical scenarios' to analyse transitions/shifts in technological systems, linking the micro-level of niches, the meso-level of regimes and the macro-level or socio-technical landscape.

Geels explores these in the case of the emerging battery-electric vehicle (BEV) niche over the last two decades in the face of the incumbent gasoline vehicles technology regime. This analysis reveals a number of specific patterns and strategies – from which it would be possible to develop scenarios based on different meso- and micro-level dynamics (which are ignored in simplistic forecasting models). Though not developing these scenarios as such, Geels articulates the steps that would be involved.

The work reported in this book seeks to capture intellectual developments in the field of SST overall, and in particular to explore conceptual advances emerging from four research centres which have been closely involved in European SST research, and which have contributed particularly to this volume: the University of Edinburgh, the Norwegian University of Science and Technology, Trondheim, the Technical University of Denmark, Copenhagen, and the University of Twente de Borderij, in the Netherlands. Each centre, and each contributor, has their own specific intellectual configuration and history which they have brought into the discussion. And debates have been carried forward for nearly two decades through collaboration and engagement as well as points of disputation and difference. This combination of engagement and of distance has been particularly fruitful in encouraging mutual learning.[7]

We hope this book can both stimulate further intellectual ferment, and make this body of findings more accessible to policymakers and a range of broader interested public.

NOTES

1. Thanks are due to Knut Sørensen, the other contributors to this book and to my colleague Wendy Faulkner for their helpful comments on this chapter.
2. Though we prefer the metaphor of social shaping, we include within this remit the significant corpus of work addressing the 'social construction' of technology (see in particular, Pinch & Bijker 1984, Elliott 1988, Bijker, Hughes & Pinch 1987, Bijker & Law 1992).
3. Thanks to Ulrik Jørgensen for suggesting this formulation.
4. There are some important exceptions here: notably Jasanoff *et al.* (1995) and the revised edition of MacKenzie & Wajcman (1999), and also, in relation to technical change at work, McLoughlin (1999).
5. For example Social Shaping of Technology figures in the European Commission 4th and 3rd Framework Programmes (in the technology policy lines of the social science research programmes) and formed part of the 'Technology' Thematic Priority adopted by the UK Economic and Social Research Council (http://www.esrc.ac.uk/th_list.html).
6. In Chapter 2, Sørensen addresses the critique often advanced of the policy applicability of qualitative, case-study based research of the type which prevails in SST. He argues that the rich empirical accounts of cases (or groups of cases) within their context, informed by theoretical agendas, do provide a basis for a creative rethinking of policy. However the relationship between research and policy is not simple. For example, though policy-makers seem to prefer quantitative 'facts', these can be at a such high level of generalisation that they provide little insight into the key policy/strategic choices being addressed. On the other hand the UK House of Commons Science and Technology Committee recently cited the use of complex case-studies as an instance of good and innovative practice. (Great Britain: House of Commons, 2000).
7. And an important element here has been the opportunity for repeated interaction over time, in particular through the various workshops organised under the European COST A4 (scientific co-operative) Action on 'the social shaping of technology' and through several joint research projects under the European Commission 4th and 3rd Framework Programmes.

BIBLIOGRAPHY

Bijker, W.E., Hughes, T.S. & Pinch, T.J. (1987), *The Social Construction of Technological Systems: New Directions in the Sociology and History of Technology*, Cambridge MA/ London: MIT Press.

Bijker, W. & Law, J. (eds) (1992), *Shaping Technology/Building Society: Studies in Sociotechnical Change*, Cambridge MA/London: MIT Press.

Callon, M. (1980), 'The State and Technical Innovation: a case-study of the electric vehicle in France', Research Policy, **9**, pp. 358–76.

Cowan, R. S. (1987), 'The Consumption Junction', in W.E. Bijker, T.P. Hughes & T. Pinch (eds), *The Social Construction of Technological Systems: New Directions in the Sociology and History of Technology*, Cambridge MA/London: MIT Press.

Cronberg, T. & Sørensen, K. H. (eds) (1995), *Similar Concerns, Different Styles? Technology Studies in Western Europe*, Proceedings of the COST A4 workshop, Ruvaslahti, Finland, 13-14 Jan 1994, COST A4, Luxembourg: Office for Official Publications of the European Communities.

Elliott, B. (ed.) (1988), *Technology and Social Process*, Edinburgh: Edinburgh University Press.

Great Britain: House of Commons (2000), *The Work of the Science and Technology Committee 1997-2000: First Special Report, session 2000-2001* (Stationery Office Dec. 2000). HC 44 ISBN 0 10 201201 6. Available at: http://www.parliament.the-stationery-office.co.uk/pa/cm200001/cmselect/cmsctech/44/4402.htm

Hughes, T. (1983), *Networks of Power*, Baltimore, MD/London: Johns Hopkins University Press.

Jasanoff, S., Markle, G. E., Petersen, J. C. & Pinch, T. (eds) (1995), *Handbook of Science and Technology Studies*, Thousand Oaks: Sage.

Law, J. (1988), 'The Anatomy of a Socio-Technical Struggle: the Design of the TSR2', in B. Elliott, (ed.), *Technology and Social Process*, Edinburgh: Edinburgh University Press, pp.44-69.

MacKenzie, D. & Wajcman, J. (eds) (1985), *The Social Shaping of Technology: How the Refrigerator got its Hum,* Milton Keynes: Open University Press

MacKenzie, D. & Wajcman, J. (eds) (1999), *The Social Shaping of Technology: Second Edition*. Milton Keynes: Open University Press.

MacKay, H. & Gillespie, G. (1992), 'Extending the Social Shaping of Technology Approach, Ideology and Appropriation' *Social Studies of Science*, **11**, pp. 685–716.

McLoughlin, I. (1999), *Creative Technological Change: The Shaping of Technology and Organisations*, London/NY: Routledge.

Pinch, T. & Bijker, W. (1984), 'The Social Construction of Facts and Artefacts: or How the Sociology of Science and the Sociology of Technology might Benefit Each Other', *Social Studies of Science*, **14**, (3), pp. 399–441.

Rip, A., Misa, T. & Schot, J. (eds) (1995), *Managing Technology in Society: The Approach of Constructive Technology Assessment*, London: Pinter.

Russell, S. (1986), 'The Social Construction of Artefacts: a Response to Pinch and Bijker', *Social Studies of Science*, **16**, (2), pp. 331-346.

Russell, S. & Williams, R. (1987), 'Opening the Black Box and Closing it Behind You: On Micro-Sociology in the Social Analysis of Technology', *Edinburgh PICT Working Paper* no.3, RCSS, Edinburgh: Edinburgh University.

Sørensen, K.H. & Levold, N. (1992), 'Tacit Networks, Heterogeneous Engineers and Embodied Technology', *Science, Technology and Human Values*, **17**, (1), pp. 13-35.

Sørensen, K.H. (ed.) (1994), *The Car and Its Environments: The Past, Present and Future of the Motor Car in Europe*, Proceedings of International Conference, COST A4, vol. 2, Luxembourg: Office for Official Publications of the European Communities.

Sørensen, K.H. (ed.) (1999), *Similar Concerns, Different Styles? Technology Studies in Western Europe: Volume II*, COST A4, Luxembourg: Office for Official Publications of the European Communities.

Winskel, M. (1998), *Privatisation and Technological Change: the Case of the British Electricity Supply Industry*, unpublished PhD thesis, Edinburgh: University of Edinburgh.

Williams, R. & Edge, D. (1996), 'The Social Shaping of Technology', *Research Policy*, **25**, pp. 865–899.

Williams, R., Faulkner, W., & Fleck, J. (eds) (1998), *Exploring Expertise*, Basingstoke: Macmillan.

2. Social Shaping on the Move? On the Policy Relevance of the Social Shaping of Technology Perspective[1]

Knut H. Sørensen

1. INTRODUCTION

Most standard textbooks on policy analysis have little or nothing to say about technology, and until recently, the technology studies literature has had little concrete engagement with policy. The main reason for the first lacuna is that technology is not a part of the toolbox of policy analysis. Technology is supposed to be affected indirectly by financial and legal instruments, but it is normally not a concern in itself. Parliamentary debates about nuclear power or genetic modification are exceptions to the rule. Normally, technology is supposed to be managed through the establishment of general institutions and standards (regulation), through public support of R&D or through other financial measures that affect supply or demand. Even in the case of public infrastructure investments, decisions are usually focused on location and budget, leaving the technical dimensions to the experts. In the understanding of mainstream economics, technology is something that is developed and used as a consequence of policy instruments, but it is not a policy instrument in itself.

One could also note that there is an unclear relationship between science policy and technology policy, to the effect that these two policy areas are confused into one, often labelled R&D policy, or technology policy is reduced to innovation policy. This means that challenges related to regulation and construction of infrastructure are left out of sight.

In much technology studies, on the other hand, policy has been interesting as a factor that may influence or shape technology, but not something that the technology studies constituency wanted to engage. The main exception is technology assessment, but this has never been considered as a proper part of the field. While efforts have been made to bridge this gap (see, e.g., Rip, Misa & Schot 1995), it is still important to discuss how and to what extent insights from technology studies may be applied constructively in technology policy. This chapter aims to contribute to this by discussing the potential policy relevance of the broad approach within technology studies designated 'social shaping of technology'. It seeks to identify interesting findings and analyse some challenges involved in attempting to transform them into a policy relevant knowledge base. Of prime importance, and a major gain from utilising technology studies in this respect, is the possibility of producing a broader agenda for technology policy. It no longer remains singularly concerned with innovation, but also with issues related to the *uses* of technology and thus to issues of regulation and infrastructure. Also, it raises questions about who should be recognised as policy-actors.

This chapter will discuss some aspects of the social shaping of technology (SST) research that are particularly relevant to analysing its potential implications for technology policy. Thus, we will be concerned with theoretical issues as well as the practical relationship between knowledge and policy-making.

2. SOCIAL SHAPING VERSUS SOCIAL CONSTRUCTION OF TECHNOLOGY: CONVERGENCE AND EXTENSIONS

The phrase 'social shaping of technology' was launched in the title of the reader by Donald MacKenzie and Judy Wajcman (1985, second revised edition 1999). It was related to the so-called strong programme of science studies, but loosely and indirectly. The SST programme, as outlined in the Introduction by the editors, was above all to criticise and transcend technological determinism – the assumption that technology had an *asocial* origin and definite, well-defined and predictable outcomes.

The initiative coincided with other efforts to develop new approaches to technology studies (see, e.g. Hughes 1983, Bijker, Hughes & Pinch 1987, Latour 1987). While these contributions displayed a variety of concepts and methods, they shared the aim of transcending technological determinism and provided a new basis for making anthropological, historical and sociological analyses of technology. Another common feature was the emphasis placed upon the study of technology during its inception, by analysing research, invention, innovation and design of new artefacts and systems. Compared with previous efforts to study technology, this was a move 'upstream', from the analysis of 'impacts', to the investigation of how technologies were constructed.

Initially, there were some differences of emphasis between the social shaping approach of MacKenzie and Wajcman on the one hand, and the social construction programme outlined by Bijker, Hughes, & Pinch (1987). To analyse social shaping meant to analyse the influence of particular political, economic and cultural interests and values, while the social construction programme was characterised by a more explicit agnosticism in terms of what should be meant by mapping the 'social'. Social shaping meant an interest in identifying different options present in the development of a given technology, asking how come some options were preferred to others: 'How did the refrigerator get its hum?,' to cite the title of a paper by Ruth Schwarz Cowan (1985). The answer should be given with reference to established social categories, like class and gender. The social constructivists had the same ambition with respect to choice of options (in their vocabulary, closure), but argued the need to keep an open mind about how closure was achieved, and by whom.

Paradigmatic demonstrations of social shaping analysis were given in David Noble's work on Numerical Control (NC) machinery (Noble 1979, 1984) and Langdon Winner's paper 'Do artefacts have politics?' (Winner 1986). Noble argued that NC technology was the outcome of coinciding interests of industry, the military and professional engineers, an amalgam of interests that he perceived as characteristic of modern capitalism. Winner provided more examples of how technologies had been fashioned according to well-defined social interests, implying that this would be typical of the development of new technologies. Artefacts were essentially politically shaped.

The last decade has made the nuances between social shaping and social construction of technology less pertinent. Both parties have modified their positions. Social shaping has become more open to the possibility that the choice between options cannot be attributed to predefined social interests (like gender, class and ethnicity), while social constructivists have become more sensitive to the importance of established powers, networks and institutions (Callon 1991, Bijker 1993, Williams & Edge 1996, MacKenzie & Wajcman 1999). Essentialism has become an invective in both camps, alongside technological determinism and linear models. Thus, we may use the label of social shaping of technology to designate a generic approach to the study of technology that remains anti-determinist and anti-linear, but less concerned with the issue of 'materialisation' of social interests. The term signifies a set of approaches that share the following set of assumptions and concerns:

- It explores the social processes related to technological change.
- Negotiations between different social groups and actors are a focal point, emphasising concepts like flexible interpretation of technology and technological controversy.
- It highlights choices between different technical options potentially available at every stage in the generation and implementation of new technologies.

To study the social shaping of technology is thus to analyse the construction of sociotechnical entities (Cronberg & Sørensen 1995, Williams & Edge 1996).

Another set of changes is related to the spaces under scrutiny. While both the social shaping and social construction perspective meant a focus on the genesis of technologies, this is now changing. The distinction between design, implementation and use is no longer perceived as clear-cut, and the field has definitely broken with the previous tendency of assuming that laboratories and other sites of design and development should be privileged locations of analysis. Increasingly, technology (or, rather, sociotechnical entities) is seen as shaped in all kinds of localities. The shaping that takes place so-to-speak outside development and design is still consequential for the outcomes in terms of social practices and cultural meanings. Development and design are important, but not decisive (Sørensen 1994, Williams & Edge 1996, Lie & Sørensen 1996).

The importance of this development in terms of the potential usefulness of the social shaping perspective should not be underestimated. First, it offers a new way of raising issues of democracy and public participation in matters related to technology. Second, this focus means that technology policy itself becomes more central as an object of social shaping research. Third, the 'consequences' of technology policies may be studied, not only in terms of invention and innovation, but also with reference to social change and social stability. This means that the vocabulary of social shaping research and of technology policy potentially has been enriched by new concepts, in particular *catalyst*, *domestication* and *entrenchment*. Let me briefly outline what this means.

The understanding of technology as having catalytic properties rather than being a force of change is not just a change of the stock of metaphors from classical mechanics to chemistry. It is a main ingredient in the development of a non-determinist and non-linear conceptualisation of technology that retains the idea that technology plays an important role in processes of social change. The challenge has been how to avoid reductionist moves, like the attribution of particular social changes to particular qualities of technologies, while also escaping the sociological trap where one is bound to explain social changes only in social terms. When technology is perceived as a catalyst, it is seen as an agent that facilitates or makes possible a destabilisation of a given social order by other actors and thus enables the opening of new options. Whether the social order is changed or not depends on the presence of actors and efficient strategies to make use of the destabilised situation. Thus, social change is neither made through new technologies nor through new social strategies or juxtapositions of structures, but rather through new sociotechnical constellations. Neither technology nor culture suffices; the result is achieved through interaction and the weaving together of material and non-material elements.

This weaving together is captured in the concept of domestication of technology. Nothing happens after the introduction of technology unless it somehow is put to work and given meaning; unless it is appropriated by social actors. It has to be transformed from being an external, *alien* element to a familiar one, embedded in social and symbolic practices. Technology has to be acquired, put to use and given meaning, it has to become a part of the relevant practical, symbolic and cognitive spaces of the actors involved (Silverstone & Hirsch 1992, Lie & Sørensen 1996, Sørensen, Aune & Hatling 2000).

The outcome of a collective process of domestication of a technology may eventually be that it becomes *entrenched*. This means that the technology is made part of a stable sociotechnical arrangement in a way that makes it increasingly difficult to do away with it. A prime example of this process is the motorcar that has been transformed from a novelty that was a luxury and a sports good of the few, to become an artefact perceived as a natural constituent of modern life. Entrenched technologies pose a particular challenge to technology policy because very often, dis-entrenchment seems to be very difficult to achieve. Thus, the space of regulation becomes quite limited and difficult to extend. On the other hand, entrenched technologies still need to be domesticated by users because new versions are made available to new people. Domestication may cause destabilisation in the sense that new meanings or new practices may be constructed, but this possibility is not yet very well understood.

3. THEORY INTO PRACTICE?

There are no well-established criteria to evaluate the policy relevance of knowledge. Both 'policy' and 'relevance' are ambiguous concepts (see, e.g., Weiss & Bucuvalas (1980)) and there is no simple recipe that allows the translation from research into practice. We have to be careful to avoid technocratic fallacies. An additional problem is that studies of technology policy have tended to be descriptive. The topic is still under-theorised and under-conceptualised.

However, there is a long tradition of analysing more generally how social science knowledge is made use of in policy work. The literature suggests at least two important findings. First, social science knowledge very often does not carry over into policy. Second, policy makers use social science knowledge for a host of different purposes, not just as input to problem-solving and decision-making (Weiss 1979). The instrumental model of knowledge utilisation, where research results are supposed to be the basis from which policy may be rationally deducted, is clearly flawed. These findings are also in concert with one of the basic tenets of social shaping of technology research, namely that it is a mistake to assume that we often find linear transformations from science to innovation.

What social shaping research may hope to contribute, in line with other social science efforts, is to help develop a knowledge base for policy. The formulation of policy is creative work, negotiating different points of view and different bodies of knowledge. However, new research-based knowledge may make a difference, to the way problems are formulated and to the tools available regarding strategies to cope with them.

Terms like knowledge creep and knowledge acquisition have been used to describe this process. On the basis of the work of Giddens (1990), one could alternatively suggest that the interaction should be understood in terms of a *reflexive* model. This means that what we perceive of as knowledge is something that is made use of as input to the process that leads to policy formulations, by providing arguments, mental models and examples (see also Næsje, chapter 10). This is not a weak and defensive argument. Decision-making is an exercise in the combination of deductive and inductive reasoning, based on many different forms of input and processed in complex ways. Knowledge, formalised as well as non-formalised, transformed, compartmentalised, etc., is a resource in this process, in two major ways. First, it helps structure the problem and discourse about the problem. Second, it helps clarify the problem and the available solutions. Both these aspects of knowledge are important to see its potential policy relevance. In the case of social shaping of technology research, a basic issue would be to what extent it helps to change the perception of the objects of technology policy or to improve the understanding of challenges, or both.

Lest this sounds overly academic and abstract, let us begin with an example: the car. All Western countries display a similar pattern in the way cars have become entrenched as the major instrument of transporting people. The now common acknowledgement that cars represent a major environmental problem in terms of pollution, noise, resources, land use, and visual qualities, has yet proved unable to destabilise this entrenchment. On the contrary, the number of cars and distances driven are still increasing.

Public regulation of car use has been dominated by financial instruments. It has been assumed that increased taxes would reduce the use of cars, either generally or within a particular local context. However, minimising costs of transport does not seem to be a central concern of the public. Over and over again, the car is established as having as many expressive as instrumental qualities. Driving is embedded in a larger system of meaning and comfort as well as a demand for flexible mobility (Sørensen & Sørgaard 1994). Thus, taxes may have an effect, but it is rather limited. People continue to drive because it is perceived as more comfortable, and because they can afford it.

Of course, taxes may be raised to a level where the burden gets sufficiently heavy to make people choose other modes of transportation. However, the economic and social side-effects of such increases (e.g. on the price of houses and land, the distribution of real income, and regional dynamics) make this move unlikely. Therefore, technical solutions like improvements of the gasoline-based car (greater fuel efficiency, use of catalysts, lighter bodies) or alternative cars (LEVs, fuel cells, LNG cars, etc.) are attractive, but they leave important problems, for example related to land use, unsolved.

Social shaping research on cars suggests different approaches to the problem of regulating car ownership and use. Since the car is such an important part of the symbolic practices in everyday life as a carrier of meaning and expression of identity, one cannot regulate it successfully by instrumental means. The problems do not reside with the car as an artefact. They emerge from the diverse set of practices where the car is involved. Thus, social shaping research could be used to change the understanding of the problem. One set of strategies would follow on the basis of the above argument, namely to consider the problem of transport in terms of quality and symbols rather than in terms of efficiency and instrumental rationality. This would imply that transport policy would need to be much more concerned about the quality and symbolic meaning of alternative modes of transport, e.g. public transport.

Another option would follow from the argument that the key concept that describes the problem is *mobility* rather than the car (see Sørensen 1999b). The issue that confronts modern industrialised societies is how come the levels of mobility have risen so much (more than tenfold in the last 50 years), and why mobility in terms of individually controlled vehicles is valued so highly. The so-called car problem, from this perspective, is basically a mobility issue. This does not make the policy challenges less pressing, but it changes the agenda by introducing different sets of arenas and tools with which to analyse problems and provide solutions. For example, the social construction of cities appears to be a very pertinent problem. Why are they commonly made to demand a high level of mobility?

Similarly, a social shaping analysis of energy consumption in Norwegian households ends up by redefining the policy issue of sustainable consumption of energy. When studied in this manner, it turns out that most people do not consciously consume energy. Energy is an invisible good, the consumption of which is generated indirectly through other activities and acts of consumption. When people construct their residences, transforming houses to homes, what they do is very consequential for their use of energy, but these consequences remain backstage (Aune 1997). Thus, to treat energy like any other good is misleading, and to use energy prices as the prime policy tool is only efficient under very particular circumstances.

A third example may be taken from ongoing research on social learning in multimedia. One striking aspect of the multimedia situation is the large number of initiatives taken by a host of actors related to newspapers, banks, governmental agencies, small media companies, consulting firms, etc. to explore the potentials of this new technology through sustained trials to find potentially interesting and/or profitable applications. Even if the efforts to organise the supply of access services are large, the – in traditional terms – strictly technological actions do not tell much about what will be gained in the future. This will be much more dependent on content providers and users (Brosveet & Sørensen 2000). If technology policy in the multimedia field ignores the latter efforts, it will be much less efficient in the promotion of multimedia technologies in the information society that is envisioned.

Many other examples could be added. To summarise what they amount to, one can identify the following policy inputs that could be extracted from SST research:

1. *The need to be concerned about technology as an object of policy, not just as a source of economic growth.* Social shaping research has identified a range of options where technology may be seen as an ingredient of broader strategies to promote specific political aims, e.g. sustainability.
2. *An extension of the space of technology policy and of the groups perceived of as policy actors.* Social shaping research has moved beyond a singular focus on research, development and design, and put much more emphasis on what is happening through implementation and use and the learning associated with these activities. Technology policy would be concerned with most areas of policy making.
3. *Redefinitions of problems and concerns*, away from individual artefacts and technologies, towards techno-cultural clusters of practices. To some extent, this means that problems are perceived in broader terms, as suggested in the multimedia example above. But redefinition can also involve a real shift in focus, for example from the car to mobility.
4. *A different view of the tools potentially available*, with increased emphasis on the need to combine policy instruments more creatively. The social shaping insistence to perceive technologies as encultured means a much greater concern for cultural aspects. This is not a rejection of economic instruments, but rather a critique of simplistic beliefs in the efficiency of such tools.

In addition, since the social shaping tradition values reflexivity as constitutive of the performance of research, it is probably also able to engage productively in reflexive exchanges with policy-makers. Perhaps ironically, one could foresee the possibility of an organised example of Giddens' (1976) double hermeneutic circle: social shaping researchers interpreting technology policy people, and *vice versa*. However, the irony may get lost when one considers the implications of such exchanges in terms of making research more interactive (Woolgar 2000) and thus socially more robust (Gibbons 1999). Arguably, openness and interactivity in the research process positively affect the social robustness of knowledge, and thus its usability in policymaking.

Having said this, one has to note that 'relevance' clearly is not a simple property of knowledge or research that can be claimed. Nor is it something that can easily be reconstructed with hindsight. 'Relevance' is a negotiated quality, and there may be different reasons to accept or dismiss the labelling of a given piece of knowledge as relevant. A first-order approximation could be that 'relevance' is related on the one hand to *availability*, and on the other hand to *applicability*, whether or not a given policy problem can be related to it or not. In addition, properties like *robustness* and *acceptability/assumed reliability* seem important. The issue of availability, which really is a knowledge transfer challenge, is of course important. SST researchers have not on the whole felt a need to make their work accessible to policy people, but some have. Probably, in this respect, social shapers are like most social scientists.

4. CHALLENGES OF/TO POLICY

One of the paradoxes of technology is in the contradictory perception of its importance. In the popular discourse, new technologies with their seemingly magical properties are seen as the major engine of change. Society is transformed, mainly because of technological development. Catchphrases like Information Society are typical of this belief. In policy as well as in academic discourses, however, technology is an off-stage phenomenon. As already mentioned, technology is not an acknowledged part of the policy toolkit, and it is definitely not seriously integrated in social theory and similar academic efforts to understand social order as well as social change.

It is no accident that the off-stage nature of technology is found both in policy and in social theory. These two discourses are related, although the relationship is distant and mediated. Intellectuals have never been really happy with technological issues (Latour 1996); they are somehow suppressed. When social shaping research claims policy relevance, at the same time it has to engage with social theory in order to support this claim. This challenge can be understood in terms of the need to conceptualise the material nature of social action. How may we integrate the material (nature, technology, and bodies) into the analysis of human societies? What is the role of technology as part of human action? How may we transcend the immanent fear of the material that we find within the social and human sciences? Even if much effort has been put into the conceptualisation of the social nature of materiality, there is still a lot of work to do. In fact, even such a basic notion as negotiation has seldom been dissected in technology studies to provide a more detailed insight into who negotiates what under which circumstance and with what outcomes (for an exception that indicates where this may lead us, see Bucciarelli (1994).

There are presently at least two streams within a broader social shaping community that can be said to work with what we could call the social theory challenge. Even if they are related, they work with different conceptual strategies. The most established is Actor Network Theory that in this respect pursues a semiotic inroad, exploring concepts like inscription, delegation, and network (Akrich 1992, Latour 1992). Partly from the same tradition, but also drawing upon other positions, is the effort to explore the use of metaphors like hybrids, cyborgs, and monsters to describe and analyse the ambiguous, but also promising, welding together of technology and culture (Latour 1991, Law 1991, Haraway 1991, 1997).

A second stream emerges from gender studies where feminists have coined the concept of mutual shaping of gender and technology, or in general terms, of culture and technology (Berg 1996). Some of the achievements are definitely inspired by Haraway's cyborg metaphor, but the analysis of 'mutual shaping' comes closer to show us the dynamics of the construction processes in question. This list could have been expanded, e.g. to include efforts of anthropologists like Emily Martin (1989, 1994). Her analysis of how knowledge from scientific communities is mediated, reproduced and transformed, adds important insights about the symbolic role of discourses in relation to both material and nonmaterial practices.

Another important paradox of technology is the reproduction of a dynamic relationship between universal development and local appropriation. Even if the same technologies are made available everywhere, and in spite of large engineering efforts of standardisation, there is no definite evidence of cultural convergence. In fact, as is demonstrated in the classical work of Thomas Hughes (1983), national, regional and even local structures may contribute to specific configurations of sociotechnical systems. This is also a basic tenet of social shaping research.

However, if sociotechnical configurations have distinct local features, this must somehow be reflected in the technology policy discourse. We could call it the national contingency problem because of its open-ended nature. We know that there is a re-embedding dynamic, to use Anthony Giddens' (1990) concept, that may introduce distinct local features, but the relationship between the universal and the local, between dis-embedding and re-embedding, cannot be determined. There is contingency at work, but no determination.

Interestingly, we can observe a similar situation when we compare social shaping research from different countries. Even if we can see similarities in the concerns and research agendas that develop, there are quite distinct national 'styles' in the way research is performed (Cronberg & Sørensen 1995, Sørensen 1999a). There are dissimilarities of research that in principle is performed on the same set of objects and drawing from the same theoretical source of inspiration. Usually, such dissimilarities would be outlined in terms of lag, misunderstandings, quality, or relative dominance of different schools of research. The underlying assumption is that, over time, convergence will take place.

However, the two collections point in a different direction. Some convergence may be observed, since the field of technology studies has become established with quite similar understandings of what constitute the important issues, but also through engagement with the same 'international' research traditions. On the other hand, national contingencies work to modify and diversify. The research problems, when we look at them in greater detail, are partly defined through political concerns that reflect different political cultures and different industrial and technological structures. They are also shaped by academic traditions. Similarly, the 'international' research traditions are translated in order to be domesticated on the existing academic terrain. Even if this terrain changes, translation is important and produces a national 'flavour' to the intellectual practice of social shaping research.

The social shaping approach may be the only effort in the study of technology that has begun to integrate this reflexive point in its own self-understanding. This should definitely be an advantage. Since there are national contingencies of technological development as well as of technology policy, the knowledge base of technology policy needs to be sensitive to such contingencies. This does not mean that social shaping research needs to be or should be local in scope and ambition. The point is to be able to continue a dynamic relationship between 'universal' and 'local' efforts of understanding, since they enrich each other, while at the same time be aware that such a dynamic is played out.

A similar insight should be teased out of yet another challenge related to the fact that social shaping research mainly has been based on detailed analysis of single cases. One must be careful not to make universal statements from these case-studies, but rather to develop theoretical agendas and conceptual frameworks that transcend the individual case. On occasion, this dominance of case-studies has been used to criticise social shaping research as unable to provide more general insights into the nature of technological development. To be useful to technology policy work, the argument goes, such lack of generalisable arguments is a serious deficiency. Consequently, one would need social shaping research not rooted in case studies.

Fortunately, there are two serious fallacies in the critique. First, the critique confuses statistical generalisation with synthesis or analytical generalisation (see Yin 1984). As indicated earlier in this chapter, social shaping research has been able to identify, through casestudies, features of technological development that are expected to be common, not to say dominant qualities. The conceptual work, leading to the suggested triangle of catalyst, domestication, and entrenchment, is one example, and many others that could be brought forward. Social shaping research – like any science of humans – cannot predict in a strict sense, but it can identify processes and phenomena that need to be observed and taken into consideration. There are regularities etc. one may look for (see, e.g., Geels, Chapter 13, in this volume).

Second, the claimed need for generalised knowledge is related to the instrumental and linear model of policy relevance, where the role of knowledge is understood as the only basis from which policy may be developed and argued. Of course, if one simply could deduce policy from knowledge, knowledge claims would need to be very robust and universal indeed. However, following from previous arguments, this sort of deductive exercise is simply not possible and this is not the way policy utilises knowledge. When we adopt a reflexive model of the relationship between knowledge and policy, the requirements placed on knowledge to be relevant are different. Of course, robustness is usually important. It currently seems as if statistically produced knowledge is held in higher esteem and given more authority than case-studies. Næsje's contribution to this volume, however, suggests that narratives close to 'urban legends' may be of prime importance in actual decision-making about policy, at least under certain circumstances. Thus, there may be a difference between the public and the private dimension of policy-making.

Having said that, the social shaping community may nevertheless be criticised for paying insufficient attention to synthetic work that cuts across various case studies. Detailed empirical work has been held in greater regard, even if theoretical concerns and interests have high visibility. Also, perhaps part of the critique about the case study orientation has been generated by a somewhat careless use of that label to designate one's research. When compared with standard case study methodology (Yin 1984), many social shaping 'cases' are really something more. Many historians are weary of the case study label. They perceive of case studies as example studies and claim that their historical analyses of particular situations transcend such categories because they provide a deeper and wider insight than single cases usually are thought to give. Maybe the historical stance should be adopted also by others?

5. WHAT IS IN A NAME? TRANSFORMATION OR DILUTION OF THE SOCIAL SHAPING OF TECHNOLOGY APPROACH?

The social shaping of technology research presented in MacKenzie & Wajcman (1985) carried radical connotations. One could observe a definite neo-Marxist and feminist influence, and the point that technology was shaped by particular social interests, was a radical one. Ideally, technology should not be socially shaped, and if it were, the shaping process should be more democratic. However, the political message from these early efforts of social shaping research was not singular and well-defined. What do we know when we know that artefacts have politics, and what do we do about it? The responses given ranged from participation strategies and beliefs in the capacities of social movements to neo-Luddite arguments. Generally, I think it is fair to say that the analysis was far better than the prescriptions, and the latter were usually missing anyway. The research was critical in the sense that it exposed how certain dominant social groups (the bourgeoisie, patriarchy) were able to shape and make use of technology to promote their interests, but it did not lend itself to any simple translation into politics. This should be understood on the basis of the predominant view of technology as a neutral instrument of social progress, a view widely shared across the political spectrum for most of the 20th century. Just to show that technology was not neutral but in fact a carrier of particular social interests was in that respect a political statement and a necessary step in order to achieve a better understanding of the technology-policy relationship.

The wider transfer of the SST concept, on which this chapter has been based, has also involved a set of translations in terms of aims, problems, and concepts. One could argue that in this way at least some of the basic original intentions have been 'betrayed', and there is some substance to that. A clear and important aim of early social shaping research was to show or, rather, expose the influence of dominant social interests in the assumed value-neutral area of development of technology. This is no longer at the forefront of the research interest. The notions of social interests are more nuanced, but also more diffuse. One is more concerned with the way interests are juxtaposed, negotiated and transformed, rather than simply the fact that there may be fundamental differences in terms of the power bases of these interests. Greater care is exercised in avoiding essentialist assumptions, regarding the character of social interests as well as the ability of technology to carry social interests in an unambiguous manner.

There are at least two influences that have shaped this outcome. One is the development within constructivist technology studies. They have demonstrated the need to be more careful in working from *a priori* assumptions about technology and social interests, and have produced contributions that show what may be gained in this way. We have now a better idea of the process through which indeterminisms of technical change are produced: we have a fragile basis of predicting 'impacts' because the interpretations of technologies are dynamic and situated, and thus inherently flexible. Not because technologies are flexible in a material sense, but because it is of great importance what the people that relate to a given technology, make out of it. Meaning and praxis is not predetermined at any stage. They are produced through controversies, negotiations, and truces, although in a contingent manner. These insights are also included in the revised version of the 1985 social shaping collection (MacKenzie and Wajcman 1999).

Given the critique from some quarters that constructivism is apolitical (Winner 1993), it may be paradoxical to argue that the second influence to transform the social shaping perspective is in fact coming out of efforts to apply that perspective to technology policy. Even if the notions of what may be political are different, it is still notable that the social shaping perspective has made a slow but definite impact in technology policy communities. In some sense, this is probably related to the reduced tendency to declare political partisanship, but there have also been more sustained efforts to explore and develop the political implication of the social shaping perspective.

This has meant an attempt to clarify the political implications of a non-determinist, non-essentialist analysis of technology and technological change. The work has brought forward interesting, but also uncomfortable, issues. What do we know when we know that technology is socially shaped, and who is able to make use of that knowledge? The eventual 'betrayal' of the original social shaping perspective is probably to be found in the answer to the latter question. What follows from non-essentialist assumptions is a loss of faith in the potential of *a priori* privileging in the design process some groups of users over others. The 1985 version of social shaping represented a belief that women and/or workers could be the ones to find a social shaping perspective most useful because it would provide them with strategically useful knowledge. The 2001 version even tries to enrol powerful groups as users, based on the belief that this means – potentially – a general form of progress. A more sensible technology policy or improved tools of technology assessment are arguably an advantage to the great majority, even if the interests to be pursued are different. The issues at stake are the construction of democratic arenas as well as practices of technology policy.

What is happening is not a loss of political argument in the social shaping perspective (or within constructivism, for that matter), but rather a loss of political innocence. The discovery of potential political usefulness is not necessarily beneficial or benign. Clearly, it represents a temptation to go pragmatic in order to gain consequence. However, it is not clear that one needs a principled rejection of such temptations. In pragmatic terms, the policy relevance is related mainly to three aspects of the social shaping approach. First, the insistence that technology is shaped in many different localities and that technology should be perceived as a process, rather than an artefact, extends the span of technology policy very considerably. While this of course complicates technology policy, it means that a lot more policy instruments may prove relevant than those traditionally directed towards the management of RTD-activities.

Second, the emphasis on users and the importance of users' activities in domesticating (or not domesticating) technologies, is in principle an agenda of empowerment of users. It may also, in combination with the first point, provide the basis for an argument for a broader conception of technology policy. Third, and in a sense more traditionally academic, we have the argument that social shaping research provides resources of reflection and thus provide a support of a more 'enlightened' technology policy. Of course, the proof of the pudding remains in the eating.

NOTES

1 An earlier version of the paper was presented at the COSTA4/TSER conference "*The Promise of Technology*", Copenhagen 2-3 Oct 1997. It has grown out the work of COST A4 and represents an effort to take stock of some of the achievements.

BIBLIOGRAPHY

Akrich, M. (1992), 'The description of technological objects', in W Bijker & J Law, (eds), *Shaping Technology/Building Society*, Cambridge, MA: MIT Press.

Aune, M. (1997), *"Nøktern" Eller "Nytende"? Energiforbruk og Hverdagsliv i Norske Huholdninger*, report 34/97, Centre for technology and society, Trondheim: University of Trondheim.

Berg, A-J. (1996), *Digital Feminism*, report 28/96 Centre for Technology and Society, Trondheim: University of Trondheim.

Bijker, W, Hughes, T. & Pinch, T. (eds) (1987), *The Social Construction of Technological Systems*, Cambridge, MA: MIT press.

Bijker, W. (1993), 'Do Not Despair: There is Life After Constructivism', *Science, Technology and Human Values*, **18**, (1), pp. 113-138.

Brosveet, J. & Sørensen, K. H. (2000), 'Fishing for fun and profit. Norway domesticates multimedia', *The Information Socity* , **16**, pp. 263-276.

Bucciarelli, L. L. (1994), *Designing Engineers*, Cambridge, MA: MIT press.

Callon, M. (1991), 'Techno-Economic Networks and Irreversibility', in J Law (ed.), *A Sociology of Monsters. Essays on Power, Technology and Domination*, London: Routledge, pp. 57-102.

Cowan, R. S. (1983), *More Work for Mother: The Ironies of Household Technology from the Open Hearth to the Microwave*, New York: Basic Books.

Cronberg, T. & Sørensen, K. H. (1995), *Similar Concerns, Different Styles?; Technology Studies in Western Europe*, COST Social Sciences, COST A4, vol. 4, Luxembourg: Office for Official Publications of the European Communities.

Gibbons, M. (1999), 'Science's new social contract with society', *Nature*, **402**, no. 6761 supp. 2 Dec. pp. C81-C84.

Giddens, A. (1976), *New Rules of Sociological Method*, London: Hutchinson.

Giddens, A. (1990), *Consequences of Modernity*, Cambridge: Polity Press.

Haraway, D. (1991), *Simians, Cyborgs and Women: The Reinvention of Nature*, New York: Routledge.

Haraway, D. (1997), *Modest_Witness@Second_Millenium.Femaleman_Meets _Oncomouse*, New York: Routledge.

Hughes, T. (1983), *Networks of Power: Electrification in Western Society, 1880-1930*, Baltimore: Johns Hopkins University Press.

Latour, B. (1987), *Science in Action*, Milton Keynes: Open University Press.

Latour, B. (1991), *We have Never Been Modern*, Cambridge, MA: Harvard University Press.

Latour, B. (1992), 'Where are the Missing Masses? The Sociology of a Few Mundane Artifacts', in W. Bijker & J. Law (eds), *Shaping Technology/Building Society*, Cambridge, MA: MIT Press.

Latour, B. (1996), *ARAMIS or The Love of Technology*, Cambridge, MA: Harvard University Press.

Law, J. (ed.) (1991), *A Sociology of Monsters: Essays on Power, Technology and Domination*, London: Routledge.

Lie, M. & Sørensen, K. H. (eds) (1996), *Making Technology our Own?: Domesticating Technology into Everyday Life*, Oslo: Scandinavian University Press.

MacKenzie, D. & Wajcman, J. (eds) (1985), *The Social Shaping of Technology*, Milton Keynes: Open University Press.

MacKenzie, D. & J. Wajcman (eds) (1999), *The Social Shaping of Technology*, Buckingham: Open University Press, 2nd edn.
Martin, E. (1989), *The Woman in the Body: a Cultural Analysis of Reproduction*, Milton Keynes: Open University Press.
Martin, E. (1994), *Flexible Bodies*, Boston: Beacon Press.
Noble, D. (1979), 'Social Choice in Machine Design', in A Zimbalist (ed.), *Case Studies in the Labor Process*, New York: Monthly Review Press.
Noble, D. (1984), *Forces of Production: A Social History of Industrial Automation*, New York: Knopf.
Rip, A., Misa T. & Schot, J. (ed.) (1995), *Managing Technology in Society. The approach of Constructive Technology Assessment*, London: Pinter.
Silverstone, R. & Hirsch, E. (eds) (1992), *Consuming Technologies: Media and Information in Domestic Spaces*, London: Routledge.
Sørensen, K. H. (1994), 'Technology in Use: Two Essays on the Domestication of Artifacts', *STS Working Paper* 2/94 Centre for Technology and Society, Trondheim: University of Trondheim.
Sørensen, K. H. (ed) (1999a), *Similar Concerns, Different Styles? Technology Studies in Europe*, COST A4 Social Science, vol. 2, Luxembourg: Office for Official Publications of the European Communities.
Sørensen, K. H. (1999b), 'Rush-Hour Blues or the Whistle of Freedom? Understanding Modern Mobility', *STS Working Paper* 3/99 Centre for Technology and Society, Trondheim: University of Trondheim.
Sørensen, K. H., Aune, M. & Hatling, M. (2000), 'Against Linearity: On the Cultural Appropriation of Science and Technology', in M. Dierkes & C. von Groete (eds), *Between Understanding and Trust*, Amsterdam: Harwood academic publishers, pp. 237-257.
Sørensen, K. H. & Sørgaard, J. (1994), 'Mobility and Modernity: Towards a Sociology of Cars', in K H Sørensen, (ed.) *The Car and its Environments: The Past, Present and Future of the Motorcar*, COST Social Sciences, COST A4 vol. 2, Luxembourg: Office for Official Publications of the European Communities.
Weiss, C. H. (1979), 'The Many Meanings of Research Evaluation', *Public Administration Review*, Sep/Oct.
Weiss, C. H. & Bucuvalas, M. J. (1980), *Social Science Research and Decision-Making*, New York: Columbia University Press.
Williams, R. & Edge, D. (1996), 'The Social Shaping of Technology', *Research Policy*, **25**, 865-899.
Winner, L. (1986), 'Do Artefacts have Politics?' in *The Whale and the Reactor: A Search for Limits in an Age of High Technology*, Chicago: University of Chicago Press, pp. 19-39.
Winner, L. (1993), 'Upon Opening the Black Box and Finding it Empty: Social Constructivism and the Philosophy of Technology', *Science, Technology, & Human Values*, **18**, (3), 362-378.
Woolgar, S. (2000), 'Social Basis of Interactive Social Science', *Science and Public Policy*, **27**, (3), pp. 165-174.
Yin, R. K. (1984), *Case Study Research: Design and Methods*, London: Sage.

3. Social Shaping of Technology: Frameworks, Findings and Implications for Policy *with Glossary of Social Shaping Concepts*

Stewart Russell and Robin Williams

1 INTRODUCTION

This chapter reviews recent developments in the rich and growing body of research and theory which has come to be known as *social shaping of technology* (SST). It is structured in a way we hope allows discussion of how specific concepts and findings from SST can be applied, and we have drawn out some of the implications of the currents and themes we identify. Our aim in this and the next chapter (Russell and Williams *Concepts, Spaces and Tools for Action: Exploring the Policy Potential of the Social Shaping Perspective*) is not to offer substantive policy advice, nor recipes for doing so, but rather to explore what is involved in putting SST insights to use.

SST is not a single well defined theory, and debate continues over which approaches should be included in the term. Here we take a fairly broad conception of its scope and of the types of work that contribute to the endeavour. At the same time, however, we select what we regard as the most important elements and try to develop a picture of how they might be integrated.[1]

SST thus includes distinctively different approaches and schemes of concepts. Its objects of study range widely across types of technology, parts of innovation processes or domains of use. Nonetheless each approach makes some claim to general applicability. There is often considerable overlap in research focus and the aspects of sociotechnical change that are emphasised. Different approaches and studies may be making similar claims about phenomena, or contrasting arguments about the best ways to understand them. Our goal then is to provide a comprehensive and comprehensible road map through this complexity: to provide some order to the arguments; to group, distinguish and compare the concepts and general claims; and to assess common ground and clarify continuing disagreements.

Though the general tenets of SST might still indicate the 'broad church' that Williams and Edge identified some years ago, we argue that there has been a noticeable convergence and common ground among SST analysts on several key issues. Theorists have been less inclined to take up polarised positions for effect. They seem more willing to blend concepts, to work with substantive findings from different frameworks within and beyond the general SST endeavour, and to explore complementaries. We try to indicate how some of the concepts and insights might be further integrated, or at least used in a consistent way.

We are not, however, trying to impose a particular model overall. Least of all do we want to end up with a bland synthesis which somehow represents a least common denominator among the variety of positions. A pluralistic approach in deploying different models and concepts can help illuminate different facets of sociotechnical change, and the remaining tensions between different views are in themselves productive. We are not in any case able to settle the many disputes – at least not to the satisfaction of the contributors! We limit ourselves therefore to suggesting possible fruitful directions.

The next two sections outline the basic idea of social shaping and review early approaches. We then identify and examine in detail themes and trends in recent SST work (sections 4 to 11). At the end of each, we indicate some of the implications for policy and intervention – at a more detailed level than the general observations we make in Chapter 4. In concluding, we comment first on the evolution of the field, and our perception of points of convergence and tension, and second on some of the issues that arise in deriving general lessons from SST research.

The proliferation of terms in SST is itself daunting. The urge to coin new ones is understandable. As researchers try to reconcile insights from new research with available concepts, they find that their connotations are inappropriate or that they fail to capture something important in the empirical material. Or they may want to distance their new framework from its competitors and predecessors. The result, however, may be overlap between different terms, a term with different meanings in different frameworks, or simply a new word for a familiar phenomenon. We have therefore appended a glossary of key social shaping concepts, both to help researchers compare and consider the contributions, and more importantly as a guide for external audiences who are less familiar with the languages.

2 THE BASIC IDEA OF SOCIAL SHAPING

The idea that technology is socially shaped is fundamentally opposed to technological determinist accounts of the nature of technology, of the relation between a society and its technologies, and indeed of the bases of social organisation and the sources of social change. Such determinist models have dominated popular and many academic discourses on technology and society. They underlie prevalent notions of economic and social development. Until relatively recently they have been implicit in most policy frameworks for technology. They depict technology as an essentially autonomous entity, which develops according to an internal logic and in a direction of its own, and then has determinate impacts on society – in effect moulds society to suit its needs. The inevitable sequence of technological advance is determined in some accounts by progressive improvements on previous versions of a technology, and in others by scientific discoveries or the application of scientific methods to discovering improvements. Either way, technological change is depicted as beyond social influence; even its adoption is often seen to be determined by a 'technological imperative'. These models implicitly deny any possibility of choice in the direction of technological development and consequent social change. In this view the scope of public policy is limited to predicting and monitoring the progress of technology along its inevitable trajectory, finding ways of speeding it up by providing the required resources and removing obstacles, and promoting the smooth adaptation of society to the changes it demands.

The reaction against technological determinism in studies of technology produced a range of arguments as to its inadequacy as explanation and its ideological function in mystifying and furthering the interests of dominant groups which benefit from technological change. This critical work asserted that technology is shaped in form and content by social forces, and that its social effects are not determined simply by the nature of the technology. Instead of technology being treated as a 'black box' in history and sociology, or as a mystery exogenous variable in economics, the approach insisted that the process and content of technological activities and products should be amenable to social investigation. There is, as Rohracher (1998) observes, 'no inherent or compelling logic of technical development' and such patterns and logics as exist are explicable socially. In terms of a political critique, since current technologies and their effects are shaped by certain social forces – to the advantage of some social interests and the detriment of others – technological development should be able to be steered according to different social goals.

From theoretical arguments and empirical demonstrations that technologies are socially shaped, arose analyses of how this shaping comes about – examining the sites and contexts of choice, and the influences steering that choice – and debates about the best way of explaining it.

There is no agreed definition of what constitutes or qualifies as an SST approach. The broadest conception takes the only criterion to be an opposition to technological determinism, in attempting to demonstrate social influence on the direction of technological change. This would admit a wide variety of contributions from history, sociology, politics and economics-based studies of technology.

In practice, the term social shaping has come to be associated with particular approaches, emphases and claims, which represent a convergence of currents from these disciplines. The work arose in particular from

- a move to contextualisation in the history of technology, attempting to overcome a traditional internalist/externalist divide;
- critiques in innovation studies of the conventional treatment of technology in economics; and
- attempts to import approaches and insights from science studies, especially the sociology of scientific knowledge.

There has been an increasing dialogue and reconciliation between early approaches, a willingness to learn from other frameworks which assume the malleable nature of technology and analyse the forces influencing its direction, and a borrowing from other areas of study of concepts and insights which can be made consistent with their principles. While it may be possible to attribute particular concepts and insights to one or other origin, it has become ever more difficult and misleading to impose a conventional notion of disciplinary boundaries on studies in SST.

SST work has been largely based on case studies of individual artefacts and systems. Its basic tenet, however, that technology is socially shaped, can be taken as applying in a broader sense: that social processes shape not only the form and features of particular technologies but also patterns and general characteristics and directions of technologies across whole areas of development and application. We shall see that some of the crucial theoretical challenges involve explaining shaping in this broader sense and reconciling concepts and insights at these different levels.

3 EARLY APPROACHES IN SST

Several key frameworks for analysis were proposed through the 1980s, drawing variously on existing strands of sociological, historical and economic work on science and technology. They have inspired much subsequent research and theorising in SST, and their influence will be seen throughout this review.[2]

A range of studies – represented in the seminal 1985 collection edited by MacKenzie and Wajcman – sought to demonstrate how technologies embody and reflect dominant social interests in their form and features. These often constituted critiques of class, gender, military or bureaucratic interests which directed and selected among technological developments, and hence of their impacts on other sections of society. They included feminist studies of the gendered development and effects of technologies, particularly in household, service and office work, and health and medical settings; labour process analyses of technological change and work organisation in industrial and commercial settings; and studies of corporate strategies to achieve technological dominance in particular markets.

The social construction of technology approach (SCOT) pioneered by Pinch and Bijker brought insights from the sociology of scientific knowledge to bear on technological development. Adopting constructivist principles from SSK, it treated technological success and failure symmetrically, insisting on the same sort of explanation for both. It followed the process by which closure is achieved among 'relevant social groups' between competing interpretations of the available technological options, so that a particular design becomes taken for granted as the essence of the technology (Pinch & Bijker 1984, Bijker, 1987, 1993 & 1995b, Bijker & Law 1992b, Bijker, Hughes & Pinch 1987).

Actor Network Theory (ANT) developed by Callon, Latour, Law and others, followed the strategies and actions of central actors – system-builders or heterogeneous engineers – as they attempt to marshal the resources necessary for the project, particularly by enrolling other actors – locking them into appropriate roles – and appropriating the right to speak for them. A technology was conceived as an emerging and increasingly stabilised network of material and non-material elements. The nature of the project and the identities and interests of the actors involved are transformed as the network takes shape. ANT also developed useful analyses of the means by which actors use stable entities, or inscribe their intentions into technologies, to obtain remotely actions they require of other groups. Proponents insisted the analysis did and should avoid prior theorisations of the social setting and of the interests and power of actors, dissolve distinctions of scale, and demonstrate the constitution of entities in the process of interaction. ANT developed a distinctive vocabulary to describe these processes and distinguish its concepts (Akrich & Latour 1992, Bowker 1987, Callon 1980, 1986a&b, 1987, 1991, Callon & Latour 1989, Callon & Law 1981, Latour 1988, Law 1986a&b, 1987a, b&c, 1992a, Law & Callon 1988.)

A strand of systems thinking from the history of technology, particularly associated with Hughes, was also influential. It also followed the attempts of central system-builders to construct the required complex of technical and social elements and compete against other nascent systems. It stressed the achievement of the successful construction of a system out of chaos, conflict, diversity and decentrality; the importance of the interconnectedness of a technological system in explaining the foci and obstacles in innovative effort;

and the strategic thinking of the system-builder (Hughes 1979, 1983, 1986a, b&c, 1987).

Programmatic statements in this early SST work tended to give the impression of distinct schools of thought and marked breaks with previous frameworks, and a consequent tendency to reject other frameworks and insights. We can see now a step back from such positioning. Moreover there was much work throughout which drew from several currents and did not fit readily into any one, or sought to stress themes or introduce dimensions of social shaping that cut across the central contentions of what are often identified as the main schools – for example, feminist work stressing the centrality of patriarchy and gender but taking a wide range of approaches from more structuralist political economy to post-modern cultural studies (Wajcman 1991).

The emphases and concerns in SST work have also exhibited differences between countries (Cronberg & Sørensen 1995a, K.H. Sørensen 1999). This is understandable given the different circumstances and features of each country's industrial, technological and political systems, and the particular preoccupations arising from those. Certainly the understandings of key concepts like technology assessment and participation depend on experimentation and experience in national contexts and differ markedly. However the field has been characterised from the outset by a high level of international collaboration, with researchers moving between countries and participating in international conferences and journals. The frameworks and concepts used in any one country are thus contingent mixtures depending on the origins and affiliations of relatively small numbers of people, and there is a limit to the usefulness of distinguishing supposed national approaches to SST.

There remain, of course, other currents in technology studies and other areas of social theory which address relevant concerns and provide important insights. We would want to reject some of the claims and arguments, and indeed some of the assumptions and principles of certain approaches.[3] But many of the insights are compatible or adaptable, and are important for a rounded understanding of technology and its social dimensions. It is instructive, for example, to consider how discussions of overarching patterns in possible future technologies of production – flexible production, post-Fordism, and so on – relate to SST arguments and findings. And although we necessarily neglect it here, much work of a more philosophical nature on technology and the human condition has a direct bearing on SST theorising.

4 THEMES AND TRENDS IN RECENT SST RESEARCH

We can trace substantial growth and development in the SST perspective over the past decade. Many more areas of technological activity and domains of application have been tackled and many more examples generated. We can now draw on a much greater variety of technologies, each showing different dynamics, contexts and forms of institutionalisation.

We have organised the following sections around what we see as the key themes and trends in this work. It is not that its character changed abruptly in the 1990s; the emphases we depict here were nascent in earlier work, and some of the arguments generally accepted now were articulated then. We identify and explore:

- the integration of insights from other currents of technology studies and more general social theory;
- further development of ideas on the situated, sociotechnical, and systemic character of technologies, including the co-production of technology and social change;
- a continuing emphasis on the contingent and unpredictable nature of sociotechnical processes and outcomes, particularly in analyses of emerging alignments among actors and their conceptions of the technology and its function;
- at the same time, continued attempts to explain regularities and constraints, and an increasing effort to reconcile or find a balance between these two emphases, particularly to allow explanation of the way technologies are or are not translated from local achievements to generic forms;
- an extension of the focus of research downstream to appropriation and use, and an increasing emphasis on the articulation between earlier and later stages in innovation processes;
- a wider conception of the terrain of innovation and of relevant groups and sites; and
- a greater emphasis on cultural aspects of technology, particularly the role of meanings and values attached to technologies, and the shaping of knowledge in and about technology.

5 INTEGRATING INSIGHTS FROM OTHER AREAS OF THEORY AND RESEARCH

SST has continued to draw, critically and with qualification, on other traditions of technology studies and diverse areas of social theory and research. We find a much increased dialogue between them, and more attempts to integrate approaches and insights. Because of the variety of influences, and the interdisciplinary background of many SST researchers, it is increasingly difficult to acknowledge all the contributing strands and sources or to draw clear boundaries. We find this interaction encouraging and fruitful.

Innovation studies based in evolutionary economics

There is a continuing relation in particular with a strong tradition of innovation studies based in evolutionary theories in economics.[4] This work is represented at a macro level by analyses of long waves, techno-economic paradigms and other patterns in innovation and economic activity; at meso levels in studies of innovation in specific industries and sectors; and in a variety of types of micro level analysis, like studies of R&D in the firm. It highlights the dynamism of economies, gives technological change a central role, and insists on treating technology as endogenous to economic processes.

In many respects the boundary between this work and SST has become less clear: their preoccupations overlap strongly and their findings are often consistent. In its focus on firms, this innovation studies work has developed accounts of the search for solutions as a bounded heuristic process, the local, tacit and cumulative character of knowledge, path dependency in directions of change, and choices and trade-offs between contradictory pressures. These seem perfectly consistent with the emphases of much SST work.

In that economic approaches are geared towards explaining order and patterning, innovation studies work at broader levels can also be useful for SST: at the macro level of patterns in innovation, of institutional environments, of the co-evolution of technology and economies, and of national systems of innovation; and at the meso level of patterns across firms and industries. Although there is much disagreement here among SST theorists, some have sought to find consistency with their own meso-level understandings of specific institutions or networks and micro analyses of development and adoption processes.[5]

There are however important points on which SST theorists remain critical of evolutionary economics based work (Rohracher 1998, McLoughlin, 1999). In concentrating on patterns and generic processes, it fails to acknowledge and explain diversity and contingency among innovation processes and firms.[6] As it focuses more on the revealed performance of organisations and systems, it tends to overlook the detailed dynamics and outcomes of particular innovation processes and treat local learning economies as black boxes. So while the techno-economic patterns it depicts can be taken as an appropriate summary of the aggregate effects of structured incentives or of routine procedures and transactions, SST in effect seeks to demonstrate how and to what extent the microeconomic mechanisms this work posits actually operate in specific cases. As Rammert emphasises (Chapter 6 in this volume), economic accounts such as supposed economic logics, attributions of rational behaviour by economic agents, and formal economic assessments, still leave space for sociological or cultural explanations of selection, interpretation and negotiation, and hence contingency and variation.

While most innovation studies frameworks would likewise reject simple metaphors from biological evolution of random variation and selection (van den Belt & Rip 1987, pp.137-138), SST analyses are better equipped to address the detailed mechanisms involved: to show how innovators interact with their environment, how successful innovation depends on their actions in anticipating and modifying selection pressures, and how they change their strategies and technologies in response to experience (Jørgensen & Karnøe, 1995, p.74). SST is also wary of treating firms and markets as fixed entities, stressing that innovation processes often cut across and transform existing entities. It typically explores the connection of developing technologies to different social groups and hence acknowledges a wider range of actual or potential influences on innovation. Focussing on the interpretation of technologies rather than assuming their properties are inherent (MacKenzie 1996a), provides a different understanding both of practitioners' shaping and of users' reception and appropriation of technologies.

Despite these criticisms, and continuing disagreement about what findings and concepts SST can accept, there is much that it can draw on in this current of innovation studies. We should not expect a complete merging of the two bodies of theory: their agendas and concerns are different, and they have strengths at different levels of generalisation. But there has certainly been greater interaction between them, a greater awareness of common research interests, and much more 'cross-fertilisation' or 'cross-border trading'. In considering implications for policy, we concur with Rohracher (1998) that we should deploy insights from the two approaches together.

Niches

The fruitful interaction of sociological and evolutionary economics ideas can be seen in the framework developed by a group of researchers at Twente and elsewhere which depicts micro level processes of *niche formation*, a meso level of *regimes*, and a macro level of *sociotechnical landscapes*. We return to consider this scheme later, but here it is useful to consider the treatment of niches (Kemp 1994, Geels & Schot 1998, Kemp, Schot & Hoogma 1998, Vergragt 1988, Weber *et al.* 1999, Geels, Chapter 13 in this volume). In a technological niche, an innovation is temporarily protected from the dominant selection criteria in existing regimes. Niches are argued to be a widespread feature of development – that is, successful innovation often requires the successful creation and management of niches to allow alignment to take place. A niche may then widen or branch through reduced costs, widened applications and markets, or changes in external conditions. Conversely the technology fails if it is unable to survive when niche conditions are removed.

The concept can also be applied to social experiments – in which groups outside conventional innovation locations are motivated to develop and apply alternative technologies by a commitment to social or environmental objectives (Verheul & Vergragt 1995). A variety of factors or routes may lead to wider adoption: a niche market, latent demand, investment or takeover by commercial interests.

Niches are seen as crucial in establishing not only individual innovations but sociotechnical change at a broader level: *niche proliferation* and a new emerging *ecology of niches* may lead to a significant shift of regime, and niche conditions allow proponents not only to work on specific developments but to promote favourable conditions in infrastructure, policy and social acceptance (Rip & Schot, Chapter 5 in this volume, Kemp, Schot & Hoogma 1998, Weber *et al.* 1999).

History of technology and history and sociology of science

SST has in part been based on contributions from contextualist history of technology. While SST theorists remain critical of a number of strands of history of technology – particularly heroic and acontextual depictions of inventors – they have usefully integrated many findings and categories from historical studies of particular developments, practitioners and contexts, regarding:

- the internal structure and hierarchy of technical communities, including differences among sub-communities, and the nature of the interfaces and interactions between them (e.g. Constant 1984);
- 'reverse salients' (Hughes 1986b) and points of dissatisfaction as foci of effort and stimuli for technological change;
- other factors and contingencies that motivate technological work and how these are mediated into decisions and actions (Vincenti 1994, 1995);
- the identification and hierarchical decomposition of problems, and the extent of novelty or tradition in that structuring;
- the integration of different forms of codified and tacit knowledge, and the relation of technical work to other functions e.g. marketing;
- commonalities and differences between organisations in terms of internal structures, practices and traditions.

Similarly observations from historians and philosophers of technology on the nature and sources of technological knowledge may provide useful insights.

Work on the complex relation of technology to science – or rather that which argues further that demarcations between them are unclear, arbitrary or obsolete – also provides important insights: the role of scientific inputs particularly in radical innovations; the parallels between scientific and technological knowledge, especially in experiment and testing; and problems with the use of scientific knowledge. SST has taken on the fundamentally changed view of science from recent sociology of scientific knowledge, particularly since many researchers in SST also have a background in science studies.

Political sciences and other areas of social theory

SST draws on many other bodies of work and themes in social theory – for example,

- debates in political economy concerning market, hierarchy and network as organising principles in economies;
- analyses of discourses and representations from cultural studies;
- sociological, historical or psychological work on scientists and engineers;
- theory of social movements; and
- elements from policy studies and organisation theory.

The field thus to some extent reflects and is dependent on changes in social theorising in general, and has its own versions of debates around, for example, gender, modernity and postmodernism. Many of the theoretical issues we raise here have parallels elsewhere. In turn we argue that SST work – by tackling what is peculiar about the institutions and knowledge forms in and around technology, and especially in stressing the sociotechnical character of social forms and questioning notions of social and material effects – has raised important challenges for the whole of social theory.

It is not surprising that SST can adopt concepts from political science, given SST's emphasis on the political nature of innovation – not simply in the sense of processes in the state or public arenas, but in the organisation and operation of all the actors and sites examined, especially at a micro level the strategies of developers and the choice of priorities and objectives. Political science offers a variety of potentially useful models – from network and self-organising systems approaches to micro-interaction and behavioural frameworks – and of substantive characterisations of political systems and state interventions. It indicates sites, actors, foci and emphases that might otherwise be neglected – from processes of forming technology and related policies within government and the effects of their outputs on innovation, to the influence of interest groups on political agendas concerning controversial technologies.[7] Among approaches drawing explicitly on political theory,

Weber and colleagues use network and systems theories (e.g. Weber 1999, Weber & Paul 1999); Hård (1993) stresses the importance of conflict dynamics; Pfaffenberger (1992) proposes the concept of *technological drama* as a frame for accounts of innovation; and de la Bruheze (1992) uses models of the negotiation of problem definitions, agenda formation, 'garbage can' and bureaucratic decision-making, as antidotes to rational linear accounts.

Work on the management of technology, like SST, has shifted away from structural to processual explanations (Clark & Staunton 1989, Green, Jones & Coombs 1996). Some recent studies have placed politics more centrally, seeing organisation as the management of different interests, and stressing negotiation, choice, power, conflict, and managers' active role in interpreting and responding to environmental signals (e.g. Thomas 1996, Cabral-Cardoso 1996, McLaughlin *et al.* 1999). Again this work has overlapped with SST and been influenced by it, and can be consistent with it and useful for it (McLoughlin 1999, Chapter 4).

6 THE SHAPE AND SOCIAL CHARACTER OF TECHNOLOGIES

Technologies as situated and sociotechnical

We can summarise much of the general thrust of SST approaches as treating technologies as socially situated, in these related senses:

- technologies are produced and used in particular social contexts, and the processes of technological change are intrinsically social rather than simply being driven by a technical logic;
- technologies function as such in an immediate setting of knowledge, use practices, skills, meanings and values, problems and purposes, and objects which they act on;
- technologies in many applications are best considered to operate as sociotechnical *systems* or *configurations*;
- technological change is always part of a *sociotechnical* transformation – technology and social arrangements are *co-produced* in the same process.

The form of technologies: features, categories and anatomies

A basic requirement for any discussion of the *shaping* of technologies is adequate categories for their *shapes* – to distinguish types of technology and to describe the form and features of particular technologies. Otherwise we risk over-generalisation and arguments at cross-purposes. Confusion arises especially when people fail to distinguish between technologies as generic capabilities, components or techniques, and technologies as specific applications or arrangements of components.

Some categories should be rejected because they represent vague or arbitrary constructions, they reify particular arrangements or features as somehow essential characteristics when a technology may be configured in different ways, and they are typically used in deterministic depictions (Kling 1992a).

Fleck's categories (1988a&b) of discrete, component, system and configurational technologies are particularly useful – though they denote ideal types, and specific technologies show elements of each. Then in identifying differences between specific systemic or configurational applications, and relating these to particular shaping processes or influences, SST analyses can use a variety of schemes for describing the anatomy or architecture of complex technologies, like Kaplinsky's scheme (1984) for manufacturing automation. These distinguish the functions replaced or the work made possible by components or sub-systems. [8]

Technologies as systems or configurations

Work has continued on the characteristics and dynamics of technological systems, both in that innovation may be conceptualised as a process of system-building, and in examining the growth, operation and reshaping of mature systems (Rohracher 1998).[9] Recent work on large systems (Mayntz & Hughes 1988, Joerges 1988, La Porte 1991, Summerton 1992 & 1994, Weber 1998, Weber 1999 & Chapter 12 in this volume) has demonstrated:

- their widely varying characteristics, particularly in the form of interdependencies and feedback between elements;
- the varied forms of knowledge and communication used in controlling and operating systems, and their limitations;
- the mechanisms by which systems fail or their performance deteriorates;
- how their systemic character, and particular forms of linkage, influence innovation processes in them (cf. Hughes 1986b & 1987);
- the problems of subjecting large systems to control or change when entrenched, and the features which provide intractability (Collingridge 1980 & 1992).

Systems analyses stress that technologies are dependent on the continued availability, reliability and compatibility of other technical components, directly related organisation, and supporting infrastructure and institutions, and that they are vulnerable to changes in those.

Clearly the usefulness of the system metaphor is easily overworked and limited if it goes no further than pointing out the ubiquity of interconnectedness; the analysis must show the character and consequences of that systemicity. Some work on technological systems has tended to assume or overstate the internal coherence and self-reinforcement of 'systems', their functionality and adjustment to their environment, and hence their stability or inertia. Thus a systems perspective may fail to anticipate potentially radical transformation of technological ensembles, like that in UK electricity systems following liberalisation and restructuring of the supply industry (Winskel 1998). Insights taken over into SST from critical organisation theories, for example, show the value of understanding conflicts and incompatibilities among system elements and operating rules and rationalities.[10] There are also problems – familiar across several disciplines – from setting conceptual boundaries to a system that are inappropriate or that produce conclusions which are artefacts of that demarcation.

In contrast, then, Fleck and others have come to identify *configurational* technologies (Fleck 1988a, 1993, 1994a&b, Fleck, Webster & Williams 1990) and to argue that information technology (IT) applications are predominantly of this sort.[11] Configurations are likewise complexes of technical elements, socially and technically mediated relations and routines, and informational and physical linkages, but they are not designed and built from scratch with dedicated components and architectures, and have no clear system-level dynamic. The concept also draws attention to the substantial innovation involved in the integration of existing technologies and routines.

The appropriateness of these metaphors – the extent of the systemic or configurational character of a technology – has to be argued for each case. Both have important implications for the processes of alignment involved in creating or changing sociotechnical ensembles. Either characteristic requires firms to work to some extent within the constraints of existing technologies if developing new components or pursuing incremental innovations, or to make alliances and bring in expertise when proposing major innovations. However, the two concepts have quite different connotations for the ways such changes can be pursued and the ease of doing so.

The co-production of technology, social organisation and impacts

Technology studies has long struggled to find appropriate ways of associating social and technological change. Conventional discourses accommodate two contradictory views: that technology can be applied straightforwardly to achieve predefined goals; and that it has widespread unintended negative impacts. Both are inadequate: the former overstates the ability of controlling groups to achieve their objectives; the latter serves to obscure and excuse the extent to which outcomes are intended and required, or could have been foreseen and avoided. SST has to transcend these: to reinstate intention, interest and responsibility, but also to acknowledge the emergent and unpredictable nature of sociotechnical change.

SST has rejected notions of technology and society as separate but interacting spheres, and coined terms that stress that technology and social arrangements develop together as part of the same process, and that technological entities are always combinations of social and technical elements (Bijker & Law 1992a, Berg & Aune 1994, Williams 1997a, Leonard-Barton 1988).[12] Technology and organisation, cultural forms, values, identities, etc. *co-evolve*, are *co-produced*, or are *mutually constitutive*. Technological change, then, is always part of a larger *sociotechnical* transformation. Hughes' metaphor (1986a) of a *seamless web*, Bijker's (1993) *sociotechnical ensembles*, *sociotechnologies* (McLaughlin *et al.* 1999), and *sociotechnical landscapes* all indicate the hybrid character of technological developments or their contexts.

To the extent that outcomes depend on intention, the elements deployed and the knowledge required to do so are *heterogeneous* (Law 1987a&b). Developers may seek at one extreme to fit an innovation into existing institutions, practices and expectations, so as to minimise uncertainty and ensure acceptance and smooth implementation. At the other extreme they may undertake extensive reshaping of conditions – creating markets, configuring users, obtaining infrastructure – to allow the successful introduction and operation of a radically new technology. Social effects thus depend on the way that particular impacts are sought or avoided by the actors involved.

Technological and social change, however, are never fully planned and predicted; they are subject to frequent setbacks and failures and emerge in the course of local struggles to produce a working technology and accommodate it in its use setting. The extent to which a technology achieves a dominant group's objectives for it, or furthers its interests, is at least in part an achievement, possibly against the actions of users and others.

SST has thus moved away from a notion of 'impacts' – simple consequences of a technology, determined by its character (Kling & Scacchi 1982) [13] – to one of changes produced in a variety of ways in the shaping of sociotechnical entities. It can retain an analysis of interests embodied in a technology – and hence assert the possibility of different design objectives and priorities (Wynne 1995) – without assuming a simple relation between interests and outcomes. Rethinking the 'impacts' of technological change in this way helps resolve some of the problems with conventional notions of 'unintended consequences' and thus has important implications for regulation and other attempts to challenge and control technologies.

The thrust of the argument so far is that the impacts which are intentionally pursued or unintentionally emerge depend on local conditions –

in contrast to the modernist view of a neutral technology having universal applicability and consequences. Clearly, however, many technologies are produced in generic forms to allow them to be applied widely. Many are transferred successfully and largely unchanged and some do contribute to a homogenisation of organisation and culture. To what extent, and how, technologies can produce a standardised domain requires an understanding both of local processes of appropriation and use, and of globalising forces – standardising scripts,[14] constraints embedded in the technology, economic pressures, uniform procedures promulgated through education and training, accompanying discourses of 'imperatives' – and of the tensions and complementarities between them. These interact in different ways in specific episodes of adoption, and patterns emerge through many such local contestations (Williams 1997a).

The material and the social

SST analyses have not only pointed to the hybrid social and technical character of technological activities, practices, knowledges and assemblages, but have also questioned distinctions between 'technical' and 'social' phenomena. They have sought terms which highlight the conventional nature of these labels and implied boundaries, and their questionable connotations.[15]

The discussion has opened out into fundamental questions about materiality and social action (McLoughlin 1997, Grint & Woolgar 1992, Kling 1992b, Law & Mol 1995):

- whether there is an irreducibly technical character to artefacts;
- how we should conceive limits to the malleability of technical forms;
- how and to what extent material components help systems to be durable and reproducible in space;
- whether artefacts have material effects not explicable as socially constructed properties attributed to them; and
- conversely whether current sociological concepts adequately treat technical mediation of social relations.

The ANT principle of treating human and non-human 'actants' symmetrically – in effect giving the latter agency – though now not widely accepted,[16] has been a particularly provocative and stimulating position on these issues. Similar questions appear, though, wherever there is a theoretical move to stress choice and active construction, and particularly when this tends toward voluntarism – as in strategic choice management theory (McLoughlin 1999). Technology first seems to slip out of the analysis altogether, as if social relations explain everything. When it is reintroduced, there is a danger that technological capabilities and characteristics are taken as inherent properties.

Technologies clearly do help order activities, meanings and relations, and it is important to retain a conception of those roles. As the ANT notion of *delegation* emphasises, artefacts can mediate relations of control and help enforce standard behaviours. The use of physical devices as instruments for controlling or standardising human behaviour has been analysed in a number of areas, notably workplace automation and military and security technologies. Other work in SST, however, stresses flexibility in the way users interact with technologies and in the meanings attributed to them. It shows users resisting or subverting prescribed patterns of use, and questions whether material artefacts can achieve determinate effects. While most SST researchers attempt pragmatically to accommodate both options, there is not yet a simple formulation that resolves this tension.

Figure 3.1 Implications of Section 6 - the sociotechnical, situated & systemic character of technologies for policy and intervention

Simply recognising that we are dealing inevitably with social influences and social criteria of choice throughout technological innovation processes – even in the early stages which are primarily the domain of technical practitioners and often assumed to be governed by purely technical logics and practices – greatly opens up the scope of possible action for policy-makers. The ramifications of this shift of perspective run throughout the policy insights we outline from here on. It points to:

ß a broader range of mechanisms and points of intervention;
ß the diverse forms of knowledge and disciplinary expertise needed for both development and intervention;
ß the key role of communication channels among groups involved in development;
ß a broader range of criteria for evaluating the success of projects & interventions.

It also implies a wider range of social groups may legitimately be involved in conception, design and development.

The complex character of much technology implies a shift of focus in management and policy, even when considering individual artefacts, to the system or configuration level and to the infrastructure supporting a technology. It means we must pay attention to the specific and varied character of the links between components and different possible means of influencing them. These vary between 'systemic' and 'configurational' technologies, with important implications when considering measures to intervene in all processes and stages of innovation:

- in considering support for new technological components – e.g. by identifying bottlenecks or 'reverse salients' which require a focus of innovative effort;
- in following the ramifications of a targeted effort or breakthrough; and
- in identifying opportunities and constraints in controlling or redirecting existing technologies where they have become entrenched.

All this points to the value of being able to respond flexibly and to redeploy resources within the complex entity as the problem focus shifts.[17] The distinction between systems and configurations, however, indicates quite different possibilities for variety and reshaping in complex technologies.

Our understanding of the co-evolution of technologies and social forms shows that treating technological development and the occurrence of 'impacts' as separate processes is severely limiting. It highlights the need to integrate policies and programmes for innovation with those for evaluation and regulation. The emergent and unpredictable nature of sociotechnical transformations points again to the value of flexibility and constant monitoring, maintaining channels of communication and arenas of debate, and avoiding disincentives to open appraisal. Strategies for combatting entrenchment imply that at least for public interests, flexibility and controllability are to be accorded a premium in evaluations of technology.

The new perspective also stresses that the changes emerging from the introduction of a technology are different in different conditions, and that the same policy instruments may likewise operate differently. Intervenors must expect and allow for changes not only in the form and features of the developing technology, but also in the interests and objectives of groups as the technology is established and diffused, as social rearrangements take effect and associated changes become apparent.

7 THE COMPLEX, CONTINGENT, LOCAL DYNAMICS OF INNOVATION

The complexity, contingency and unpredictability of innovation

Recent research and theorising in SST has produced a wealth of accounts of development processes across a wider range of technologies and contexts. It has yielded many more detailed models of the events, dynamics and types of player in technological development processes, and many more terms for the specific phenomena identified. This much expanded coverage has demonstrated the variety of:

- divisions and articulations of design, development, testing, and other activities in different industries, organisations and technology areas (e.g. Downey 1995, Law 1992b);
- loci of innovative activity;
- practices, cultures and divisions of labour;
- influences on innovation practices and decisions;
- industry structures and relations with users.

SST has tended to stress the dependence of technological outcomes on these local and detailed features.

SST represents a critique of the linear models of innovation that until recently underlay much technology policy discourse (Russell & Williams, Chapter 4 in this volume, Tait & Williams 1999) – not only the simple assumption that technologies are conceived for a purpose, built and introduced, and thereby straightforwardly achieve it, but also in the depiction of a sequence of distinct stages: typically invention, development, design, testing, refinement, implementation and diffusion. Even the 'linear plus' models identified in some more recent policy initiatives (Tait & Williams 1999) remain inadequate as descriptions of innovation and a misleading basis for policy: they still fail to capture the overlap, interaction and different ordering of activities, and the variety of sources and inputs and the multiple relationships thus entailed.

In contrast, the revelation in SST research of complex influences, unpredictable courses of development, multiple sites of innovative activity, and in particular, extensive innovation during configuration and appropriation, has led to the overarching narrative frame of an *innovation journey* and notions of *distributed innovation*. Development often shows a meandering and branching path so that the emerging entity is quite different to that anticipated and planned. Technologies do not advance steadily along a self-evident trajectory of improvement, nor are they transferred unchanged across settings. As ANT analyses stress, as more actors become involved in the development, so it diverges from the original concept. There may be a complete transformation of the idea even before it attains material form (Irwin & Vergragt 1989). The final outcome as implemented may be quite different to that conceived at the outset. Further cycles of innovation may follow as the technology is adopted and diffused.

Alignment and stabilisation: building networks or constituencies

SST has substantially developed its analysis of the ways technological developments are stabilised into particular forms, and the accompanying processes of alignment – the orientation of actors to contribute, the shaping of the development, and the modification of the actors and their interests on the way. In variants of ANT it involves the formation of a network through the enrolment of actors and the translation of their interests. In the theory of sociotechnical constituencies, it involves building an alliance of interests around a project at an inter-organisational level (Molina 1995, 1997 & 1999).[18] In the SCOT framework it is the process of achieving closure among relevant social groups over the meaning of an artefact.[19] Disco and van der Meulen (1998b) stress the variety of modes of *coordination* involved in establishing and maintaining such alignments.

All these analyses tend to stress the protracted process of alignment, the difficulty of achieving technological change and the extent to which it is an *accomplishment* (Badham, Couchman & McLoughlin 1997). Technologies emerge as unreliable strategic vehicles for even those nominally directing innovation. Information and communications technology (ICT) systems in particular are widely seen as having 'failed to meet the expectations generated in organisations by technologically-driven visions' (Williams 1997d). Outcomes are much less determinate and much more of a compromise than, for example, early labour process studies would have predicted; industrial technologies are no longer viewed as simply reflecting management objectives and values.

Some form of network theory remains popular for analysing alignment, allowing us to conceptualise not only the emerging stability and resilience of a new arrangement, but the resistance of existing arrangements to change. The notion of network highlights a mode of coordination of activities distinct from markets and hierarchies (Freeman 1991, Rammert, Chapter 6 in this volume). Representations of networks can form a basis for examining: features across different cases and their correlation with innovation outcomes; discourses, information flows, patterns of communication or negotiations at interfaces between organisations; different sorts of link which are mobilised; and the sources of inputs to innovation (Conway 1995, Steward, Conway &

Overton 1995). They can provide useful insights into the limits of change in existing alignments and the way in which these may need to be broken up to reorient a field (Elzen, Enserink & Smit 1996). Thus when networks are new, small and struggling to get started, they can and must adapt, so that the form of technology can and must change. As the network grows the entity becomes more stable, especially as it takes on a material as well as conceptual form, as supporting infrastructure is put in place, and as identities become more firmly defined in relation to it (Sørensen, Chapter 2 in this volume).[20]

Analyses of alignment often highlight the political and conflictual character of development processes – the existence of choice, the different power of actors, and the need to coordinate diverse activities and cultures (Jørgensen & Sørensen, Chapter 7 in this volume). Groups try to direct the innovation as they see fit, closing down options, black-boxing some elements and reopening others.[21] ANT in particular, concentrating on the machinations of system-builders, identified various strategies and objectives, and later studies have greatly expanded the repertoire.[22] Recognising the importance of appropriate markets, infrastructure and regulation for the success of a development, and their position in relation to competing and supporting technologies, developers seek collaboration in, for example, developing standards, building demand, and sharing costs and risks.

These processes can display complex mixtures of competition and cooperation. In IT, change is so rapid and technologies so complex that no single actor controls development of an entire field. There is fierce competition between firms over particular markets but at the same time a concerted effort at a broader level to produce a unified vision of an ensemble of technologies. The effect may be to set controlled conditions in which competition takes place – perhaps the standardisation of one aspect of the system while allowing a proliferation of products in others. The phenomenon points to a multi-layered model of how interests in the sector interact, and a careful dissection of the advantages and disadvantages to any one of them of particular strategies (Williams 1997d).

The emerging sociotechnical entity thus attains some degree of obduracy or irreversibility. For network analyses, this results from increasingly stable interdependencies (Callon 1991). Other frameworks accept explanations from elsewhere in innovation studies: mechanisms of *entrenchment*, *lock-in* or *path dependence* in which features of the technologies or their settings provide positive externalities or tend to exclude competing technologies (David 1986, Collingridge 1980 & 1992, Arthur 1988 & 1992, Knie 1992). Such stabilisation mechanisms can be a major obstacle to change. The exclusionary effects of entrenched technological forms may require strategies to protect embryonic competitors e.g. by creating *niches* (Rip & Schot, Chapter 5 in this volume, Geels, Chapter 13 in this volume).

Clearly then, the stabilisation and entrenchment of a technology, however we explain it, are crucial to successful innovation. They also, however, entail foreclosing on certain options and thus giving up flexibility to deal with changing requirements. Premature decisions or unduly specific visions can lock developers onto particular paths which can prove disastrous if circumstances change (Collingridge 1992, Howells & Hine 1993).

We must also examine the converse – the way formerly stable arrangements are reopened and trajectories destroyed – and the tensions and balance between the two tendencies. Work following from the SCOT approach, for example, has refined our understanding of closure around particular interpretations of the technology and its function, of test results or other appraisals, or of symbolic meanings in its appropriation (Bijker 1995b, McLaughlin *et al.* 1999). It implies that closure is a matter of degree, does not occur definitively at one point, and is more or less temporary (Rosen 1993). It is not so much settling on a common meaning as a workable 'meeting of different positions and perspectives' (McLaughlin *et al.* 1999, Chapter 8). Fundamental questions about the direction and suitability of the technology may be reopened later; controversies can be revived; meanings and evaluations can be challenged again; and boundary objects can be unpacked. The same observation applies to stabilisation more generally: stability is relative and temporary.

Indeed some analysts question the usefulness of the stabilisation metaphor for other than relatively isolated artefacts. They argue that certain technologies never display an identifiable envelope or trajectory, but are subject to continual significant reformation in local circumstances and perpetual shifting in their overall character (Grint & Woolgar 1992). The substantial change in many technologies as they are appropriated, or the way they are launched as vague concepts and only given definite form as they are applied, implies that we cannot depict a single stabilisation point before their launch. Such stabilisation as does occur in one technology may produce a 'cascading' process of reconfiguration in other areas (Fleck, Webster & Williams 1990), and stimulate innovation in competing technologies. This work finds lessons in examples, like multimedia technologies, of extreme fluidity and uncertainty, rapid change, conscious experimenting, extensive innovation in use, and an absence of central control. The degree of stability or fluidity, then, has to be demonstrated and explained for particular areas of development and particular times.

Standardisation

A number of SST and similar studies focus on standardisation as an important form of alignment and as a process of competition among visions or early variants of a technology (Law 1986b, 1987b&c, Fleck 1988b, pp.41-45, Berg & Schumny 1990, Cowan 1992, Schmidt & Werle 1992 & 1998, Egyedi 1996, Cawson, Haddon & Miles 1995, Williams 1997d). Standardisation takes various forms: from prior agreement of the required characteristics of a technology through public fora in *de jure* standard-setting, to emergent *de facto* standardisation around a specific component, operating procedure or form of infrastructure. Motives and pressures vary between settings, but the large costs of having a technology become obsolete encourage prior negotiation and collaboration.

Standards generally serve to cut diversity and to create order and compatibility, but there is usually a complex matrix of standards at different levels – in IT, for example, basic component, application, platform and system – which provide a mix of constraint and autonomy. Thus the precise role, effects and significance of a standard can vary. It may in some respects be exclusive, and thus for a particular firm or consortium the basis of a monopoly position. A standard however usually relates only to one component or aspect of operation, or a particular interface among machines, materials and humans. It may simply ensure interoperability. Any one standard still typically permits different ways of building technologies around it; technologies may be designed to accommodate competing standards; and translation devices can be created to mediate between them. Standards clearly influence the application and use of ITs, but they need not, and generally cannot, be made to encompass those uses.

Studies of standardisation have examined the strategies:

- of developers to fix a standard that suits them;
- of others to reinforce it, or to undermine it – to establish a different standard or to maintain diversity;
- of other parties, like suppliers of complementary components and services, to negotiate its forms and terms;
- and of regulators or other public agencies to impose or resist standardisation.

Standards 'games' or 'wars' have thus become complex, and the fluidity and dynamism in some areas of IT in particular makes any form of central control difficult and likely to fail (Williams 1999).

Visions

Visions are important resources by which developers and proponents – and opponents – seek to mobilise and shape expectations and commitments around new technologies. A variety of other terms have been coined: *scenarios*, *design concepts*, *product spaces*, *poles of attraction*, *framing objects* (Sørensen, Chapter 2 in this volume, Geels & Schot 1998, Rip & Schot, Chapter 5 in this volume, Williams *et al.* 2000, Jørgensen & Sørensen, Chapter 7 in this volume). A vision includes a view of not just the form and features of the technology, but also its functions and benefits, and a new sociotechnical order in its domain of application. It therefore entails particular assumptions about markets, users and use practices, supplier-user relations, costs, and so on – and hence a model of the wider social context, and values and commitments concerning desirable futures.[23]

Visions serve many purposes within and outside the developing organisation (Sørensen, Chapter 2 in this volume, Geels & Schot 1998, Molina 1999):

- in orientating potential partners, suppliers of inputs and complementary products, intermediaries, users and wider constituencies;
- in persuading them of the feasibility and desirability of the innovation, especially in promising to solve a recognised problem or fulfil a common objective;
- in gaining commitments to participate, and building alliances or networks;
- in guiding and giving coherence to work within a firm or project;
- in testing and refining ideas for the technology and its use in a less costly way than experimental release;
- in warding off competitors and pursuing standards;
- in raising user expectations, creating demand, and shaping advantageous wider discourse; and
- in enlisting supportive state policies.

A vision typically deploys metaphors which simplify choice. It is likely to start out vague – perhaps to the advantage of the developer – and then as other groups are enrolled it shifts and sharpens. It becomes an important element of the *protected space* for development, allowing participants to suspend questioning of its basic premises.

Thus key moves effectively foreclosing options in advance are made in the 'virtual space' of visions before there is significant material development (Jørgensen & Sørensen, Chapter 7 in this volume). Visions may be contested, or they may stimulate alternative proposals. For consumer products in particular, intensive activity takes place among groups of producers while product ideas are negotiated and evaluated (Cawson, Haddon

& Miles 1995, Chapter 3); there are flexible interpretations and competing designs in a *product space* within one producer unit, and different and overlapping product concepts proposed by rival developers in the industry, or cutting across related industries.

Visions are particularly significant – and explicit – for groups developing alternative technologies outside mainstream sites of innovation (Verheul & Vergragt 1995). The social visions which inspire initial development, however, may become submerged in other agendas. A technology may be supported initially because of the social objectives identified with it and because it is consistent with particular values and practices. But these are not necessary and inseparable features of it. As the technology is modified, as other interests shape its form and uses, as it is absorbed into mainstream production and use and economic logics takes over, it may achieve quite different goals, benefit different interests, and have different effects, to those envisaged.

Figure 3.2 Implications of Section 7 - the complexity, contingency and dynamism of innovation processes for policy and intervention

SST studies have put paid to the decision-maker's dream of effective central control of technological direction. Interventions must address the difficulties of attaining technological objectives, with frequent failures, unintended outcomes and unstable consequences. Contingencies in innovation processes make it difficult to produce generally valid prescriptions; policies and instruments may need to be tailored for specific technologies, industries and areas of application. A catalogue of characteristic 'constituency structures' and 'innovation journeys' might be compiled for various generic technology and sector combinations, but much depends on specific contexts. Though this more complex view of technology decision-making seems daunting, SST has produced crucial insights for guiding the more subtle and varied interventions needed to stimulate or redirect development:

1. While simple prescriptions for technology management now seem futile, a *processual* view points to important tactics in building new technologies, sociotechnical constituencies and markets, and the circumstances in which these should be effective. Analyses of alignment processes provide plenty of lessons in how to facilitate them – or if need be obstruct them – and judge their success.[24] The work reinforces some general requirements for design and development like aligning offerings to international standards and catering for diverse markets.
2. There may be crucial points in innovation processes at which outcomes are more malleable; participants more receptive; and intervention most effective – and conversely phases at which intervention is more difficult or simply too late.[25]
3. The meandering character of innovation journeys, Rip (1995) argues, translates into advice to developers and managers not to go into a 'head-on attack'. Letting the development take a 'sociotechnical detour' from the planned course may avoid powerful opposition or a fatal obstruction and enable at least a substantial part of their original aims to be realised.[26]
4. Formulating a project and creating favourable conditions entail balancing the advantages of a definite vision against allowing for later choice and change. Although maintaining flexibility may be costly, too particular a vision, a premature decision between options, or too narrow a focus on specific technologies, can itself prove disastrous if circumstances and perceived user requirements shift (Collingridge 1992, Howells & Hine 1993).
5. Flexibility may best be maintained not within one project or firm but at the level of an industry, by providing conditions for multiple developments which compete later. Attempts to define and support a *single* path of development at a policy level are likely to be misguided.[27]
6. To help avoid premature alignment around unrealistic visions, it is valuable to involve a wider range of actors [Figure 3.5], and provide for diverse views or counter-intuitive ideas. Public policy objectives may of course warrant introducing a wider range of assumptions and objectives into the articulation of a vision than those of the developer.
7. SST work suggests direct confrontation with entrenched technologies is likely to fail. In contrast, relatively modest changes of circumstance may significantly alter relationships between actors and the technical options chosen [Figure 3.3].

8 CONTINUITY AND PATTERNING

Structure and constraint in innovation

While one current of SST work has stressed the contingency and unpredictability of sociotechnical change, and the plurality of actors involved, another has continued to focus on structure and constraint rather than dynamics and choice. This work has sought concepts and forms of analysis which attempt to capture the regularities and patterning in technological development, and the cognitive and institutional structures and mechanisms that produce them – that favour specific outcomes and exclude others.

Concepts used to describe regularity in the development of technological devices or systems include:

- *trajectories* of change according to particular criteria of assumed improvement;
- *envelopes*, *avenues* or *corridors*, depicting boundaries of form within which variation takes place;
- *product cycles* in which the type of innovative activity varies according to the stage of the stage reached.

The terms *paradigms*, *frames*, *traditions*, *styles* or *regimes* convey with varying emphases the social, political, legal, cultural or cognitive means by which those directions and constraints are produced and reproduced.[28] These notions stress the routine and rule-bound character of much technical work, and suggest that options are ruled out before they are conceived – a perspective supported by empirical studies of design processes demonstrating that typically few alternative concepts are ever evaluated (Cawson, Haddon & Miles 1995).

At a more specific level, analysts identify elements of a frame or paradigm: *guiding principles* (Elzen, Enserink & Smit 1996), *norms* (Hameri 1994), *roles*, *product standards*, *production routines*, *use practices*, *instrumentalities*, *dominant problem definitions*, *do-able problems* (Fujimura 1987), *focusing devices* (Rosenberg 1976), *guide-posts* (Sahal 1985), *modes of satisficing*, and other forms of rule or local regularity in practice which make up the larger patterns.

These concepts attempt to account for stable paths of development which cannot be attributed to conscious choice or formal evaluation. They indicate ways in which practitioners cope with multiple uncertainties and complexity of choice. They also represent means by which broader social influences are mediated into technological development, and economic and political selection criteria are embedded in technological practice. They can therefore in turn contribute to explaining patterns and characteristics of technological forms across industries or areas of technology.

These two aspects of SST work have developed in tension, and have overlain deeper theoretical differences – particularly over the use of agency and structure as explanatory resources, and over the relevance of analyses of the contexts in which innovation takes place.[29] This has of course been a core debate in social theory, but the issues have been particularly contentious in technology studies, where it is argued that explaining major sociotechnical transformations requires radical theoretical innovations.

Depictions of regularities face several criticisms: that allocating innovations to categories is contestable and *post hoc*; that the extent of patterning is easily overstated and in many areas huge variation and rapid change mean that no trajectory or envelope can be identified (Fleck 1988b, Fleck, Webster & Williams 1990, Bijker & Law 1992b&c); and that such concepts imply determinism or automaticity. The social structural explanations favoured in this work have been accused of reductionism (Bijker 1993) – that they read off outcomes from a limited range of unsuitably blunt social categories, disregarding intervening influences, the local complexity of identity and interest, and their transformation in processes of alignment.

On the other side, the stress on fluidity and contingency has raised concerns:

- that such analyses cannot explain such patterns and continuities in sociotechnical practice and outcomes as can be observed;
- that the malleability of sociotechnical arrangements is overstated and the possibilities available to system builders are seen as unrestricted;
- that the accounts are ultimately limited to complex descriptions; and
- that outcomes are seen as products solely of interactions among groups, or relations within a network – that the accounts cannot show how patterns in technologies reflect broader social structures, contribute to their maintenance, or may lead to their transformation.

Thus some early SST work was criticised (as noted in section 3) for putting too much emphasis on local or micro levels while either failing to situate this action in a broader context or denying the need to do so; and for its voluntarism, depicting society as exhaustively constituted of active choices, while implicitly adopting specific contestable notions of social structure (Wynne 1995; cf. Russell 1986, Rammert 1997, Russell & Williams 1987).

While the debates are no longer as heated and the positions as polarised, the basic theoretical questions remain:

- whether we can or should apply prior conceptions of social systems, and how these systems shape or pattern local action;
- whether we need to, or can, reconcile analyses at different levels of social aggregation, or whether, for example, analyses of network-building obviate distinctions of scale;

- how we can best theorise the processes by which some locally produced outcomes are translated into substantial changes across a wider domain, and others fail to do so.

Balancing regularity and contingency

Much recent SST work has in different ways aimed to get the best of both explanatory worlds and balance these emphases: retaining the complexity or fluidity of accounts of interactions, and showing the extent of contingency and acknowledging diversity or uniqueness; but at the same time accounting for regularity and continuity. While it acknowledges the role of technological change in bringing about transformations of some part of the sociotechnical terrain, this work sees existing social systems influencing particular technological outcomes. Thus it insists that patterning in technology does not come from an aggregation of random or wholly contingent decisions and outcomes, but is structured by its context (Rammert 1997, Wyatt 1998).[30]

A large body of work in SST, and several of the approaches represented in this book, now assert the importance of analysis at micro-, meso- and macro-levels. It aims to find:

- different sets of concepts and categories appropriate to each level;
- consistent forms of explanation at each; and
- means of explaining the coupling between levels.

The work varies considerably, however, in the theoretical schemes it has borrowed from or reworked, the concepts it uses, the processes and structures it focuses on, and the substantive accounts of institutions and processes of change it accepts.[31]

In practice this work is able to switch comfortably between the language of agency and that of structures or systems. But the reconciliation has been sought not only pragmatically in handling case studies, but also in theoretical formulations which transcend the conflicting positions. We sketch some of them here.

Whether it is reasonable to identify a trajectory or envelope, say, remains a matter of evidence and judgement, and the extent of patterning is understood to vary between areas of technology. Acknowledging a trajectory need not imply that it will continue – that the concept has any predictive value. Any apparent momentum, of a trajectory or an individual project, has to be explained in terms of mechanisms of maintenance or entrenchment (Bijker & Law 1992b&c). Patterns of practice and outcome are thus not seen as immutable: we must account for the processes by which they emerge, and are institutionalised, challenged, defended, modified or overthrown.

Nor need the analysis be deterministic and static. Context is not a fixed 'backdrop' against which action takes place, nor does it form fixed limits within which there is supposedly free choice.[32] The analysis must still examine the extent to which organisations need active managing, patterns of belief need argument and reassertion, and institutions are challenged and changed in the process of their reproduction. It can still be consistent with ideas of co-production – allowing for existing structures being subverted or overturned without having to forego any notion of prior structuring.[33]

None of this structuring of locality need be depicted as determinate or easily predictable: there are always complex chains of influence through the layers of context. Actors interpret rules and draw on resources in different ways; they engage with economic, political and discursive systems in the process of recursively recreating and modifying them, rather than their actions being determined by them. It may still be justifiable, however, to bracket some effects of social systems as structures confronting actors with different sets of constraints and opportunities.

A study should examine economic, political and cultural structures and dynamics, and can draw selectively on general analysis from the corresponding discipline, but without treating them as separate 'factors' or somehow institutionally separate spheres. It has to acknowledge that technologies are implicated in all of them – that systems at all levels are inescapably sociotechnical, rather than social forms simply providing 'context' for technologies.

Clearly changes in the broader sociotechnical terrain, brought about through processes unconnected with the particular technology in focus, may be fundamental in shaping it – that is, we cannot understand everything concerning the origins and development of a technology in terms of local action (Russell & Bunting 1997). Dramatically changing contexts are a challenge to any macro level analysis used in SST studies (Sørensen & Levold 1992).

It is the meso level where there has been a meeting between approaches examining how sociotechnical entities are built as groups interact, and those seeking a more refined institutional analysis starting from broad features of a social formation. Some early SST theorising tended either to neglect the specific character of the meso level institutions important for technology, or to import questionable assumptions about them (Sørensen & Levold 1992). SST researchers differ widely in the concepts and substantive analyses they use in this middle ground, the features they give prominence, and the institutions they bracket as forming a relatively stable context or a coherent locus of technology. They draw, for example – with reservations and adaptations – on general analyses of *national innovation systems* or *styles*[34] (Lundvall 1988, 1992 & 1998, Nelson 1993, Wijnberg 1994, Green, Walsh & Richards 1998, Guerrieri & Tylecote 1998), of firms and markets (Dosi 1988, Dosi & Orsenigo 1988), and of political and legal systems as they affect policy and regulation. Mappings of inter-organisational networks are useful where relations are diverse.

Dutch researchers have developed one such multi-level SST framework (Rip & Kemp 1998, Rip & Moors 1999, Rip & Schot, Chapter 5 in this volume, Geels, Chapter 13 in this volume): at the macro level it depicts a *sociotechnical landscape*; at the meso level, a mosaic of *regimes* at different stages which set crucial conditions for innovation; and at the micro level, local search practices and processes in technological projects, and the creation of *niches* – though it might equally well accommodate analyses of alignment in other terms. The group has been examining regime change in terms of the coincidence and coupling of changes at different levels, and particularly how clusters of niche developments could be fostered to achieve a substantial shift of regime towards environmental sustainability (Kemp 1994, Schot, Hoogma & Elzen 1994). Niches are conceptualised in relation to surrounding conditions set by relevant regimes, and activities like standardisation can be analysed as being both required by a regime and contributing to it (Deuten, Rip & Jelsma 1997).

Other theorists have suggested concepts to connect patterns and rules of micro-level technological practice and choice to broader social structures, for example: *orientation complexes* (Weingart 1984), *codes* or *encoding* (Mollinga & Mooij 1989, Feenberg 1991, Mackay & Gillespie 1992) and *structural filters* (Rammert 1997). In specific cases of technological development these concepts suggest how influences on practice or choice are mediated, though the notions still have to be given specific form: chains of influence have to be traced convincingly, and the existence of particular sets of rules and their institutionalisation in particular forms have to be demonstrated. The usefulness of the concepts, however, is perhaps more in explaining how patterns of practice help reproduce dominant institutions and disadvantage technological options which would challenge them.

One broad strand of SST theorising, then, has sought ways of reconciling and articulating different levels of analysis. Another, by contrast, has retained a strong scepticism towards the very idea, maintaining that the notion of levels has been used loosely; that it interferes with explanation of how sociotechnical systems and the interests around them are constituted through interaction; and that significant sociotechnical transformations, where local achievements are translated into something much more widely significant, are precisely those in which previous hierarchies are overturned (Sørensen 1998a). Thus much SST work continues to follow processes of network building on the lines of ANT, without presuppositions about the character, interests and power of participants.

Figure 3.3 Implications of Section 8 - analyses of stability, continuity and patterning for policy and intervention

Focussing on the patterns and continuities of sociotechnical practice which tend to produce particular outcomes, rather than on the purposive processes of choice, negotiation and alignment involved in innovation, draws attention to an altogether different set of considerations for policy-makers and other intervenors. A multi-level analysis of the sort outlined here can be particularly fruitful.

This emphasis, and concepts like paradigm, routine and style, demonstrate how limits to technological choice are built in from the start – how particular objectives and options are favoured and others are not even articulated. Reconstructing the implicit rules and exclusions, and the premises on which they are based, should show interveners how they could contest existing assumptions, introduce other criteria and considerations, and explicitly formulate alternative possibilities for development.

Understanding the layers of influence that shape a development, and the attempts of key actors to change conditions in several of those, points correspondingly to the need for intervention at a range of levels or distances from the immediate situation of the technological activity. A minor adjustment to a technological development, an incremental innovation along an existing trajectory may be pursued through a change in relatively detailed practices or immediate conditions – intervening at the micro level of the niche conditions – without requiring the more substantial changes in the broader setting and incentives for innovation that might be called for with more radical technological change.

Head-on confrontation with an entrenched technology will rarely be successful. In the short-term, a new challenger will be unlikely to prevail. Strategic niche management may be needed to allow the initial development of a competing technology. More importantly, relatively modest changes in conditions could destabilise and reorient existing regimes by inducing a reconfiguration of actors – as illustrated by the work on the development of cleaner production (Clayton, Spinardi & Williams 1999).

Substantial re-alignment of a regime may thus come from a steady build-up of innovations challenging the established pattern. It may nonetheless require direct intervention to bring about favourable changes in the broader sociotechnical landscape – to a range of sociotechnical institutions and infrastructure. Like the modifications of electricity markets and regulatory regimes to encourage renewable energy and conservation technologies, these changes may represent modifications to more fundamental restructuring being pursued for other reasons.

Likewise, confronting institutions head-on in a concerted attempt to restructure them may be politically unfeasible and produce unwanted outcomes. More piecemeal, gradual and indirect strategies may be more effective, opening up opportunities for progressively more significant change to conditions. The experience of niche management demonstrates that rapid shifts may be possible as positive reinforcements take effect, as costs drop or as critical market sizes are surpassed.[35]

9 APPROPRIATION AND USE

Extending the focus of SST downstream

Recent SST work has given much greater attention to:
- processes of acquisition and appropriation in specific contexts;
- the extent of innovation as technologies diffuse and are used;
- the relations and strategies of developers and users in shaping the locus and control of innovative activity;
- the articulation of supply and demand sides;
- the connection between innovation and the public evaluation and political steering of technology (this last set of issues is taken up in section 10).

Again these concerns have long been present in technology studies:
- in calls for research on consumption (Cowan 1987);
- in studies of implementation in work organisations (Rhodes and Wield 1994a);
- in a focus on incremental innovations and their cumulative significance (Rothwell & Gardiner 1985);
- in arguments on the local character of technological knowledge;
- in notions like *innofusion* (Fleck 1988a) which criticise depictions of adoption and use as passive processes, and of diffusion as spreading static forms of device or technique;
- in feminist analyses of the renegotiation of identities and work practices around new technologies (Cowan 1976 & 1983, Cockburn 1981 & 1985); and
- in SCOT analyses of user groups and their influence on design.

Recent work, however, has greatly developed and consolidated this focus.

Selection and acquisition

SST work has started to draw on general sociological work on consumption.[36] While it is important to recognise the differences between consumer technologies and technologies in industrial and other settings, the differences can be overstated, and analyses of the two can learn much from each other. The boundaries between business and consumer markets in ICTs are no longer distinct, and much 'consumption' is for work in the home, a site of production and consumption of goods and services (Silverstone and Hirsch 1992).[37]

Acquisition is much more complex than the conventional picture of replacement by a self-evidently better technology. Users, or those purchasing technology for organisations, are guided and constrained by complex mixes of acknowledged objectives, priorities, criteria and perhaps regulatory constraints, as well as values, pressures, images and associations. Purchasers in all areas more or less overtly balance cost with a variety of attributes – functionality, labour-, cost- and time-saving, convenience, complexity, skill level, quality, safety and reliability. However, the requirements, criteria, assessments, and weightings given to each, are variable and contestable. Choice is both informed and constrained by knowledge, experience, availability and relations with suppliers. And even in organisational settings, and within explicit strategies for technology acquisition and management, choice is determined less by formal evaluations than is projected, and more by tradition, style, symbolism and other cultural influences. In organisational settings, of course, groups using a technology are involved to varying extents in its selection, but frequently not at all.

Users are not limited to choosing between available options in a product class, but may envisage altogether different technologies for performing a task. More consumers appear to be questioning the sources of commodities and the impacts of production processes. New markets are allowing them to choose on grounds other than price, functionality and quality, and in some cases to have an influence over the production route. This is in part an effect of wider social concerns – for example, animal welfare in farming or chemical testing, or sustainability in energy production. It is also reinforced by producers and retailers encouraging niche markets and having to educate consumers about the differences.

Appropriation and use

Beyond acquisition lies a process of domestication (Lie & Sørensen 1996): embedding a technology in practice, getting it to work adequately, making it usable, making sense of it, and evaluating it. This appropriation takes place at two levels: an individual or local level of assimilating the technology, adjusting practices around it, and according it value – entailing, as Williams (1997b) points out practical, symbolic and cognitive dimensions[38] – and broader social processes of accommodating the technology and judging its acceptability.

While the ease of introducing and adopting technologies varies greatly, in all settings it is a more or less protracted struggle. It is probably simplest for incremental changes and discrete new technologies, and massively complicated for major systems or configurations. It is not simply a technical process of adjusting the technology, nor in managerial terms one of straightforwardly marshalling resources and organising work around it. It is complex and inherently political: it depends on the initial roles, commitments, identities, knowledge and expectations of a range of groups and individuals, and entails change to those.

In work settings appropriation depends on interactions across various divides: management and workforce, professional and occupational groups, gender divisions of labour, functional divisions in an organisation, and perhaps different organisations. The attempts of managers and implementers to enforce standard practice and limit flexibility are typically in tension with the need to involve users in getting the technology to work through local adaptation.

User groups vary in their power:

- to choose the technology,
- to acquire the skills and authority to use it in different ways and at different levels,
- to fit it into work routines or reshape them,
- to adapt or modify it, fix problems, override functions or by-pass its outputs,
- to influence its evaluation and interpretation,
- to resecure their identity and status – in gender and occupational or professional terms – and their sense of order and certainty around it,
- and perhaps to subvert or reject it.

This is partly because of the varying efficacy of the technology and its embedded scripts in enforcing intended patterns of use, but also because of the internal politics of the organisation and the different forms of regulation and self-regulation of behaviour to which users are subject (McLaughlin *et al.* 1999).

Users thus actively create a role for a technology and an evaluation of it. Even for consumer goods, the picture is not one of passive consumption or automatic acceptance of their inscribed meanings and use patterns (McLaughlin *et al.* 1999). Users take more or less informed and conscious decisions to fit technologies into their detailed practices and their modes of living and working, or to modify those to accommodate the technologies. This appropriation is often experimental or iterative.

Usability and usefulness are not inherent qualities of the technology but are developed by users over time, out of different purposes and needs, using different criteria, and through different experiences (McLaughlin *et al.* 1999). Even to the extent that the technology is eventually stabilised and integrated, and its routines of use come to appear natural – in Weingart's term (1984), are *trivialised* – user groups may continue to differ in their judgements about its relevance, functioning and value.

These processes are all related to the external environment: political and cultural structures, such as industrial relations regimes or gender divisions of labour in workplaces or households, affect opportunities for change; political resources overlap the boundaries of the setting; and identities and meanings are referred to wider discourses and allegiances. Users are influenced not only by explicit arguments about the value and benefits of the technology, but also by prevalent general discourses – especially those asserting the rationality, efficacy and neutrality of technology. These reflect particular assumptions about its properties and use, and thereby advantage particular groups and interpretations (McLaughlin *et al.* 1999).

Innovation in use

SST work thus stresses the extent of innovation that emerges in implementation and use (Rhodes & Wield 1994a) as technologies and settings co-evolve. Shaping activity is not confined to design and development: rather than being launched in a finished state and thereafter available for straightforward use, technologies are introduced, intentionally or otherwise, in a form that requires further work. Significant reshaping takes place as technologies are configured, appropriated and used in particular applications, and as they diffuse across different settings and different domains of use.[39] There is often a cyclical interaction as suppliers learn from early introductions and adapt their offerings to suit new groups of users.

SST analyses are thus able to apply many concepts and insights from their studies of design and development, albeit in a more complex terrain (Bijker 1992, Kline & Pinch 1996). To differing extents former stabilisations of design, and former closures of contested visions or evaluations, are reopened – that is, the technology may undergo several phases of *recontextualisation* (Rip & Schot, Chapter 5 this volume, Schumm & Kocyba 1997). As McLaughlin *et al.* (1999, p.198) point out, however, from the point of view of those users far removed from the strategic sites of design, the process may be less one of restabilisation of the technology and its organisational setting, than multiple, local and more or less temporary *accommodations* in differing ways for different groups as the technology is incorporated, more or less comfortably, as part of their routines, roles and ways of thinking.

Again insights into systemic or configurational technologies are crucial in analysing innovation in use. Such technologies evolve in a learning process involving selective restructuring and replacement of elements, and devising strategies to cope with unforeseen problems or incompatibilities. User shaping is especially important in configurational technologies because, as Fleck (1988b) observes, requirements are not so much set internally by the system itself as explicitly decided by the implementer for that application.

Supplier and user strategies and industry structures

Studies of innovation in use have demonstrated the varied strategies of suppliers towards downstream shaping, the range of responses of users, and the complex industry structures which develop from these moves and in turn influence them.

Suppliers adopt a variety of approaches towards uncertainty about user behaviour and in response to the unpredictability and negotiability of technological change in implementation. The key axis of choice is specificity: at one extreme to provide versions fitted to the characteristics and requirements of specific users, and at the other to produce a standard product. Differentiation attempts to ensure the adoption and successful use of a product by fitting it as closely as possible to a perceived setting. It therefore requires a detailed understanding of users and local circumstances. Users are of course highly diverse, even within one locality and area of application. With mass market products, suppliers can either attempt to prefigure particular patterns of use and thus *configure* the user (Woolgar 1991a, Akrich 1992b & 1992c, Akrich & Latour 1992), or allow for flexibility – configuration, customisation or modification by users or intermediaries. The second approach acknowledges the limits of their understanding of user settings and requirements and of how these will change. In practice we find combinations of the two, with suppliers choosing which features to fix and which to leave flexible.[40]

While many technologies require some form of site-specific adaptation, local knowledge is crucial in IT applications, and standard products which do not acknowledge diversity among users may fail. The history of computer-aided production management (CAPM), for example, points to the widespread failure of early attempts to introduce standard packages where there was little uniformity in user organisations; either the packages required extensive customisation or organisational practices had to be adapted to the package – or both – in a long process of reconfiguration (Webster & Williams 1993, Fleck 1994b, Clausen & Williams 1997a). The experience of devising new configurations may later allow more generic systems to be offered, but there remains a danger of reintroducing inflexibilities (Fleck 1994b, Williams 1999).

Users, or those responsible for technology acquisition in organisations, vary widely in their capacity to evaluate the solutions on offer and in their responses to suppliers' strategies. User approaches in software development for factory and office automation, for example, range from complete in-house development to accepting commercial packages. In between comes cooperative development of tailor-made solutions involving supplier, user and perhaps intermediaries – a process from which both users and suppliers can gain expertise (Tierney & Williams 1990, Brady, Tierney & Williams 1992). For consumer technologies, the trade off is often between expensive variation or flexibility requiring adaptation, and a cheap and convenient standard product. Similarly for some industrial technologies, system-builders may find sufficient flexibility in how they can configure largely standard components.

Industry structures in many areas are thus much more complex than is captured by a simple notion of suppliers, users and markets. For some technologies, because of the complexity of configuring and adapting them for use in different contexts, and because of the different strategies of developers and users approaching this implementation, there have developed chains of intermediaries which shape a basic technology or configure components towards precise user needs. These intermediaries – retailers, IT professionals, managers, organisers of trials, and pioneer users within adopting organisations – play a key role combining universal technical knowledge with local knowledges of the organisational and cultural context of use (Williams *et al.* 2000). They not only adapt and allow the diffusion of the technology, but also help users understand possibilities and formulate requirements, and mediate between users and suppliers (e.g. Windrum 2000).

Each of these chains is subject to a variety of pressures towards vertical integration or separation. Activity is often segmented into self-contained organised markets containing only certain supplier-user chains and involving significant cooperation along them (Lundvall 1988, Freeman 1988, Clausen & Koch, Chapter 8 in this volume). At the level of an industry providing a major technology, the combination of more or less standard components in configurations for different uses may also contribute to a complex supply structure with complex patterns of specialisation, differentiation, competition and cooperation.

In these ways industry structures both result in part from developers' and users' choices, and in turn influence them, and mapping them is crucial to understanding the shaping of technologies.[41]

Articulation of supply and demand sides

What we term the articulation of supply and demand sides covers the range of ways a technology comes to be aligned with its required functions and its wider roles and significance. Again SST work is shedding new light on a familiar set of questions: what stimulates innovation, how developers envisage the users and uses of their technologies, and how options are formulated, objectives set, and conceptions of need and functionality mediated in technical practice.[42]

For developers the problem is that 'user requirements' are not pre-existing and finite characteristics that can simply be captured by the system designer. Users often cannot specify what they want or assess what technologies might be capable of. Requirements constantly shift in the light of new possibilities and experiences. Users can reinterpret intended meanings and functionalities in unpredictable ways. And beyond individual user responses, of course, innovations often face a process of formal admission through regulatory regimes and of wider social debate and acceptance.

The nature and efficacy of supplier-user linkages varies greatly according to the area of technology, the domain of application, the radical or incremental character of the innovation, and the significance of its potential impacts. Discrete technologies and minor variations in consumer products perhaps get closest to pure reliance on markets. Although one finds increasingly elaborate methods of gauging and stimulating consumer demand, Cawson, Haddon and Miles (1995) judge that it is rare for consumer IT products to be derived from explicit consideration of a want or need. Rather, opportunities are conceived by technical staff in terms of particular trajectories, and their market potential is then judged against a limited range of priorities and values derived from their image of consumption.

In other contexts, particularly technologies for industrial and commercial use, producers rely on a variety of means to guide development, to judge demand and acceptability, to create favourable conditions for uptake, and for radical innovations to create a market. All entail much closer connections with users than a simple market nexus. In some cases, retailers, consumers' associations, user clubs, advocacy groups, or intermediate adapters of the technology, perform the role of representing the market to the supplier, and to some extent *vice versa*. In some cases political, professional or other collective representation exists among users and other groups potentially affected. Suppliers may pilot technologies in different ways with selected adopters. In some areas, users submit explicit specifications, though even this often does not obviate further informal user involvement (Fleck 1988b, p.20). Developers can be expected to be ambivalent about direct user involvement in design: while it gives them a more direct way of understanding user requirements, it may also threaten their control over formulating the mode of use (Hatling & Sørensen 1998).

The success of these communication and positioning activities varies greatly, and is crucial to the successful introduction of a technology. Clearly some sort of mutual articulation of supply and demand occurs for all developments, but as Rip (1995) suggests, it is generally haphazard and in some areas huge investments are made before demand and acceptability has been adequately assessed.

This improved understanding of appropriation and use, and of their articulation with design and development, are encapsulated in the notion of *social learning* (Rip *et al.* 1995a, Wynne 1995, Sørensen 1996, Williams *et al.* 2000). It is both an analytical concept for envisaging the overall process and a normative principle to get managers and policy makers to recognise its necessity and requirements: it denotes processes in which developers, implementers and users, linked in diverse ways in networks, learn from experience and interaction. Social learning is therefore seen not just in cognitive terms but as necessarily social and political (Sørensen 1996).

Figure 3.4 Implications of Section 9 - extension of focus to include appropriation and use for policy and intervention

In extending the innovation focus 'downstream' to technology appropriation and use, SST adds weight to a long-standing message from evolutionary economics about the importance of 'coupling' between technology supply and markets. SST highlights questions not addressed by broad brush evolutionary approaches: the *detailed processes* involved; what is learnt, and how it is promoted by particular conditions. There are no simple prescriptions here for developers and intervenors. However a better understanding of the transformation and alignment processes in technology appropriation and use is essential for any attempts to facilitate or influence them.

The appropriation focus draws attention to the *diversity* of users, with their specific and changing expectations, and their *active role* in developing practices, concepts of use and meaning around artefacts. Suppliers and promoters of technologies need a better understanding of user culture and context and how they shape individual user decisions and broader cultural and political processes of acceptance.[43] Innovators should address demand and acceptability systematically and early in design and development.[44] SST demonstrates various ways to 'anticipate on contextualisation' (Rip & Schot, Chapter 5 in this volume). Its general insights translate readily into a checklist for embedding technology that project managers need to consider.[45] The advice may sound basic and self-evident, but there is ample evidence that large investments are made without this foresight. Recognising the diversity of users and their values can also feed directly into niche strategies for mobilising markets.[46]

For policy-makers concerned to encourage a particular avenue of innovation, the clear lesson is that support for R&D alone may be ineffective; encouraging adoption and social learning may be more productive. This suggests the need:

- to map and probably influence the evolving 'distributed innovation network' of suppliers, intermediaries and users, their strategies, incentives and interactions.
- to assist potential users in evaluating and introducing the technology.
- to foster localised expertise among users, or support the roles of intermediaries.

To encourage local learning and explicit experiment, technology demonstration programmes in particular should be seen as an opportunity not merely for display but also for productive supplier-user interaction: for experiments in configuring and adapting offerings and matching user requirements with technological capabilities.

Policy issues, however, go beyond facilitating the successful introduction of specific technologies and include wider aspects of sociotechnical change. Public scrutiny and regulatory oversight may need to encompass predicted or emerging second-order effects. This draws other interests into the process. A key role for public agencies will thus be fostering dialogue and cooperation among the parties (Rip, 1995).

Successful innovation implies a process of mutual shaping between new technologies and the regimes in which they are embedded – including policies for promoting innovation, and regulation – both formal legislative requirements and informal rules. Understanding how these conditions are formed and how they might be reshaped – for example how 'proper' uses of a technology or its imputed physical and moral risks are defined – bear critically on the acceptance of new technology offerings.

10 A WIDER CONCEPTION OF RELEVANT ACTORS AND OF THE TERRAIN OF TRANSFORMATION

Decentring specific actors and technologies

Several SST approaches in the 1980s examined innovation processes through the strategies of central actors and as primarily shaped by them.[47] The tendency in recent SST work has been to take a broader terrain from the start – a sector, system, arena or other part of the sociotechnical landscape – and a multi-actor and often multi-level scope. In this way it facilitates the analysis of distributed innovation (Rammert, Chapter 6 in this volume). It can follow more easily an innovation journey where there is no dominant driving actor, or in which the locus of activity passes through – or even reconstitutes – several organisations. It allows examination of multiple related strands of development where activity does not centre on any one artefact – where there are competing agendas or systems, or where emergent technologies represent a confluence or separation of different streams of innovation (Cawson, Haddon & Miles 1995, Chapter 6). Along with the extension to appropriation and use, the effect of these shifts of perspective is therefore to decentre the developer and to some extent specific artefacts – to move away from mapping networks according to actors' direct relationship with particular technologies.[48] The complexity of the terrain, differences between domains and localities, and the effects on innovative activity of changes and policies elsewhere, mean it is vital to look beyond the immediate setting of a specific innovation.

This broader perspective is particularly suited to complex systems involving different kinds of expertise. The development of interactive TV, for example, requires knowledge about ICT delivery systems, broadcasting, public and private services, and consumers' capabilities, interest and willingness to pay (Nicoll 1999). Increasingly the knowledge required is not available within a single organisation, but is achieved through collaboration, producing sometimes extensive and changing networks. The analyses are also better able, at least in principle, to account for the exclusion or marginalisation of certain interests.

Reconceptualising the transformation terrain

Broadening the perspective like this then raises the question of how best to bound the scope of a study: to take manageable slices of a complex terrain or process, or to bracket off parts of the context, in ways which allow access to, and perhaps have the effect of highlighting, the most important parts of sociotechnical transformations. Clearly the choice of simplifying device has to be justified and its dangers kept in mind.

Jørgensen and Sørensen (Chapter 7 in this volume) propose the overarching concept of *arenas of development*, to capture key participants and features of particular contexts of innovation. It allows subsidiary concepts and accounts of specific processes to be organised fruitfully within it: its proponents identify *modes of performance* undertaken within the arena – like constructing competing visions and networks, and establishing user configurations – and the strategies associated with each. Examining how the arena is constituted can capture the constraints which operate on those processes, the inclusion or exclusion of groups, the privileging of particular procedures or discourses, and so on. Generally it has been used to accommodate and organise network analyses, drawing in some cases explicitly on ANT concepts (O. Sørensen 1999). Analysis starts, however, with a mapping of the groups and organisations involved, and traces relations and interactions between them, rather than building up the network through the perspective of any one actor.[49]

Clausen and Koch (Chapter 8 in this volume) use the terms *spaces* and *occasions* to denote important loci of transformation – times and settings where crucial choices are made. For software, for example, they select spaces of particular industry segments of suppliers and users, grouped around distinctively different technological forms; internal dynamics of adopting firms, focussing on the interplay between their structures of rules, norms and principles, and a variety of forces for change; and design for mass production and user customisation. The crucial occasions in each are times of fluidity in relations internal or external to the space – like the entry of a new player or a dramatic change to product markets – but at which decisions are made which profoundly affect later options.

Rammert (Chapter 6 in this volume) proposes *technological project* as a suitable unit of analysis – a reasonably bounded and self-contained system of organising interactions between cultural patterns and technological potential. Vergragt (1988), among others, advocates identifying *critical events* or episodes in an innovation process where actors are compelled to take a major decision about the course of development.[50]

Moving outward: innovation, infrastructure and regulation

While much early SST work concentrated on design and development, there is now a move not only 'downstream' to use and appropriation, but also 'outwards' to incorporate analyses of infrastructure and of regulation – or means of political control more broadly conceived (Weber & Paul 1999). Clearly changing industry and market structures, physical and organisational infrastructures, environmental and health regulation, taxation, and competition and industrial policies, all set fundamental conditions affecting the development and adoption of technologies. Numerous other areas of economic and social activity and policy have indirect and often inadvertent effects as well. Yet typically the connections between these areas have been neglected in policy analysis, and little has been done to coordinate interventions in them.

In this broader perspective we can examine how, with the increasing importance of public acceptance and concern about technological risks, developers try to establish conditions favourable to their promulgation: technical infrastructure and skills, a supportive policy context, visions of their use, and criteria for their assessment. Conversely, it is equally important for understanding how some technologies remain marginalised: we need to analyse the sociotechnical context into which they would have to fit, and in which a favourable niche would have to be opened up (Russell & Bunting 1998).

Regulation to control technologies and their impacts has largely been retrospective, with options limited where a technology is well developed and perhaps well entrenched by the time it is scrutinised. Attempts at control are then easily depicted as obstructive and burdensome (Irwin and Vergragt 1989). The separation of responsibility for innovation and regulation in state institutions reflects that of the technology supply and implementation phases in technological change (Rip, Misa & Schot 1995c), and is reflected in the limited overlap between the fields of innovation and risk studies in academic work.

Rather than a matter of speeding up or slowing down development along a fixed path, SST work suggests the control of technology should be seen as intervening in different ways in the complex terrain of influences on innovation with the effect of sending it off in a different direction. It is widely accepted that regulation may promote or inhibit innovation, and work on innovation-forcing points to the possibility of using regulation to stimulate and steer innovation (Boden 1994). SST work has started to investigate the circumstances in which that can work and the likely or actual effects of specific measures (Irwin & Vergragt 1989, Cramer *et al.* 1990, Groenewegen & Vergragt 1991, Braun & Wield 1994, Weber 1999). Some pollution regulations, for example, have had powerful effects in forcing technological and industrial change and fostering inherently cleaner production methods; others, however, produce minimal compliance and 'end-of-pipe' fixes (Clayton, Spinardi & Williams 1999).

Paradoxically perhaps, it was initially the shift in technology studies upstream from a preoccupation with 'impacts', that led to a proposed programme of intervention in technological development – perhaps the best developed example of applied SST work: constructive technology assessment (CTA) (Schot 1991, 1992, 1998, Schot & Rip 1996, Rip, Misa & Schot 1995a, Rip 1999).[51] Its basis is an acknowledgement that the development of a technology and the creation of its social and environmental effects is an extended and simultaneous process. It attributes the failure of traditional technology assessment to a poor understanding of those dynamics. Further, it tries to overcome the dilemma identified by Collingridge (1980): that attempts to control a technology early in its development fail because its final form and effects cannot be predicted; but that technology is not amenable to control later because it becomes entrenched. Thus CTA rejects the separation of promotion and control – analytically and politically, temporally and institutionally. It advocates integrating an iterative assessment into design and development as well as implementation phases. It envisages continuous interaction between policy makers or regulators and developers. It acknowledges that development necessarily involves multiple actors and that effective assessment and redirection requires input from them. It is consistent with a perspective that stresses reflexivity and social learning.

Figure 3.5 Implications of Section 10 - a wider conception of relevant actors and the terrain of transformation for policy and intervention

The growing emphasis in SST analyses on diverse actors and a wider terrain of sociotechnical transformation reinforces our view of the profound changes needed in conceptions of policy intervention and technology management, away from top-down planning and control – directed by the state or managed within an individual firm – towards a network model. Thus, despite enduring differences in policy style and tradition, between for example *laissez faire* and *dirigiste* models, we can see across the OECD nations a move away from the state acting as a direct controller of technology decisions, towards a role as organiser of collaboration and knowledge flows between firms, research institutes and other players (Williams *et al.* 2000). At the same time the globalisation of technology supply forces nations to seek niche opportunities in a dynamic context. In contrast to the 1980s concept of planned and orderly indigenous development across the board – this new strategy is described by Brosveet and Sørensen (1997) as 'fishing not farming'.

We see a process of experimentation with modes of intervention and new kinds of regulatory intervention such as fiscal measures – proposed as an efficient way of mobilising the behaviour of widely dispersed players – as well as goal-orientated regulatory regimes. Though appearing weaker than direct control, these measures can have significant effects on the dynamics of innovation. Conversely SST studies have highlighted unintended and undesired consequences of more direct types of purposive intervention. Policymakers must engage with a variety of mechanisms and contend with their respective advantages and disadvantages. Controversy continues about the usefulness and appropriateness of different approaches and instruments, in a context where systematic knowledge is still lacking.

One of the more established developments, however, concerns the important role of the state as an intermediary [Figure 3.4], either directly or more commonly as a provider of the resources that allow other intermediaries to sustain the interorganisational collaborations needed for successful technological innovation. Templates of existing successful networks or regime structures, carefully used, should help the more successful construction or improvement of new development alliances (Rip 1995). It should be possible to learn from the experience of existing often *ad hoc* 'macro-actors'.

It is also clearly important to allow actors to form links which cut across social and institutional barriers to remove hindrances to exchanges of views (Clausen & Koch, Chapter 8 in this volume). Interventions to broaden involvement and facilitate learning [Figure 3.6] must deal sensitively with the diverse cultures which will be brought together, and address the needs identified in work on public participation in scientific and technological debates: for information and education, for appropriate arenas and procedures, and crucially for ways of enabling communication among different groups of technical specialists and between them and non-specialists.

11 CULTURE AND KNOWLEDGE IN TECHNOLOGY

There has been an increased stress in recent SST work on aspects of technology which concern culture and knowledge:

- meanings and values attached to technologies, and the importance of that symbolism and evaluation in selection and appropriation;
- the role of cultural influences among technical practitioners, and from other institutions, in technological development;
- knowledge of the functioning and performance of technologies, and knowledge in formal and informal evaluations of their social and environmental impacts.

Variants of sociology of knowledge, discourse analysis and ideology critique, are used in analysing cultural forms around technologies and their social functions. From anthropology and ethnography, SST work has taken on the importance both of immersion in the social world of the research objects and at the same time of distance from them. Debates on reflexivity have emphasised the need for analysts to reflect on their own categories and assumptions. Concepts from elsewhere have proved useful, particularly:

- *boundary work* – the processes by which groups try to impose categories and distinctions in order to control participation in scientific and technological debates and activities (Gieryn 1995, Shackley & Wynne 1996, Jasanoff 1987);
- *boundary objects* – entities at least nominally common to several actors' discourses, enabling them to discuss an issue and perceive a shared interest (Fujimura 1992, Grin & Van de Graaf 1996).

Analysing culture in technology

There is wide agreement that culture has to be treated not as a uniform and fixed background but as complex socially constructed systems, intimately interwoven with political, economic and other sociotechnical forms, though usefully separated from them analytically. There remain, however, differences in explaining how cultural forms are constructed – particularly in the emphasis on dynamics or structure. Again we find the most fruitful work seeking to combine both.

Cultural shaping of technology takes place at two levels: the private construction of technical practice and significance; and the public construction of accepted uses and rules of using technologies in different arenas (Rammert, Chapter 6 in this volume).

At the second level it has been shown that innovation systems, policy processes and styles of regulatory interventions are strongly dependent on political cultures at an institutional or national level (Jasanoff 1986, Vogel

1986). Clearly, however, cultural analysis in SST has to avoid the common tendency to focus exclusively on those aspects which distinguish national cultures or styles of work. It should encompass other levels, institutions and groupings – cultures specific to technical disciplines, occupations, genders, localities, firms, movements, etc. – and allow particularly for local diversity in meanings and practices among users. Much of the pioneering work here has come from a gender perspective, concerned to explore the consequences of gender assumptions embodied in design visions and about users and use settings; gender differences in the reception of technologies; the way in which gender identities and technical practice are mutually constitutive; gendered constructions of skill and worth in work; and the way these practices and meanings are renegotiated in the face of change.

Among other influences shaping cultural engagement with technology are general discourses concerning technology, especially deterministic and utopian discourses stressing the neutrality, inevitability or rationality of technological change.

A cultural analysis is crucial for understanding processes of innovation within technical arenas and communities, and the mediation of wider social influences into technical practice. SST work has continued to examine how shared patterns of thought and practice – visions; traditions, paradigms or styles; evaluations of need, performance and effects; and aesthetic appreciations, non-material goals and rewards – are institutionalised, challenged and renegotiated. Culture is implicated both in creativity and the generation of opportunities, and in restricting potential openness and variety.

Cultural influences can usefully be understood as both:

- 'filling in' where choices are underdetermined by economic and political mechanisms, to the point where culturally specific routines, heuristics and modes of learning dominate (Rammert, Chapter 6 in this volume); and,
- mediating other forms of determination, for example, by embedding economic and political assumptions and criteria in apparently technical practice, in informing conceptions of users and use settings, and by shaping rationalities and problem definitions in formal evaluations.

These insights apply not just to developers but also to users interacting with and modifying technologies: rather than users adjusting their interpretations around a fixed material and organisational form, there is a further shaping of technology attributable to cultural patterns and dynamics. Assessments of usefulness, for example, are constructed not only discursively but in the course of practical engagement with the technology and perhaps its modification. Some illuminating work has focussed on the culture of groups which are particularly active in reinterpreting technologies as they put them to unconventional uses – such as technology hobbyists and computer hackers.

In the case of ICTs, cultural analysis is particularly fruitful because they are sites for creating, or media for conveying, cultural products. Cultural analysis of the content of the media is well established, but the mutual influence between content and technological systems and artefacts themselves is becoming more complex in, for example, multimedia technology (Williams 1997b, Williams *et al.* 2000).

Knowledge forms in technology

In analysing the social construction of technological knowledge, SST work continues to draw on sociology of scientific knowledge,[52] particularly in exploring:

- the development of explanations of how technical practices work or technologies function;
- the way evidence and judgements about performance are negotiated; and
- the contingent character of criteria of practicability, adequacy, efficacy, efficiency, success, etc.

Studies of testing in particular have shown how test results and their significance may be contested by opening questions about theoretical or practical aspects of the testing set-up (MacKenzie 1989, 1990 & 1996a, A. Webster, 1991, p.45, Pinch, Ashmore & Mulkay 1992, Pinch 1993). Users and others affected by the technology, of course, may make altogether different assessments of adequacy and performance when the technology is introduced.

SST also draws on a wide range of work examining the construction and deployment of knowledges involved in selecting and judging technologies – more or less formal evaluations in more or less public arenas. This work, for example on the politics of regulatory science, and on risk discourses and assessments, relates the discourses and specific claims to the groups involved and the context of the evaluation.

Knowledge has of course become an important focus in most currents of innovation studies, technology management and technology policy studies.[53] SST work has largely accepted useful distinctions developed here between categories of knowledge – codified, tacit, experiential, locally contingent, firm-specific, etc. In turn SST has contributed to analyses of their dynamics and their significance for understanding attempts to embody knowledge in technologies, by virtue of its distinctive focus on the negotiated and contingent content of knowledge.[54] There are also important questions in the broader political economy of knowledge which affect the social shaping of technologies – especially its distribution and ownership, and their effect on priorities and directions of development. Again SST work both draws selectively on these analyses, and can contribute by showing the importance of treating the substance of technological knowledge and countering the tendencies to reduce notions of knowledge to one of information.

Figure 3.6 Implications of Section 11 - increasing attention to culture and knowledge in technology for policy and intervention

SST work has developed a strong and fruitful concentration on the symbolic aspects of technologies and their relation to wider cultural patterns, on discourses on technology in public and policy arenas as well as in development and use, and on the way these cultural forms help shape sociotechnical outcomes. This work can contribute much to improving our understanding of policy and strategy discussions, and points directly to ways of structuring and facilitating them.

1. The use of narratives and metaphors is central to our attempts to make sense of future worlds of which we have no direct experience and convince others to support or oppose these visions. Awareness of how we create and deploy these meanings, and how they affect discussion and acceptance of options will help policymakers, strategists and wider publics (Naesje, Chapter 10 in this volume).
2. Technologies are understood in profoundly different ways by different players – developers, engineers, managers, policymakers, potential consumers and other publics.[55] Successful development may require facilitating communication amongst players with different traditions and commitments.
3. Involving users in design is not just about improving the physical functions of the technology; it should help developers & intervenors understand the contexts and practices of use and anticipate the symbolic significance of artefacts for potential users which is crucial in their market acceptance or rejection.
4. Understanding and promoting public acceptance of technology entails taking seriously different perceptions of technologies. SST and related studies of risk communication show that different publics may approach their evaluations of technologies with very different purposes and criteria from those assumed by developers and technical experts – in short, with quite different frameworks of meaning. Moreover, lay assessments of a technology, based on experience and understanding of local conditions, may be more realistic and finely tuned than the idealised or over-generalised models of remote experts and decision-makers.
5. These moves are justified not simply in terms of a commitment to democratic involvement. Insights from SST emphasise the need for developers or managers to test the robustness or sensitivity of decisions and guard against tendencies towards overoptimistic claims or forecasts, and for policymakers or regulators concerned to ensure disinterested appraisal, and thus also point to the value of:
   - a range of alternative assessments from different viewpoints,
   - open scrutiny and evaluation of developers' proposals and official assessments,
   - careful and more sceptical use of a variety of sources of expert advice, and
   - provision for challenges to the assumptions underlying decisions.[56]
6. This wider involvement calls for arenas in which diverse rationalities and concerns can be expressed and considered, rather than treated as deviations to be denigrated or corrected. Decision-makers may need to recognise, and intervene to counter, attempts by particular groups to establish exclusive authority or expertise and downgrade the contributions of other groups. Failure to involve affected groups and address their concerns early may lead later to outright rejection and resistance through other channels.

12 CONCLUDING DISCUSSION

Confluences and tensions

In examining recent contributions to the SST endeavour, we have depicted considerable convergence and common ground:

- a move away from polarised positions, radically novel theoretical claims and programmatic statements;
- substantial reconciliation between what were sharply divided and often antagonistic camps:
- a willingness to explore overlaps and complementarities;
- the pragmatic use of a variety of theoretical resources in case studies;
- more attempts to find theoretical syntheses;
- drawing on concepts and substantive findings from other areas of technology studies and broader social theory;
- trying to find a balance between what were exclusive emphases, or accepting that different characteristics may predominate in different areas of technology and domains of use, rather than assuming that one pattern holds for all.

We have examined some of the attempts to find resolutions on key axes:

- between explanations in terms of agency and structural determination: by using some form of interactive or structurational model that acknowledges the recursive construction of sociotechnical institutions and patterns through human practice, and the possibility of change to them, so that we have to see neither social structures as a fixed backdrop, nor everything as constructed afresh and without constraints through purposive action;
- between emphases on fluidity and stability in sociotechnical forms: so that the question shifts to the characteristic extent of each in different areas of technology, to the variation between cases and between moments in an innovation process, to the way each is produced, and to how stabilisation in one area may destabilise another;
- between emphases on contingency and patterning in outcomes: in particular so that we can try to understand the tensions and complementarities between localising and standardising forces in the application of technologies – how these interact in particular episodes and how patterns emerge;
- between the need to use categories and analytical boundaries and to recognise their constructed and conventional nature: so that we can for different purposes treat boundaries as useful and justifiable analytical devices, as contingent and contestable constructs of participants, as reflecting real institutional separations, or as created by actors in the

process of network building and in their strategies of inclusion and exclusion.

Many of these matters have long been debated within the field, and some, like the structure-agency debate, are central issues in social science as a whole. We cannot expect a final resolution on many of them. There has however been a sustained attempt to grapple with the complexities involved, and this has, we argue, been very fruitful. A range of schemes and concepts have emerged, and our review has found points of synergy and interplay between them.

There remain, of course, plenty of disagreements over emphasis and forms of explanation. Some reflect quite fundamental theoretical commitments. For example, we outlined two major synthetic frameworks represented in this book: the scheme of *landscape/regime/niche* which represents a strong blending of sociological and evolutionary economics approaches, and the model of *arenas of development* which more comfortably accommodates depictions of network-building. Each of them subsumes and usefully organises a range of other concepts and insights, and resolves some of the problems of earlier more narrowly defined efforts. The divergence between them shows the extent of differences across SST work. These differences and tensions, however, can be productive rather than an obstacle. Different models and concepts can illuminate different aspects of the same case, and specific points of conflict indicate important research questions and form some of the most interesting sites for future theoretical development.

We believe the developments we have identified reflect the growing intellectual strength and confidence of the field in terms of a greater willingness to engage both with different explanatory traditions within SST and more broadly, and with technology practitioners and policy-makers. The emergence of broadly acceptable theoretical formulations and substantive insights should make both the explanation of its significance to outsiders and the work of practical application that much easier.

On the other hand, while there is a strong sense of maturation of the field of SST, there is less sense of 'closure'. We note here the continued dynamism of technological innovation and the involvement of new technologies in ever more domains of social life. The 'normalisation' of technology as part of everyday life in some ways seems to break down the barriers between the concerns of technology studies and other realms of social enquiry. On the other hand, new technological forms and knowledges seem to bring new threats and new strangenesses. The sociotechnical landscape is becoming more varied and complex. Accordingly we can expect the continued emergence of new explanatory schema, concepts and tools.

Generalising from SST studies

A key requirement for policy-makers and practitioners seeking to draw on SST work is some guidance as to how, and under what circumstances, empirical insights can be applied more broadly. The SST approach has characteristically been built on the basis of case studies, involving detailed qualitative analysis. This approach has been a necessary response to the conditions in which technology studies grew up: the neglect of technology in much social science, the prevalence of technological determinism, and the inadequacy of existing sociological categories for explaining the intricacies of technological change. It probably also reflects the interdisciplinary character of a field defined in terms of a set of issues, and the corresponding interdisciplinary background of many of the researchers.

The mention of 'case studies' carries the negative connotation for many people interested in the applicability of social science knowledge that there is little possibility of valid generalisation. SST does rely on 'thick description' as a means of approaching complexity and has accumulated a large body of such cases. There is something of a trade-off between depth and breadth of analysis given the exigencies of research funding. SST work seldom involves larger-scale surveys across a range of organisations or technologies using common standardised research instruments – though there have been recent attempts to undertake more systematic comparative work.[57] Nor has it generally tried to make correlations between conditions and outcomes as other traditions of innovation studies have done.[58] Moreover SST work has taken as a major theoretical theme the *specificity* of processes in different areas of technology and different domains of application, and the dependence of outcomes on contingent factors. This emphasis on the particularities of historical and social settings is in part a reaction to the over-generalisation typical of early diffusion studies or in economic analyses of firms as sites of innovation. All these features, then, present serious questions about how we should derive a general understanding.

We argue nonetheless that we can obtain robust and useful general insights from SST work. The strong case study basis does not leave the work solely descriptive. Deriving applicable lessons does not have to involve some inductive process which asserts that the findings of limited studies apply across the board. Certainly there are dangers of taking one example as paradigmatic without adequate justification, and of transferring insights from one set of conditions to another. Examples of this may occur in SST theorising, but it is not inherent in the form of theory development, and the increasing diversity of areas studied makes it less likely.

The ways we assess the significance of research findings are obviously more complex and varied for qualitative and historical analyses such as SST than for large-scale surveys. They entail interactions between theory and empirical material and between SST and more general social theory. There have been some differences in how analysts have approached this according to their framework; that is, different theoretical bases entail different views of what constitutes valid explanation, argument and evidence. But typically, analysts identify and justify the salient features of a case and its context which form the basis of a class of cases which can be treated similarly. General claims are then evaluated in standard ways: against a range of empirical evidence, allowing for the theory-dependence of observation and other forms of data collection; and in terms of consistency of explanation at different levels of abstraction. As more areas of technology and application are examined and compared, clearly we can assess the usefulness of frameworks and the validity of concepts and claims across a wider range of sites than the ones on which their proponents have concentrated.

Even a single study, if well analysed, can provide rich insights for intervention in that particular case or in similar cases and conditions. An explanation of the ways in which choices were made, or of the conditions which produced the particular outcome, should allow an analyst reasonably to argue what would have been required to produce a different result: how close the development was to being taken down a different route, what changes or interventions would have shifted it, and how difficult those would have been. If, for example, we understand the position of an actor and their degree of indifference to different potential outcomes, it may be possible to persuade them that their interests lie in a different direction.

In this respect, the rich accounts emerging from SST research may provide insights which are *more* useful for practitioners and policy-makers, as they bear directly on the available choices and constraints. Rather than accepting the negative interpretation of the limitations of qualitative research – often by quantitative analysts – we need to turn the question around: other approaches which *appear* to provide more readily applicable knowledge – for example industrial economics analyses – are in the main unable to explain the limits to generalisation from their empirical findings. Large-scale survey approaches run risks, particularly of inappropriately grouping different phenomena on the basis of common features which are not the most significant. Such quantitative approaches may of course be valid in contexts where technology artefacts and practices have become stabilised (MacKenzie 1992). But they are ill-equipped to explore how the new world may turn out to operate rather differently from what simple extrapolation of current practices and expectations would indicate. SST in contrast provides guidelines and concepts to explore both continuity and change.

Though we argue in the next chapter that the direct application of SST insights to technology policy and other forms of intervention in technological change is not yet well developed, the work in this book demonstrates the potential for a much greater contribution. One key area of work will be analyses of the shaping of policy integrally with the shaping of technology. Another, we hope, will be built on a growing dialogue between SST researchers and decision-makers, technology practitioners and other external audiences.

NOTES

1 We have had to strike a balance here between giving a full presentation of the variety of approaches in SST and the range of arguments on contentious issues, and allowing ourselves to examine them critically and comment on their strengths and weaknesses. We have excluded certain strands of work in technology studies, explicitly or implicitly. Beyond that basic boundary, though, it is not our intention to endorse the detail of particular contributions in this book or elsewhere.

2 We do not accept contemporary depictions that implied these frameworks necessarily superseded, were superior to and broke completely with what had come before. Some accounts also inaccurately equate SST with particular approaches – for example, using 'social constructivism' interchangeably with 'social shaping'.

3 Thus we would want to distance SST from some continuing currents of work on technology – e.g. conventional impact analysis, innovation studies based on linear models, and traditional technological forecasting.

4 For useful summaries and commentaries on evolutionary economics, and its relationship to SST, see van den Belt & Rip 1987, Dosi 1988 & 1990, Saviotti & Metcalfe 1991, Blume 1993, Chapter 2, MacKenzie 1992, Burns & Dietz 1992, Metcalfe 1995, Rohracher 1998, McLoughlin 1999, Williams & Edge 1996, Green *et al.* 1999. Quasi-evolutionary models are also often referred to as neo-Schumpeterian.

5. Rohracher (1998) argues there need be 'no serious theoretical conflict' between the approaches, though he points on the basis of citations to minimal overlap between the groups of theorists. He lists additional themes which can be common to both approaches – especially the systemic nature of technology; the importance of diffusion; and the role of users.

6 Evolutionary economics has itself been interested in differences between contexts and cases. Typically however, it starts with models of general behaviour of agents and then differentiates among them, either drawing attention to sector-specific features or using typologies of capabilities and characteristics. Much of the activity is orientated to finding correlations between firm or market characteristics and innovative activity, rather than qualitative models of mechanisms and processes.

7 For a systematic review and evaluation of political science and organisation theory resources, see Weber & Paul 1999. On network and self-organising systems approaches see also the SEIN project <http://www.uni-bielefeld.de/iwt>. Weber notes that innovation studies in Germany have drawn on political systems theory more extensively and use systems language much more readily than SST work in e.g. the UK (see e.g. Görlitz & Druwe 1990, Görlitz 1994). See also Rüdig 1989.

8 Kaplinsky 1984, and see the elaboration by Fleck 1988b. Kaplinsky's analysis of automation considers the spheres of manufacturing, design and coordination, and identifies the activities which comprise them – including transfer and control as well as transformation of inputs. It distinguishes three forms of automation: intra-activity, intra-sphere, and inter-sphere. The distinctions allow the mapping of varied and uneven patterns of automation within and across industries.
Here we can also incorporate Rammert's (1997) distinctions between forms of *technicisation* – transformation of practice into a *techno-structure*: *habitualisation*, *algorithmisation*, and *mechanisation*, respectively the use of routines, sign systems and objects. See also Fleck's protocol (1988b) for a *technology audit*.

9 It is now widely recognised that all systemic forms of technologies are sociotechnical. The term *technological systems* tends to be reserved for those displaying close technical coupling of material components.

10 Similarly a metaphor of *ecologies* stresses co-existence, interdependence and mutual adaptation, without the connotation of necessary systemicity.
11 Fleck (1994b) argues that robotics, expert systems, office automation, materials requirement programming (MRP), total quality management (TQM), automated warehousing, just-in-time (JIT) systems, computer integrated manufacturing (CIM), can all be fruitfully analysed as configurations. See Kling & Scacchi 1982, pp.36-38, for similar arguments.
12 For corresponding notions in evolutionary economics of the co-evolution of techno-economic entities and their institutional context, see e.g. Dosi 1991. The language of (mutual) shaping between technology and society or of co-evolution does not imply a smooth and harmonious relationship. Instead SST highlights the tensions and contradictions surrounding innovation, e.g. between the different dynamics surrounding technology supply and appropriation and conflicts between diverse players and goals in innovation (Williams, Slack and Stewart 2000).
13 The criticism can be made (Bijker 1993) that this was the implication of some early SST approaches – not only that a technology reflects the interests of the group which shaped its development, but that these interests are unproblematically furthered by its deployment.
14 In conceptualising the way in which technologies may produce change across time and space and particularly in analysing user involvement in technology development, it can be productive to follow the notion of technology as *text*. (Hill 1988 & 1992, Woolgar 1991b, Akrich 1992b & 1992c, Akrich & Latour 1992). Developers' strategies involve influencing not only the practical use of the technology, but also attitudes towards it: its desirability, confidence in its capabilities, and satisfaction that it fulfils needs or objectives. Developers thus try to *inscribe* use and meaning in the technology; the *script* is intended to enforce a particular reading or form of use, in effect to limit users' reinterpretation or reconfiguring. It is important not to assume that scripts always work as intended and that delegation is assured (Johnson/Latour 1988, Latour 1992, Law & Bijker 1992). See also Mackay & Gillespie (1992) on ideological encoding – embedding 'preferred forms' of use.
15 The categories and boundaries can be analysed as constructs which serve rhetorical purposes (Rachel & Woolgar 1995) – e.g. in depicting the failure of a project as resulting from interference by the 'political' in the 'technological' (MacKenzie 1996a).
16 Collins and Yearley (1991) argue that this reversed the gains made by a symmetrical sociological analysis of knowledge forms; ANT accounts reverted to actors' accounts of the properties and behaviour of physical entities rather than seeing these as requiring social explanation.
17 The argument follows Collingridge's work (1980) identifying features of technological systems and their contexts which engender flexibility or inflexibility.
18 See also Green's model (1991, 1992), based on biotechnology, of the interacting development, from concept to material form, of the technology and its functionality, firm structure and market knowledge.
19 While the analysis of closure is fruitful, it is widely accepted this is only one mechanism by which directions of development might be settled. Consensus over the meaning and value of a technology is unlikely; it will retain different meanings for the groups which use it or are affected by it (McLaughlin *et al.* 1999, Grin & van de Graaf 1996, Bijker & Law 1992b).
20 Again the connotations of *network* can be contentious: the loose and varied nature of links between actors around e.g. multimedia technologies leads Williams *et al.* (2000) to talk of sociotechnical *constellations*.
21 For example, Vergragt 1988. Law (1992b) analyses the strategies of groups allotted different functions in design and development; each seeks to influence or restrict the actions of those that follow.
22 For example, Jørgensen and Sørensen's list (Chapter 7 in this volume) of *modes of performance* and strategies within a development arena.
23 As Rip & Schot (Chapter 5 in this volume) observe, a vision is necessary because the technology is as yet a 'monstrosity', 'full of promise but not able to perform very well'. A trajectory can likewise be interpreted as a participant's rhetorical resource – utopian or dystopian.
24 Some frameworks for understanding alignment, like Molina's theory of sociotechnical constituencies and a 'diamond of alignment', do readily provide some basic indications of whether a project can be expected to succeed and what needs to be done to effect

alignment, and for cases like the Europrocessor (1998), specific recommendations are derived.

25 Rip & Moors (1999) & Rip & Schot (1999) identify three main loci: at the formation of a protected space; at the point of first introduction or demonstration; and as niches start branching and piling up. This identification forms an important premise of the programme of strategic niche management. These three points indicate strategies respectively of modulating promise-requirement cycles and introducing social visions; of modulating introduction, typically so that the experiment is conducted on a wider social basis than simply market demand; and of modulating the cumulative processes that can lead to regime change. Network analyses similarly indicate phases in which networks are forming or reforming – where they may need help in consolidation, be amenable to steering, or be vulnerable to disruption – and other phases when they are stable and resistant to interference.

26 We need to develop techniques for envisaging and allowing processes involving such branching and discontinuities (Rip, 1995), like scenario building rather than extrapolation of existing trends (Geels & Schot 1998, Geels, Chapter 13 in this volume), and back-casting from future imagined requirements instead of starting from currently unfolding technologies. The complexity also points to a need for improved mapping techniques for following developments in real time and facilitating understanding. Work on visual representations of innovation networks is one practical example to emerge from innovation studies (Steward, Conway & Overton 1995).
The convoluted character of a typical innovation journey, and the variety of inputs to an innovation as different actors are enrolled in its development, have further implications similar to those from earlier studies of the sources of innovation and the science-technology relation: monitoring and judging the achievements of the overall project are less than straightforward, and likewise assessing the contribution made by particular inputs of, say, basic research, or distinguishing the effects of policy interventions or regulations from other influences. Measures and criteria for evaluating the effectiveness of funding, investment, policy instruments or other form of intervention need to take this complexity into account.

27 Rammert (this volume) thus stresses the need to recognise and foster 'technodiversity'.
Path-dependencies compound the problems of 'picking winners'. See Schneider (2000) on Minitel in France: an exemplar first of public support assisting the development of an information market, but later of rigidity in that it could not readily be opened up to internet use. Such cases have lessons for aligning developments to global *de facto* standards and providing more 'future-proofing' – reverse compatibility, gateways and migration paths.
Clearly public agencies already intervene in some areas at industry or sector level to influence processes of competition and collaboration; see Fransman (1991) on the Japanese system of controlled competition under MITI.

28 *Paradigm* – in the broader of its two meanings as a mental tool-kit characteristic of a particular technical community, rather than an exemplar – has been particularly controversial (Dosi 1982, Gutting 1984, Clark 1987, Fleck, Webster & Williams 1990). Besides its allegedly deterministic connotations, it is argued to focus too much on cognitive aspects and too little on social and economic influence (Geels & Schot 1998). A strong argument can be made that neither is necessary: the original formulation was intended to explain how trajectories are maintained in the absence of explicit decisions at every point and to allow analysis of how political and economic influences were incorporated in the assumptions and practices of technical practitioners (Dosi 1982 & 1988). The concept of paradigm nonetheless has been widely by-passed in favour of *regimes* – a 'broader more socially embedded version' of paradigms (Geels & Schot 1998) – or *frames* (Bijker 1995b). Overlapping concepts are used in other traditions of innovation studies – E.g. *cognitive model* or *judgement system* (Burns & Dietz 1992).

29 The extent of apparent regularity or contingency is only in part a feature of the area studied: it is also a product of the approach, concepts and focus. As in political science more generally we find a corresponding variety of notions in SST approaches of contested concepts like power and interests. For a review see E.g. Russell 1991.

30 Patterning may include divergences in practice as well as consistency: see E.g. trends in Computer Numerical Control machine tools in European countries attributed in part to market characteristics and to national conditions and histories (Sorge *et al.* 1983, Andersen 1988, Fleck, Webster & Williams 1990, Hartmann *et al.* 1994).

31 For an early formulation see Weingart 1984. Several frameworks draw on Giddens' theory of structuration: E.g. Russell & Williams 1988, Asheim 1990, Orlikowski 1992, Rammert 1997, 1998.
32 Nor does it entail a reversion to an externalist view in which the 'social' stops at some boundary around 'technology', though it for certain purposes it may be justifiable to depict a division between context and content – to regard the context as the summary effects of stable institutions or external networks. cf. Frickel 1996.
33 Proponents of multi-level frameworks accept that sociotechnical institutions represent a previously negotiated order but argue there is little point and much to be lost in limiting ourselves to a grammar of action. They insist that we can use different but consistent concepts appropriate to different levels – that different scales of phenomenon are not usefully handled by the same concepts (Russell & Williams 1987, Russell & Bunting 1997).
34 The character of a national innovation system depends E.g. on aspects of sectoral composition, labour markets, venture capital provision, professional and research organisation, education and training, intellectual property regime, and government industry and innovation policies. An understanding of the organisational setting of innovation in particular industries or technological areas, however, requires a much more specific analysis, in which national characteristics may not be the most relevant.
35 The fundamental longer term challenge (Rip 1999), rather than simply encouraging particular technological objectives or preconceived forms of development, is to design processes and arrangements which enable tractability – including providing the conditions for niches.
36 This and the following sub-section draw heavily on two major studies of the appropriation of ITs in different contexts: McLaughlin *et al.* 1999 on the introduction of management information systems in organisations, and Cawson, Haddon & Miles 1995 on consumer IT products. See also Silverstone & Hirsch 1992, Mackay & Gillespie 1992, Sørensen 1996, Lie & Sørensen 1996.
37 Feminist work has analysed the introduction of domestic technologies and adaptation to them in terms of influences within the home – particularly gender and generational divisions of labour – and outside – like the availability of other forms of provision and the pattern of paid employment.
38 In examining difficulties in getting a technology to work and of judging whether it works, useful parallels can be drawn with the work of sociologists and historians of science on the replication of experiments. Understanding localised processes of embedding technologies and the shaping of practices involved in their operation has, as Wynne (1988b) shows, profound implications for the analysis of accidents and control of risks.
39 Wynne (1988b) and Fleck (1988b) take the argument further by stressing the experimental character of some major technologies.
40 McLoughlin (1999, p.112) points out that the need to match technological capabilities to users, whether by meeting user requirements in the technology, or by reconfiguring the user, has stimulated extensive research in marketing, ergonomics, psychology and organisational sociology.
41 For example, Brady, Tierney & Williams 1992 and Williams 1999 on IT software. There are difficulties in separating analytically different 'industries' when there is so much convergence and overlap, in say ICTs, between computing, software, DP, telecommunications and consumer electronics.
42 Early innovation studies often framed this in terms of technology-push versus market-pull, but the issues clearly go far beyond the simple market mechanisms assumed in classical economics. The interactions are better handled in concepts like the *promise-requirement cycle* (Rip & Schot, Chapter 5 in this volume). In quasi-evolutionary terms there is 'anticipation on selection' and a 'nexus between variation and selection' which becomes institutionalised in a variety of ways.
43 The work on innofusion, for example, indicates alternative technology supply strategies between capturing the specificity of user requirements and supplying generic components which users can configure – although, as Williams *et al.* (2000) note, we do not yet know which strategy works in which circumstances.
44 New kinds of mechanism for engaging directly with users in ICTs, for example, include alpha and beta testing and the proliferation of information superhighway trials and experiments.

45 Deuten, Rip & Jelsma (1997) develop suggestions in the three areas of:
- integration into relevant industry and market contexts;
- admissibility according to existing and likely regulation;
- acceptance by the public.

46 The success of some green marketing and recycling initiatives demonstrates a willingness among some users to make choices on other criteria than price and technical performance, and in particular to pay a premium or undertake some effort or inconvenience in favour of environmental values. See E.g. Sonneborn & Russell (1999) on 'green power' schemes to retail electricity from renewable energy sources, and Hansen (1999) on organic food.

47 For criticism of ANT accounts on this aspect see E.g. Singleton & Michael 1993, Summerton 1997, and of SCOT see E.g. Russell 1986, Winner 1993.

48 Summerton (1997) argues that the 'heroicism' and 'managerialism' of narratives of major change based on the exploits of individuals are fundamentally misleading; no one actor is capable of creating a new regime or a major change in regime.
A possible danger in this theoretical shift is that we lose sight of the technology and revert to dealing only with social entities and processes. It is crucial to keep the stress on the sociotechnical nature of landscapes, institutions, networks or processes.

49 Rammert (Chapter 6 in this volume) and Hage & Hollingsworth (2000) use a similar term *arenas of negotiation* and identify various types.

50 Rammert (Chapter 6 in this volume) and Freeman (1991) draw attention in the notion of a *circle of uncertainty* to the conditions of high risk and poor information under which such critical decisions may be made.

51 See also Rammert's similar concept of *reflexive technology development*.

52 We do not need here to broach the ontological arguments into which ideas of the social construction of knowledge often lead: it is sufficient to accept that knowledge forms are variable and contested, contextually dependent and exert real influences. The concentration here on technology should not be taken to imply a clear separation between spheres of 'science' and 'technology'. SST work acknowledges a complex relation between forms of explanation and practical intervention, and a spectrum of scientific-technological activities, organisations and fields.

53 For example, Senker 1995, Williams, Faulkner & Fleck 1998. The practical side of these concerns is driving the burgeoning area of knowledge management (Jørgensen, 1999).

54 Especially Collins 1990 on artificial intelligence.

55 See Grin & van de Graaf (1996, pp.84-85) for a useful general schema of the different focuses of managers, technologists and policy-makers. As they imply, legitimate intervention does not require complete agreement and fully shared meanings, only a degree of congruence over the meaning of a policy outcome.

56 Several commentators have suggested that controversy should be regarded not as a hindrance to effective policy, but as necessary and productive – in effect as informal technology assessment (Rip 1995, Cambrosio & Limoges 1991)

57 See McLaughlin *et al.* (1999) for a demonstration of the power of a group of case studies showing different features but tackled in the same qualitative framework. See also Weber, Chapter 12 in this volume.

58 Extensive studies in the 1960s and 70s which attempted to map the sources of innovation in basic research, or to make systematic comparisons to isolate variables in strategies or conditions of successful or failed innovation, demonstrated how difficult such large scale work is. SST may offer more theoretically informed insights about the scope and risks of such extrapolation (see Sørensen, Chapter 2 in this volume).

BIBLIOGRAPHY

Aichholzer, G. & Schienstock, G. (eds) (1994), Technology Policy: Towards an Integration of Social and Economic Concerns, Berlin: de Gruyter.

Akrich, M. (1992a), 'Beyond Social Construction of Technology: the Shaping of People and Things in the Innovation Process', in M. Dierkes & U. Hoffmann (eds), *New Technology at the Outset*, Boulder CO: Westview, pp. 173-190.

Akrich, M. (1992b), 'The De-Scription of Technical Objects', in W.E. Bijker & J. Law (eds), *Shaping Technology/Building Society*, Cambridge MA: MIT Press, pp. 205-224.

Akrich, M. (1992c), 'User Representations: Practices, Methods and Sociology', in W.E. Bijker & J. Law (eds), *Shaping Technology/Building Society*, Cambridge MA: MIT Press, pp. 167-184.

Akrich, M. & Latour, B. (1992), 'A Summary of Convenient Vocabulary for the Semiotics of Human and Non-Human Assemblies', in W.E. Bijker & J. Law (eds), *Shaping Technology/Building Society*, Cambridge MA: MIT Press, pp. 259-264.

Allen, P.M. (1988), 'Evolution, Innovation and Economics', in G. Dosi *et al.*(eds), *Technical Change and Economic Theory*, London: Pinter, pp. 95-119.

Andersen, H.W. (1988), 'Technological Trajectories, Cultural Values and the Labour Process: the Development of NC Machinery in the Norwegian Shipbuilding Industry', *Social Studies of Science*, **18**, pp. 465-482.

Arthur, W.B. (1988), 'Competing Technologies, Increasing Returns and Lock-in by Historical Events', *Economic Journal*, **99**, pp. 116-131.

Arthur, W.B. (1992), 'Competing Technologies: an Overview', in G. Dosi *et al.* (eds), *Technical Change and Economic Theory*, London: Pinter, pp. 590-607.

Asheim, B.T. (1990), 'Innovation Diffusion and Small Firms: Between the Agency of the Lifeworld and the Structure of Systems', in N. Alderman *et al.* (eds), *Technological Change in a Spatial Context*, Berlin: Springer-Verlag, pp. 37-55.

Badham, R. (1986), 'Technology and Public Choice: Strategies for Technological Control and the Selection of Technologies', *Prometheus*, **4**, (2), pp. 288-307.

Badham, R. (1994), 'From Socio-Economic to Socially Orientated Innovation Policy', in G. Aichholzer & G. Schienstock (eds), *Technology Policy: Towards an Integration of Social and Economic Concerns*, Berlin: de Gruyter, pp. 25-66.

Badham, R. (1995), 'Managing Socio-Technical Change: a Configuration Approach to Technology Implementation', in H. Benders, J. de Haan & D. Bennett (eds), *The Symbiosis of Work and Technology*, London: Taylor & Francis, pp. 77-94.

Badham, R., Couchman, P. & McLoughlin, I. (1997), 'Implementing Vulnerable Sociotechnical Change Projects', in I. McLoughlin & M. Harris (eds), *Innovation, Organisational Change and Technology*, London: International Thomson Business Press, pp. 146-169.

Berg, J.L. & Schumny, H. (1990), *An Analysis of the Information Technology Standardization Process*, Amsterdam: Elsevier/North Holland.

Berg, A.-J. (1994a), 'A Gendered Socio-Technical Construction: the Smart House', in C. Cockburn & R. Fürst-Dilic (eds), *Bringing Technology Home:*

Gender and Technology in a Changing Europe, Buckingham: Open University Press.

Berg, A.-J. (1994b), 'Technological Flexibility: Bringing Gender into Technology (or Was it the Other Way Round)?', in C. Cockburn & R. Fürst-Dilic (eds), *Bringing Technology Home: Gender and Technology in a Changing Europe*, Buckingham: Open University Press.

Berg, A.-J. & Aune, M. (1994), *Domestic Technology and Everyday Life: Mutual Shaping Processes*, COST A4, Luxembourg: Office for Official Publications of the European Communities.

Berg, A.-J. & Lie, M. (1995), 'Feminism and Constructivism: Do Artifacts have Gender?', *Science, Technology and Human Values*, **20**, (3), pp. 332-351.

Berg & Schumny 1990

Berg, M. (1997), 'Of Forms, Containers and the Electronic Record: Some Tools for a Sociology of the Formal', *Science, Technology and Human Values*, **22**, (4), pp. 403-433.

Berg, M. (1998), 'The Politics of Technology: On Bringing Social Theory into Technological Design', *Science, Technology and Human Values*, **23**, (4), pp. 456-490.

Berry, M.M.J. & Taggart, J.H. (1994), 'Managing Technology and Innovation: a Review', *R&D Management*, **24**, (4), pp. 341-353.

Bijker, W.E. (1987), 'The Social Construction of Bakelite: Towards a Theory of Invention', in W.E. Bijker, T.P. Hughes & T.J. Pinch (eds), *The Social Construction of Technological Systems*, Cambridge MA: MIT Press, pp. 159-187.

Bijker, W.E. (1992), 'The Social Construction of Fluorescent Lighting, or How an Artefact Was Invented in its Diffusion Stage', in W.E. Bijker & J. Law (eds), *Shaping Technology/Building Society*, Cambridge MA: MIT Press, pp. 75-102.

Bijker, W. (1993), 'Do Not Despair: There is Life After Constructivism', *Science, Technology and Human Values*, **18**, (1), pp. 113-138.

Bijker, W. (1995a), 'Sociohistorical Technology Studies', in S. Jasanoff *et al.* (eds), *A Handbook of Science and Technology Studies*, New York: Sage, pp. 229-256.

Bijker, W. (1995b), *Of Bicycles, Baekelite and Bulbs: Towards a Theory of Sociotechnical Change*, Cambridge MA: MIT Press.

Bijker, W.E., Hughes, T.S. & Pinch, T.J. (eds) (1987), *The Social Construction of Technological Systems: New Directions in the Sociology and History of Technology*, Cambridge MA: MIT Press.

Bijker, W.E. & Law, J. (eds) (1992a), *Shaping Technology/Building Society*, Cambridge MA: MIT Press.

Bijker, W.E. & Law, J. (1992b), 'General Introduction', in W.E. Bijker & J. Law (eds), *Shaping Technology/Building Society*, Cambridge MA: MIT Press, pp. 1-14.

Bijker, W.E. & Law, J. (1992c), 'Do Technologies Have Trajectories?', in W.E. Bijker & J. Law (eds), *Shaping Technology/Building Society*, Cambridge MA: MIT Press, pp. 17-20.

Bloomfield, B. & Vurdubakis, T. (1994), 'Boundary Disputes: Negotiating the Boundary between the Technical and the Social in the Development of IT Systems', *Information Technology and People*, **7**, (1), pp. 9-24.

Blume, S. (1993), *Insight and Industry: on the Dynamics of Technological Change in Medicine*, Cambridge MA: MIT Press.

Boden, M. (1994), 'Shifting the Strategic Paradigm: the Case of the Catalytic Converter', *Technology Analysis and Strategic Management*, **6**, (2), pp. 147-160.
Borg, K. (1999), 'The "Chauffeur" in the Early Auto Era: Structuration Theory and the Users of Technology', *Technology and Culture*, **40**, (4), pp. 797-832.
Bowker, G. (1987), 'A Well-Ordered Reality: Aspects of the Development of Schlumberger, 1920-39', *Social Studies of Science*, **17**, pp. 611-655.
Brady, T., Tierney, M. & Williams, R. (1992), 'The Commodification of Industry Applications Software', *Industrial and Corporate Change*, **1**, (3), pp. 489-514.
Braun, E. & Wield, D. (1994), 'Regulation as a Means for the Social Control of Technology', *Technology Analysis & Strategic Management*, **6**, pp. 259-272.
Brosveet, J. & Sørensen, K.H. (1997), *Fishing for Fun and Profit: Norway Domesticates Multimedia*, STS Working Paper 1/97, Centre for Technology and Society, Trondheim: University of Trondheim.
Burns, T.R. & Dietz, T. (1992), 'Technology, Sociotechnical Systems, Technological Development: an Evolutionary Perspective', in M. Dierkes & U. Hoffmann (eds), *New Technology at the Outset*, Boulder CO: Westview, pp. 206-238.
Cabral-Cardoso, C. (1996), 'The Politics of Technology Management: Influence and Tactics in Project Selection', *Technology Analysis and Strategic Management*, **8**, pp. 47-58.
Callon, M. (1980), 'The State and Technical Innovation: a Case Study of the Electric Vehicle in France', *Research Policy*, **9**, pp. 358-376.
Callon, M. (1986), 'The Sociology of an Actor-Network: the Case of the Electric Vehicle', in M. Callon, J. Law & A. Rip (eds), *Mapping the Dynamics of Science and Technology: Sociology of Science in the Real World*, Basingstoke: Macmillan.
Callon, M. (1987), 'Society in the Making: the Study of Technology as a Tool for Sociological Analysis', in W.E. Bijker, T.P. Hughes & T.J. Pinch (eds), *The Social Construction of Technological Systems*, Cambridge MA: MIT Press, pp. 83-103.
Callon, M. (1991), 'Techno-Economic Networks and Irreversibility', in J. Law (ed.), *A Sociology of Monsters: Essays on Power, Technology and Domination*, London: RKP, pp. 132-161.
Callon, M. (1993), 'Variety and Irreversibility in Networks of Technique Conception and Adoption', in D. Foray & C. Freeman (eds), *Technology and the Wealth of Nations*, London: OECD, pp. 232-268.
Callon, M. *et al.* (1991), 'Tools for the Evaluation of Technological Programmes: an Account of the Work Done at the Centre for the Sociology of Innovation', *Technology Analysis and Strategic Management*, **3**, (1), pp. 3-41.
Callon, M., Larédo, P. & Rabeharisoa, V. (1992), 'The Management and Evaluation of Technological Programs and the Dynamics of Techno-Economic Networks: the Case of the AFME', *Research Policy*, **21**, pp. 215-236.
Callon, M. & Latour, B. (1981), 'Unscreening the Big Leviathon: How Actors Macrostructure Reality and How Sociologists Help Them to Do So' in K. Knorr-Cetina & A.V. Cicourel (eds) *Advances in Social Theory and Methodology: Towards an Integration of Micro and Macro Sociologies*, London: RKP, pp. 277-303.

Callon, M. & Law, J. (1989) 'On the Construction of Sociotechnical Networks: Content and Context Revisited' Knowledge and Society, **9**, pp. 57-83

Cambrosio, A. & Limoges, C. (1991), 'Controversies as Governing Processes in Technology Assessment', *Technology Analysis and Strategic Management*, **3**, (4), pp. 377-396.

Cawson, A., Haddon, L. & Miles, I. (1995), *The Shape of Things to Consume: Delivering Information Technology into the Home*, Aldershot: Avebury.

Clark, K.B. (1985), 'The Interaction of Design Hierarchies and Market Concepts in Technological Evolution', *Research Policy*, **14**, pp. 235-251.

Clark, N. (1987), 'Similarities and Differences Between Scientific and Technological Paradigms', *Futures*, **19** pp. 26-42.

Clark P. & Staunton, N. (1989), *Innovation in Technology and Organization*, London: Routledge.

Clarke, A. & Montini, T. (1993), 'The Many Faces of RU486: Tales of Situated Knowledges and Technological Contestations', *Science, Technology and Human Values*, **18**, (1), pp. 42-78.

Clausen, C. & Jensen, P.L. (1993), 'Action-Oriented Approaches to Technology Assessment and Working Life in Scandinavia', *Technology Analysis and Strategic Management*, **5**, (2), pp. 83-97.

Clausen, C. & Koch, C. (2001) 'Spaces and Occasions in the Social Shaping of Information Technologies: the Transformation of IT Systems for Manufacturing in a Danish Context', Chapter 8 in this volume.

Clausen, C. & Williams, R. (eds) (1997a), *The Social Shaping of Computer-Aided Production Management and Computer-Integrated Manufacture*, COST A4, Luxembourg: Office for Official Publications of the European Communities.

Clausen, C. & Williams, R. (1997b), 'The Social Shaping of Computer-Aided Production Management and Computer-Integrated Manufacture', in C. Clausen & R. Williams (eds), *The Social Shaping of Computer-Aided Production Management and Computer-Integrated Manufacture*, COST A4, Luxembourg: Office for Official Publications of the European Communities, pp. 1-27.

Clayton, A., Spinardi, G. & Williams, R. (1999), *Policies for Cleaner Technology: a New Agenda for Government and Industry*, London: Earthscan.

Cockburn, C. (1981), 'The Material of Male Power', *Feminist Review*, **9**, pp. 41-58.

Cockburn, C. (1983), *Brothers: Male Dominance and Technological Change*, London: Pluto.

Cockburn, C. (1985), *Machinery of Dominance: Men, Women and Technical Know-How*, London: Pluto.

Cockburn, C. & Ormrod, S. (1993), *Gender and Technology in the Making*, London: Sage.

Collingridge, D. (1980), *The Social Control of Technology*, London: Pinter.

Collingridge, D. (1992), *The Management of Scale: Big Organizations, Big Decisions, Big Mistakes*, London: Routledge.

Collins, H.M. (1990), *Artificial Experts: Social Knowledge and Intelligent Machines*, Cambridge MA: MIT Press.

Collins, H.M. & Pinch, T. (1998), *The Golem at Large: What You Should Know about Technology*, Cambridge: Cambridge University Press.

Collins, H.M. & Yearley, S. (1991), 'Epistemological Chicken', in A. Pickering (ed), *Science as Practice and Culture*, Chicago: Chicago University Press, pp. 301-326.

Constant, E.W. (1984), 'Communities and Hierarchies: Structure in the Practice of Science and Technology', in R. Laudan (ed.), *The Nature of Technological Knowledge: Are Models of Scientific Change Relevant?*, Dordrecht: Reidel.

Conway, S. (1995), 'Informal Boundary-Spanning Networks in Successful Technological Innovation', *Technology Analysis and Strategic Management*, **7**, (3), pp. 327-342.

Coombs, R., Saviotti, P. & Walsh, V. (eds) (1992), *Technical Change and Company Strategies*, London: Academic Press.

Coombs, R., Green, K., Richards, V. & Walsh, V. (eds) (1998), *Technological Change and Organisation*, Aldershot: Edward Elgar.

Cowan, R.S. (1976), 'The "Industrial Revolution" in the Home: Household Technology and Social Change in the Twentieth Century', *Technology and Culture*, **17**, pp. 1-23.

Cowan, R.S. (1983), *More Work for Mother: the Ironies of Household Technology from the Open Hearth to the Microwave*, New York: Basic Books.

Cowan, R.S. (1987), 'The Consumption Junction', in W.E. Bijker, T.P. Hughes & T.J. Pinch (eds), *The Social Construction of Technological Systems: New Directions in the Sociology and History of Technology*, Cambridge MA: MIT Press.

Cowan, R. (1992), 'High Technology and the Economics of Standardization', in M. Dierkes & U. Hoffmann (eds), *New Technology at the Outset*, Boulder CO: Westview, pp. 279-300.

Cramer, J. *et al.* (1990), 'Stimulating Cleaner Technologies through Economic Instruments: Possibilities and Constraints', *UNEP Industry and Technology Review*, **13**, (2), pp. 46-53.

Cronberg, T. & Sørensen, K. (eds) (1995a), *Similar Concerns, Different Styles? Technology Studies in Western Europe*, COST A4, Luxembourg: Office for Official Publications of the European Communities.

Cronberg, T. & Sørensen, K. (1995b), 'Similar Concerns, Different Styles? a Note on European Approaches to the Social Shaping of Technology', in T. Cronberg & K. Sørensen, (eds), *Similar Concerns, Different Styles? Technology Studies in Western Europe*, COST A4, Luxembourg: Office for Official Publications of the European Communities, pp. 1-23.

David, P. (1975), *Technical Choice, Innovation and Economic Growth: Essays on American and British Experience in the Nineteenth Century*, Cambridge: Cambridge University Press.

David, P. (1986), 'Understanding the Economics of QWERTY: the Necessity of History', in W.N. Parker (ed.), *Economic History and the Modern Economist*, Oxford: Oxford University Press, pp. 30-49.

Davies, A. (1996), 'Innovation in Large Technical Systems: the Case of Telecommunications', *Industrial and Corporate Change*, **5**, (4), pp. 1142-1180.

de Bresson, C. & Amesse, F. (1991), 'Networks of Innovators: a Review and Introduction to the Issue', *Research Policy*, **20**, (5), pp. 363-379.

de la Bruheze, A.A.A. (1992), *Political Construction of Technology: Nuclear Waste Disposal in the United States*, 1945-1972, Enschede: Universiteit Twente.

Deuten, J.J., Rip, A. & Jelsma, J. (1997), 'Societal Embedment and Product Creation Management', *Technology Analysis and Strategic Management*, **9**, (2), pp. 219-236.

Dierkes, M. & Hoffmann, U. (1992), 'Understanding Technological Development as a Social Process: an Introductory Note', in M. Dierkes & U. Hoffmann (eds), *New Technology at the Outset*, Boulder CO: Westview, pp. 9-13.

Dierkes, M., Hoffmann, U. & Marz, L. (1996), *Visions of Technology: Social and Institutional Factors Shaping the Development of New Technologies*, Berlin: Campus.

Disco, C. & van der Meulen, B. (eds) (1998a), *Getting New Technologies Together: Studies in Making Sociotechnical Order*, Berlin: Walter de Gruyter.

Disco, C. & van der Meulen, B. (1998b), 'Introduction', in C. Disco & B. van der Meulen (eds), *Getting New Technologies Together: Studies in Making Sociotechnical Order*, Berlin: Walter de Gruyter, pp. 1-13.

Disco, C. & van der Meulen, B. (1998c), 'Getting Case Studies Together: Conclusions on the Coordination of Sociotechnical Order', in C. Disco & B. van der Meulen (eds), *Getting New Technologies Together: Studies in Making Sociotechnical Order*, Berlin: Walter de Gruyter, pp. 323-352.

Dosi, G. (1982), 'Technological Paradigms and Technological Trajectories', *Research Policy*, **11**, pp. 147-162.

Dosi, G. (1988), 'The Nature of the Innovative Process', in G. Dosi *et al.* (eds), *Technical Change and Economic Theory*, London: Pinter.

Dosi, G. (1990), 'The Nature of the Innovative Process', in C. Freeman & L. Soete (eds), *New Explorations in the Economics of Technological Change*, London: Pinter.

Dosi, G. (1991), 'Perspectives on Evolutionary Theory', *Science and Public Policy*, **18**, (6), pp. 353-361.

Dosi, G. *et al.* (eds) (1988), *Technical Change and Economic Theory*, London: Pinter.

Dosi, G. & Orsenigo, L. (1988), 'Coordination and Transformation: an Overview of Structures, Behaviour and Change in Evolutionary Environments', in G. Dosi *et al.* (eds), *Technical Change and Economic Theory*, London: Pinter, pp. 13-37.

Downey, E. (1995), 'Engineering Studies' in S. Jasanoff *et al.*, *Handbook of Science and Technology Studies*, London: Sage, pp. 167-188.

Drejer, A. (1996), 'Frameworks for the Management of Technology: Towards a Contingent Approach', *Technology Analysis and Strategic Management*, **8**, pp. 9-20.

Duncker, E. & Disco, C. (1998), 'Meaningful Boundaries: Symbolic Representations in Heterogenous Research and Development Projects', in C. Disco & B. van der Meulen (eds), *Getting New Technologies Together: Studies in Making Sociotechnical Order*, Berlin: Walter de Gruyter, pp. 265-289.

Edge, D. (1988), *The Social Shaping of Technology*, Edinburgh PICT Working Paper no. 1, RCSS, Edinburgh: University of Edinburgh.

Edge, D. (1994), 'The Social Shaping of Technology', in N. Heap *et al.* (eds), *Information Technology and Society*, London: Sage.

Egyedi, T. (1996), *Shaping Standardisation: A Study of the Standards Process and Standards Policies in the Field of Telematic Services*, Delft: Delft University of Technology.

Elliott, B. (ed.) (1988), *Technology and Social Process*, Edinburgh: Edinburgh University Press.

Elzen, B., Enserink, B. & Smit, W.A. (1990), 'Weapon Innovation: Networks and Guiding Principles', *Science and Public Policy*, **17**, (3), pp. 171-193.

Elzen, B., Enserink, B. & Smit, W.A. (1996), 'Socio-Technical Networks: How a Technology Studies Approach May Help to Solve Problems Related to Technical Change', *Social Studies of Science*, **26**, pp. 95-141.
Elzen, B. & Mulder, K. (1995), 'What can Technology Assessment (not) Learn from Technology Studies? an Analysis Using Case Studies on Electric Vehicles and Biodegradable Plastics', paper to *Third International ASEAT Conference Managing Technology into the 21st Century*, Manchester, 6-8 Sep 1995.
Feenberg, A. (1991), *Critical Theory of Technology*, New York: Oxford University Press.
Fleck, J. (1988a), *Innofusion or Diffusation*?, Edinburgh PICT Working Paper no. 7, RCSS, Edinburgh: Edinburgh University.
Fleck, J. (1988b), *The Development of Information Integration: Beyond CIM*?, Edinburgh PICT Working Paper no. 9, RCSS, Edinburgh: Edinburgh University.
Fleck, J. (1991), *Innovation During Implementation: Configurations and CAPM*, Edinburgh PICT Paper no. 34, RCSS, Edinburgh: Edinburgh University.
Fleck, J. (1993), 'Configuration: Crystallising Contingency', *International Journal of Human Factors in Engineering*, **3**, (3), pp. 15-36.
Fleck, J. (1994a), 'Continuous Evolution: Corporate Configurations of Information Technology', in R. Mansell (ed.), *The Management of Information and Communication Technologies: Emergent Patterns of Control*, London: ASLIB/Association of Information Management.
Fleck, J. (1994b), 'Learning by Trying: the Implementation of Configurational Technology', *Research Policy*, **23**, pp. 637-652.
Fleck, J. (1997), 'Contingent Knowledge and Technology Development', *Technology Analysis and Strategic Management*, **9**, (4), pp. 383-397.
Fleck, J. & Howells, J. (1997), *Defining Technology and the Paradox of Technological Determinism*, Department of Management Studies Working Paper, Uxbridge: Brunel University.
Fleck, J., Webster, J. & Williams, R. (1990), 'Dynamics of Information Technology Implementation: a Reassessment of Paradigms and Trajectories of Development', *Futures*, **22**, pp. 618-640.
Forrest, J. (1991), 'Models of the Process of Technological Innovation', *Technology Analysis and Strategic Management*, **3**, (4), pp. 439-453.
Fransman, M. (1991), 'Controlled Competition in the Japanese Telecommunications Equipment Industry: the Case of Central Office Switches', in C. Antonelli (ed.), *The Economics of Information Networks*, Amsterdam: Elsevier, pp. 247-271.
Freeman, C. (1988), 'Introduction', in G. Dosi *et al.*(eds), *Technical Change and Economic Theory*, London: Pinter, pp. 1-12.
Freeman, C. (1991), 'Networks of Innovators: a Synthesis of Research Issues', *Research Policy*, **20**, (5), pp. 499-514.
Freeman, C. (1996), 'The Greening of Technology and Models of Innovation', *Technological Forecasting and Social Change*, **53**, pp. 27-39.
Frickel, S. (1996), 'Engineering Heterogeneous Accounts: the Case of Submarine Thermal Reactor Mark-I', *Science, Technology and Human Values*, **21**, (1), pp. 28-53.
Fujimura, J. (1987), 'Constructing Do-Able Problems in Cancer Research: Articulating Alignment', *Social Studies of Science*, **17**, pp. 257-293.

Fujimura, J. (1992), 'Crafting Science: Standardized Packages, Boundary Objects, and "Translation"', in A. Pickering (ed.), *Science as Practice and Culture*, Chicago: Chicago University Press, pp.168-211.
Geels, F. & Schot, J. (1998), 'Reflexive Technology Policies and Sociotechnical Scenarios', working paper for conference *Constructing Tomorrow: Technology Strategies for the New Millenium*, Bristol Business School, 14-15 Sep 1998.
Geels, F. (2000), 'Towards Sociotechnical Scenarios and Reflexive Anticipation: Using Patterns and Regularities in Technology Dynamics', Chapter 13 in this volume.
Genus, A. (1993), 'Political Construction and Control of Technology: Wave-Power Renewable Technologies', *Technology Analysis and Strategic Management*, **5**, (2), pp. 137-149.
Giere, R. (1993), 'Science and Technology Studies: Prospects for an Enlightened Postmodern Synthesis', *Science, Technology and Human Values*, **18**, (1), pp. 102-112.
Gieryn, T.F. (1995), 'Boundaries of Science', in S. Jasanoff *et al.* (eds), *Handbook of Science and Technology Studies*, New York: Sage.
Görlitz, A. & Druwe, U. (1990), *Politische Steuerung und Systemumwelt*, Pfaffenweiler: Centaurus.
Görlitz, A. (1994), *Umweltpolitische Steuerung*, Baden-Baden: Nomos.
Graham, I. (1999), 'Electronic Markets: the Emergence of Technological Systems and their Materiality in the Lifeworld', paper to conference *Bringing Materiality (Back) into Management*, Gilleleje, 21-23 Oct 1999.
Green, E., Owen, J. & Pain, D. (eds) (1993), *Gendered by Design? Information Technology and Office Systems*, London: Taylor & Francis.
Green, K. (1991), 'Shaping Technologies and Shaping Markets: Creating Demand for Biotechnology', *Technology Analysis and Strategic Management*, **3**, (1), pp. 57-76.
Green, K. (1992), 'Creating Demand for Biotechnology: Shaping Technologies and Markets', in R. Coombs, P. Saviotti & V. Walsh (eds), *Technical Change and Company Strategies*, London: Academic Press, pp. 164-184.
Green, K., Jones, O. & Coombs, R. (1996), 'Critical Perspectives on Technology Management: an Introduction', *Technology Analysis and Strategic Management*, **8**, (1), pp. 3-7.
Green, K., Walsh, V. & Richards, A. (1998), 'Differences in "Styles" of Technological Innovation: an Introduction', *Technology Analysis and Strategic Management*, **10**, (4), pp. 403-405.
Green, K., Hull, R., McMeekin, A. & Walsh, V. (1999), 'The Construction of the Techno-Economic: Networks vs. Paradigms', *Research Policy*, **28**, (7), pp. 777-792.
Grin, J. & van de Graaf, H. (1996), 'Technology Assessment as Learning', *Science, Technology and Human Values*, **21**, (1), pp. 72-99.
Grint, K. & Woolgar, S. (1992), 'Computers, Guns, and Roses: What's Social about Being Shot?', *Science, Technology and Human Values*, **17**, (3), pp. 366-80.
Grint, K. & Woolgar, S. (1997), *The Machine at Work: Technology, Work and Organization*, Cambridge: Polity.
Groenewegen, P. & Vergragt, P. (1991), 'Environmental Issues as Threats and Opportunities for Technological Innovation', *Technology Analysis and Strategic Management*, **3**, (1), pp. 43-55.

Guerrieri, P. & Tylecote, A. (1998), 'Cultural and Institutional Determinants of National Technological Advantage', in R. Coombs *et al.* (eds), *Technological Change and Organization*, Aldershot: Edward Elgar, pp. 180-209.

Gutting, G. (1984), 'Paradigms, Revolutions and Technology', in R. Laudan (ed.), *The Nature of Technological Knowledge: Are Models of Scientific Change Relevant?*, Dordrecht: Reidel, pp. 47-65.

Haddon, L. (1992), 'Explaining ICT Consumption: the Case of the Home Computer', in R. Silverstone & E. Hirsch (eds), *Consuming Technologies: Media and Information in Domestic Spaces*, London: Routledge, pp. 82-96.

Hage, J. & Hollingsworth, R. (2000), *Idea Innovation Networks: a Strategy for Integrating Organizational und Institutional Analysis*, Working Paper TUTS-WP-5-2000, Berlin: Technical University Berlin.

Hameri, A.-P. (1994), 'Using Technical Norms in Explaining Technological Action', *Technovation*, **14**, (8), p. 505.

Hansen, A. (1999), 'Anticipated Consumer Demand or Technology Push: the Sources for Development of Organic and Functional Foods in the Dairy Industry', paper to conference *Bringing Materiality (Back) into Management*, Gilleleje, 21-23 Oct 1999.

Hård, M. (1993), 'Beyond Harmony and Consensus: a Social Conflict Approach to Technology', *Science, Technology and Human Values*, **18**, (4), pp. 408-432.

Hård, M. (1994), 'Technology as Practice: Local and Global Closure Processes in Diesel Engine Design', *Social Studies of Science*, **24**, (3), pp. 549-585.

Hård, M. & Jamison, A. (eds) (1998), *The Intellectual Appropriation of Technology: Discourses on Modernity, 1900-1939*, Cambridge MA: MIT Press.

Hartmann, G. *et al.* (1994), 'Computerized Machine Tools, Manpower Consequences and Skill Utilization: a Study of British and West German Manufacturing Firms', in E. Rhodes & D. Wield (eds), *Implementing New Technologies: Innovation and the Management of Technology*, Oxford: Blackwell, pp. 308-316.

Hatling, M. & Sørensen, K. H. (1998), 'Social Constructions of User Participation', in K.H. Sørensen (ed.), *The Spectre of Participation: Technology and Work in a Welfare State*, Olso: Scandinavian University Press, pp. 171-188.

Hawkins, R., Mansell, R. & Skea, J. (eds) (1995), *Standards, Innovation and Competitiveness: The Politics and Economics of Standards in Natural and Technical Environments*, Aldershot: Edward Elgar.

Hill, S. (1988), *The Tragedy of Technology*, London: Pluto.

Hill, S. (1992), 'Cultural Power of the Technology "Text"', paper to *4S/EASST Joint Conference on Science, Technology and Development*, Göteborg, 12-15 Aug 1992.

Howells, J. (1995), 'A Socio-Cognitive Approach to Innovation', *Research Policy*, **24**, pp. 883-894.

Howells, J. & Hine, J. (eds) (1993), *Innovative Banking: Competition and the Management of a New Network Technology*, London: Routledge.

Hughes, T.P. (1978), 'Inventors: the Problems They Choose, the Ideas They Have and the Inventions They Make', in P. Kelly *et al.* (eds), *Technological Innovation: a Critical Review of Current Knowledge*, San Francisco: San Francisco Press, pp. 166-182.

Hughes, T.P. (1979), 'The Electrification of America: the System-Builders', *Technology and Culture*, **20**, pp. 124-161.
Hughes, T.P. (1983), *Networks of Power: Electrification in Western Society, 1880-1930*, Baltimore MA: Johns Hopkins University Press.
Hughes, T.P. (1986a), 'The Seamless Web: Technology, Science, etcetera, etcetera', *Social Studies of Science*, **16**, pp. 281-292.
Hughes, T.P. (1986b), 'Reverse Salients and Critical Problems', paper to conference *Technology and Social Change*, Edinburgh, June 1986.
Hughes, T.P. (1986c), 'The System Builders', in B. Elliott (ed.), *Technology, Innovation and Change*, Centre of Canadian Studies, Edinburgh: Edinburgh University pp. 17-28.
Hughes, T.P. (1987), 'The Evolution of Large Technological Systems', in W.E. Bijker, T.P. Hughes & T.J. Pinch (eds), *The Social Construction of Technological Systems*, Cambridge MA: MIT Press, pp. 51-82.
Hughes, T.P. (1994), 'Technological Momentum', in M.R. Smith & L. Marx (eds), *Does Technology Drive History? the Dilemma of Technological Determinism*, Cambridge MA: MIT Press, pp. 101-113.
Hughes, T.P. (1995), 'Shaped Technology: an Afterword', *Science in Context*, **8**, (2), p. 451.
Irwin, A. & Vergragt, P. (1989), 'Re-Thinking the Relationship Between Environmental Regulation and Industrial Innovation: the Social Negotiation of Technical Change', *Technology Analysis and Strategic Management*, **1**, (1), pp. 57-70.
Jasanoff, S. (1986), *Risk Management and Political Culture*, New York: Russell Sage Foundation.
Jasanoff, S. (1987), 'Contested Boundaries in Policy Relevant Science', *Social Studies of Science*, **17**, pp. 195-230.
Joerges, B. (1988), 'Large Technical Systems: Concepts and Issues', in R. Mayntz & T.P. Hughes (eds), *The Development of Large Technical Systems*, Frankfurt-am-Main: Campus Verlag, pp. 9-36.
Johnson, J. / Latour, B. (1988), 'Mixing Humans and Non-Humans Together: the Sociology of a Door-Closer', *Social Problems*, **35**, (3), pp. 298-310.
Jørgensen, J. & Karnøe, P. (1995), 'The Danish Wind-Turbine Story: Technical Solutions to Political Visions?', in A. Rip, T. Misa & J. Schot (eds), *Managing Technology in Society: the Approach of Constructive Technology Assessment*, London: Pinter, pp. 57-82.
Jørgensen, U. (1999), 'Can Knowledge be an Object of Management?', paper to conference *Bringing Materiality (Back) in to Management*, Gilleleje, 21-23 Oct 1999.
Jørgensen, U. & Sørensen, O. (2001), 'Arenas of Development: a Space Populated by Actor-Worlds, Artefacts, and Surprises', Chapter 7 in this volume.
Jørgensen, U. & Strunge, L. (2001), 'Restructuring the Power Arena in Denmark: Shaping Markets, Technology and Environmental Priorities', Chapter 11 in this volume.
Karnøe, P. (1996), 'The Social Process of Competence Building', *International Journal of Technology Management*, **11**, (7/8), p. 770.
Kaplinsky, R. (1984), *Automation: the Technology and Society*, Harlow: Longman.
Kash, D.E. & Rycroft, R.W. (1993), 'Two Streams of Technological Innovation: Implications for Policy', *Science and Public Policy*, **20**, (1), pp. 27-36.

Kemp, R. (1994), 'Technology and the Transition to Environmental Sustainability: the Problem of Technological Regime Shifts', *Futures*, **26**, (10), p. 1023.
Kemp, R., Schot, J. & Hoogma, R. (1998), 'Regime Shifts to Sustainability through Processes of Niche Formation; the Approach of Strategic Niche Management', *Technology Analysis and Strategic Management*, **10**, (2), pp. 175-195.
Kline, R. & Pinch, T. (1996), 'Users as Agents of Technological Change: the Social Construction of the Automobile in the Rural United States', *Technology and Culture*, **37**, (4).
Kling, R. (1992a), 'Audiences, Narratives and Human Values in Social Studies of Technology', *Science, Technology and Human Values*, **17**, (3), pp. 349-365.
Kling, R. (1992b), 'When Gunfire Shatters Bone: Reducing Socio-Technical Systems to Social Relationships', *Science, Technology and Human Values*, **17**, (3), pp. 381-385.
Kling, R. & Scacchi, W. (1982), 'The Web of Computing: Computer Technology as Social Organization', *Advances in Computers*, **21**, pp. 1-85.
Knie, A. (1992), 'Yesterday's Decisions Determine Tomorrow's Options: the Case of the Mechanical Typewriter', in M. Dierkes & U. Hoffmann (eds), *New Technology at the Outset*, Boulder CO: Westview, pp. 161-172.
Landström, C. (1998), 'National Strategies: the Gendered Appropriation of Household Technology', in M. Hård & A. Jamison (eds) *The Intellectual Appropriation of Technology: Discourses on Modernity, 1900-1939*, Cambridge MA: MIT Press, pp. 163-188.
La Porte, T.R. (ed.) (1991), *Social Responses to Large Technical Systems: Control or Anticipation*, Dordrecht: Kluwer.
Latour, B. (1988), 'The Prince for Machines as Well as for Machinations', in B. Elliott (ed.), *Technology and Social Process*, Edinburgh: Edinburgh University Press, pp. 20 – 43.
Latour, B. (1991), 'Technology is Society Made Durable', in J. Law (ed.), *A Sociology of Monsters: Essays on Power, Technology and Domination*, London: RKP, pp. 103-131.
Latour, B. (1992), 'Where are the Missing Masses? the Sociology of a Few Mundane Artifacts', in W.E. Bijker & J. Law (eds), *Shaping Technology/Building Society*, Cambridge MA: MIT Press, pp. 225-258.
Laudan, R. (1984), 'Cognitive Change in Science and Technology', in R. Laudan (ed.), *The Nature of Technological Knowledge: Are Models of Scientific Change Relevant?*, Dordrecht: Reidel, pp. 83-104.
Law, J. (1986a), 'The Anatomy of a Socio-Technical Struggle: the Design of the TSR2', in B. Elliott (ed.), *Technology and Social Process*, Edinburgh: Edinburgh University Press, pp. 44-69.
Law, J. (1986b), 'On the Methods of Long-Distance Control: Vessels, Navigation and the Portuguese Route to India', in L. Law (ed.), *Power, Action and Belief: a New Sociology of Knowledge?*, London: RKP.
Law, J. (1987a), 'The Structure of Sociotechnical Engineering: a Review of the New Sociology of Technology', *Sociological Review*, **35**, pp. 404-425.
Law, J. (1987b), 'On the Social Explanation of Technological Change: the Case of Portuguese Maritime Expansion', *Technology and Culture*, **28**, pp. 227-252.
Law, J. (1987c), 'Technology and Heterogeneous Engineering: the Case of Portuguese Expansion', in W.E. Bijker, T.P. Hughes & T.J. Pinch (eds), *The*

Social Construction of Technological Systems, Cambridge MA: MIT Press, pp. 111-134.

Law, J. (1991) (ed.), *A Sociology of Monsters: Essays on Power, Technology and Domination*, London: RKP.

Law, J. (1992a), 'Notes on the Theory of the Actor-Network: Ordering, Strategy and Homogeneity', *Systems Practice*, **5**, pp. 379-394.

Law, J. (1992b), 'The Olympus 320 Engine: a Case Study in Design, Development, and Organizational Control', *Technology and Culture*, **33**, pp. 409-440.

Law, J. & Callon, M. (1988), 'Engineering and Sociology in a Military Aircraft Project: a Network Analysis of Technological Change', *Social Problems*, **35**, pp. 284-297.

Law, J. & Bijker, W. 1992, 'Postscript: Technology, Stability and Social Theory', in W.E. Bijker & J. Law (eds), *Shaping Technology/Building Society*, Cambridge MA: MIT Press, pp. 291-308.

Law, J. & Mol, A. (1995), 'Notes on Materiality and Sociality', *Sociological Review*, **43**, pp. 275-294.

Leonard-Barton, D. (1988), 'Implementation as Mutual Adaptation of Technology and Organization', *Research Policy*, **17**, pp. 251-267.

Leyten, J. & Smits, R. (1996), 'The Role of Technology Assessment in Technology Policy', *International Journal of Technology Management*, **11**, (5/6), pp. 688-702.

Lie, M. & Sørensen, K.H. (eds) (1996), *Making Technology our Own? Domesticating Technology into Everyday Life*, Oslo: Scandinavian University Press.

Lundvall, B.-A. (1988), 'Innovation as an Interactive Process: from User-Producer Interaction to the National System of Innovation', in G. Dosi *et al.* (eds), *Technical Change and Economic Theory*, London: Pinter.

Lundvall, B.-A. (1992), *National Systems of Innovation: Towards a System of Innovation and Interactive Learning*, London: Pinter.

Lundvall, B.-A. (1998), 'Why Study National Systems and National Styles of Innovation?', *Technology Analysis and Strategic Management*, **10**, (4), pp. 407-421.

Lutz, B. (1992), 'Technology Research and Technology Policy: Impacts of a Paradigm Shift', in M. Dierkes & U. Hoffmann (eds), *New Technology at the Outset*, Boulder CO: Westview, pp. 14-30.

Mackay, H. & Gillespie, G. (1992), 'Extending the Social Shaping of Technology: Ideology and Appropriation', *Social Studies of Science*, **22**, pp. 685-716.

MacKenzie, D. (1988), 'The Problem with "the Facts": Nuclear Weapons Policy and the Social Negotiation of Data', in R. Davidson & P. White (eds), *Information and Government: Studies in the Dynamics of Policy-Making*, Edinburgh: Edinburgh University Press, pp. 232-251.

MacKenzie, D. (1989), 'From Kwajalein to Armageddon? Testing and the Social Construction of Missile Accuracy', in D. Gooding, T.J. Pinch & S. Schaffer (eds), *The Uses of Experiment: Studies in the Natural Sciences*, Cambridge: Cambridge University Press, pp. 409-436.

MacKenzie, D. (1990), *Inventing Accuracy: an Historical Sociology of Ballistic Missile Guidance*, Cambridge MA: MIT Press.

MacKenzie, D. (1992), 'Economic and Sociological Explanation of Technological Change', in R. Coombs, P. Saviotti & V. Walsh (eds), *Technical Change and Company Strategies*, London: Academic Press.
MacKenzie, D. (1996a), 'How Do We Know the Properties of Artefacts? Applying the Sociology of Knowledge to Technology', in R. Fox (ed.), *Technological Change: Methods and Themes in the History of Technology*, Amsterdam: Harwood Academic, pp. 247-264.
MacKenzie, D. (1996b), *Knowing Machines: Essays on Technological Change*, Cambridge MA: MIT Press.
MacKenzie, D., Rüdig, W. & Spinardi, G. (1988), 'Social Research of Technology and the Policy Agenda: an Example from the Strategic Arms Race', in B. Elliott (ed.), *Technology and Social Process*, Edinburgh: Edinburgh University Press, pp. 152-180.
MacKenzie, D. & Wajcman, J. (eds) (1985), *The Social Shaping of Technology: or How the Refrigerator got its Hum*, Buckingham: Open University Press.
MacKenzie, D. & Wajcman, J. (eds) (1999a), *The Social Shaping of Technology*, 2nd edition, Buckingham: Open University Press.
MacKenzie, D. & Wajcman, J. (1999b), 'Introductory Essay: the Social Shaping of Technology', in D. MacKenzie & J. Wajcman (eds), *The Social Shaping of Technology*, Buckingham: Open University Press, pp. 3-27.
Mangematin, V. & Callon, M. (1995), 'Technological Competition, Strategies of the Firms and the Choice of First Users: the Case of Road Guidance Technologies', *Research Policy*, **24**, pp. 441-458.
McLaughlin, J., Rosen, P., Skinner, D. & Webster, A. (1999), *Valuing Technology: Organisations, Culture and Change*, London: Routledge.
McLoughlin, I. (1997), 'Babies, Bathwater, Guns and Roses', in I. McLoughlin & M. Harris (eds), *Innovation, Organisational Change and Technology*, London: Thomson International Business Press, pp. 207-221.
McLoughlin, I. (1999), *Creative Technological Change: the Shaping of Technology and Organisations*, London: Routledge.
Mayntz, R. & Hughes, T.P. (eds) (1988), *The Development of Large Technical Systems*, Frankfurt am Main: Campus.
Metcalfe, S. (1994a), 'Evolutionary Economics and Technology Policy', *Economic Journal*, **104**, pp. 931-944.
Metcalfe, S. (1994b), 'Evolution, Technology Policy and Technology Management', *Prometheus*, **12**, (1), pp. 29-35.
Metcalfe, S. (1995), 'Technology Systems and Technology Policy in an Evolutionary Framework', *Cambridge Journal of Economics*, **19**, pp. 25-46.
Metcalfe, S. & Boden, M. (1992), 'Evolutionary Epistemology and the Nature of Technology Strategy', in R. Coombs, P. Saviotti & V. Walsh (eds), *Technical Change and Company Strategies*, London: Academic Press, pp. 49-71.
Miles, I., Cawson, A. & Haddon, L. (1992), 'The Shape of Things to Consume', in R. Silverstone & E. Hirsch (eds), *Consuming Technologies: Media and Information in Domestic Spaces*, London: Routledge, pp. 67-81.
Misa, T. (1992), 'Controversy and Closure in Technological Change: Constructing "Steel"', in W.E. Bijker & J. Law (eds), *Shaping Technology / Building Society: Studies in Sociotechnical Change*, Cambridge MA: MIT Press, pp. 109-139.
Misa, T. (1994), 'Retrieving Socio-Technical Change from Technological Determinism', in M.R. Smith & L. Marx (eds), *Does Technology Drive*

History? the Dilemma of Technological Determinism, Cambridge MA: MIT Press, pp. 115-141.

Molina, A. (1992a), 'Integrating the Creation, Production and Diffusion of Technology in the Design of Large-Scale and Targeted European IT Programmes', *Technology Analysis and Strategic Management*, **4**, (3), pp. 299-309.

Molina, A. (1992b), 'Europe's IT Strategy Moves Closer to the Users: the Open Microprocessor Systems Initiative (OMI)', *Science and Public Policy*, **19**, (3), pp. 145-156.

Molina, A. (1993), 'In Search of Insights into the Generation of Techno-Economic Trends: Micro- and Macro-Constituencies in the Microprocessor Industry', *Research Policy*, **22**, (5/6), pp. 479-506.

Molina, A. (1994a), 'The Generation of Large-Scale Capability-Building Initiatives: Extending the Sociotechnical Constituencies Perspective', in R. Mansell (ed.), *The Management of Information and Communication Technologies: Emergent Patterns of Control*, London: ASLIB/Association of Information Management, pp. 90-120.

Molina, A. (1994b), 'Understanding the Emergence of a Large-Scale European Initiative in Technology', *Science and Public Policy*, **21**, (1), pp. 31-41.

Molina, A. (1995), 'Sociotechnical Constituencies as Processes of Alignment: The Rise of a Large-Scale European Information Technology Initiative', *Technology and Society*, **17**, pp. 385-412.

Molina, A. (1997), 'Insights into the Nature of Technology Diffusion and Implementation: the Perspective of Sociotechnical Alignment', *Technovation*, **17**, pp. 601-626.

Molina, A. (1998), 'The Nature of "Failure" in a Technological Initiative: the Case of the Europrocessor', *Technology Analysis and Strategic Management*, **10**, (1), pp. 23-40.

Molina, A. (1999), 'Understanding the Role of the Technical in the Build-up of Sociotechnical Constituencies', *Technovation*, **19**, (1), p. 1.

Mollinga, P. & Mooij, J. (1989), *Cracking the Code: Towards a Conceptualisation of the Social Content of Technical Artefacts*, Occasional Paper 18, Technology Policy Group, Milton Keynes: Open University.

Naesje, P. (2001), 'Governing Measures: User-Stories and Heat Pump Subsidies', Chapter 10 in this volume.

Nelson, R. (1993), *National Innovation Systems: a Comparative Study*, Oxford: Oxford University Press.

Nelson, R.R. & Winter, S.G. (1977), 'In Search of a Useful Theory of Innovation', *Research Policy*, **6**, pp. 36-76.

Nelson, R.R. & Winter, S.G. (1982), *An Evolutionary Theory of Economic Change*, Cambridge MA: Bellknap Press.

Nelson, T. (1994), 'The Co-Evolution of Technology, Industrial Structure, and Supporting Institutions', *Industrial and Corporate Change*, **3**, pp. 47-63.

Newman, S.E. (1999), 'That Obscure Object of Design: Object-Centered Sociality and the Design of Mediations in Software Engineering', paper to conference *Sociality/Materiality: the Status of the Object in Social Science*, Brunel University, Uxbridge, 9-11 Sep 1999.

Nicoll, D. W. (1999), 'Users as Currency: Technology and Marketing Trials', in *The Social Shaping of Multimedia*, eds. R. Slack, J. Stewart & R. Williams, COST A4, Luxembourg: Office for Official Publications of the European Communities, pp. 55-72.

Orlikowski, W.J. (1992), 'The Duality of Technology: Rethinking the Concept of Technology in Organizations', *Organization Science*, **3**, (3), pp. 398-427.

Pacey, A. (1983), *The Culture of Technology*, Oxford: Blackwell.

Pavitt, K. (1984), 'Sectoral Patterns of Technological Change: Towards a Taxonomy and a Theory', *Research Policy*, **13**, (6), pp. 343-373.

Pavitt, K. (1987), 'The Objectives of Technology Policy', *Science and Public Policy*, **14**, (4), pp. 182-188.

Pelaez, E. (1990), *What Shapes Software Development?*, Edinburgh PICT Working Paper no. 10, RCSS, Edinburgh: Edinburgh University.

Perrin, J. & Vinck, D. (1996), *The Role of Design in the Shaping of Technology*, COSTA4, Luxembourg: Office for Official Publications of the European Communities.

Pfaffenberger, B. (1992), 'Technological Dramas', *Science, Technology and Human Values*, **17**, pp. 282-312.

Pickering, A. (1993), 'The Mangle of Practice: Agency and Emergence in the Sociology of Science', *American Journal of Sociology*, (99), pp. 559-589.

Pinch, T.J. (1988), 'Understanding Technology: Some Possible Implications of Work in the Sociology of Science', in B. Elliott (ed.), *Technology and Social Process*, Edinburgh: Edinburgh University Press.

Pinch, T.J. (1993), '"Testing – One, Two, Three ... Testing!": Toward a Sociology of Testing', *Science, Technology and Human Values*, **18**, (1), pp. 25-41.

Pinch, T.J. (1996), 'The Social Construction of Technology: a Review', in R. Fox (ed.), *Technological Change: Methods and Themes in the History of Technology*, Amsterdam: Harwood Academic, pp. 17-36.

Pinch, T.J. & Bijker, W. (1984), 'The Social Construction of Facts and Artefacts: or How the Sociology of Science and the Sociology of Technology Might Benefit Each Other', *Social Studies of Science*, **14**, pp. 399-444.

Pinch, T.J. & Bijker, W. (1986), 'Science, Relativism and the New Sociology of Technology: Reply to Russell', *Social Studies of Science*, **16**, (2), pp. 347-360.

Pinch, T.J., Ashmore, M. & Mulkay, M. (1992), 'Technology, Testing, Text: Clinical Budgeting in the UK National Health Service', in W.E. Bijker & J. Law (eds), *Shaping Technology/Building Society*, Cambridge MA: MIT Press, pp. 265-289.

Rachel, J. & Woolgar, S. (1995), 'The Discursive Structure of the Socio-Technical Divide: the Example of Information Systems Development', *Sociological Review*,**43**, pp. 250-273.

Radnor, M. (1992), 'Technology Acquisition Practices and Processes', *International Journal of Technology Management*, **7**, pp. 113-135.

Radosevic, S. (1991), 'Technology Policy for the 90s: 25 Tips for a Policy Maker', *Science and Public Policy*, **18**, (4), pp. 251-258.

Rammert, W. (1997), 'New Rules of Sociological Method: Rethinking Technology Studies', *British Journal of Sociology*, **48**, (2), p. 171.

Rammert, W. (1998), *Ritardando and Accelerando in Reflexive Innovation or How Networks Synchronize the Tempi of Technological Innovation*, Working Paper TUTS-WP-7-2000 Technical University Berlin.

Rammert, W. (2001), 'The Cultural Shaping of Technologies and the Politics of Technodiversity', Chapter 6 in this volume.

Rauner, F., Rasmussen, L. & Corbett, M. (1988), 'The Social Shaping of Technology and Work: Human-Centred CIM Systems', *AI and Society*, **2**, pp. 47-61.

Rhodes, E. & Wield, D. (eds) (1994a), *Implementing New Technologies: Innovation and the Management of Technology*, Oxford: Blackwell.

Rhodes, E. & Wield, D. (1994b), 'Technology, Implementation Theory and the Implementation Process', in E. Rhodes & D. Wield (eds), *Implementing New Technologies: Innovation and the Management of Technology*, Oxford: Blackwell, pp. 79-95.

Rip, A. (1995), 'Introduction of New Technology: Making Use of Recent Insights from Sociology and Economics of Technology', *Technology Analysis and Strategic Management*, **7**, (4), pp. 417-431.

Rip, A. (1998), 'The Dancer and the Dance: Steering on/of Science and Technology', in A. Rip (ed.), *Steering and Effectiveness in a Developing Knowledge Society*, Utrecht: Uitgevereij Lemma, pp. 27-49.

Rip, A. (1999), 'Contributions from Social Studies of Science and Constructive Technology Assessment', scoping paper for ESTO project on *Technological Risk and the Management of Uncertainty*, Apr 1999.

Rip, A. & Kemp, R. (1998), 'Technological Change', in S. Rayner & E.L. Malone (eds), *Human Choice and Climate Change, vol. 2*, Columbus OH: Battelle Press, pp. 327-399.

Rip, A., Misa, T. & Schot, J. (eds) (1995a), *Managing Technology in Society: the Approach of Constructive Technology Assessment*, London: Pinter.

Rip, A., Misa, T. & Schot, J. (1995b), 'Constructive Technology Assessment: a New Paradigm for Managing Technology in Society', in A. Rip, T. Misa & J. Schot (eds), *Managing Technology in Society: the Approach of Constructive Technology Assessment*, London: Pinter, pp. 1-12.

Rip, A., Misa, T. & Schot, J. (1995c), 'Epilogue', in A. Rip, T. Misa & J. Schot (eds), *Managing Technology in Society: the Approach of Constructive Technology Assessment*, London: Pinter, pp. 347-354.

Rip, A. & Moors, E. (1999), 'A Technology-in-Society Perspective', in W. Dolfsma, F. Geels & R. Kemp (eds), *Management of Technology Responses to the Climate Change Challenge: Theoretical Elaboration of the Co-evolutionary 'Technology-in-Society' Perspective*, Center for Clean Technology and Environmental Policy CSTM, Enschede: University of Twente.

Rip, A., and J. Schot (1999): 'Anticipation on Contextualization: Loci for Influencing the Dynamics of Technological Development,' Sauer, D., and C. Lang (Hrsg.): Paradoxien der Innovation. Perspektiven sozialwissenschaftlichter Innovationsforschung, Frankfurt/New York: Campus Verlag, 1999. Revised edition as Chapter 5 in this volume.

Rip, A & Schot, J. (2001), 'Identifying Loci for Influencing the Dynamics of Technological Development', Chapter 5 in this volume.

Rip, A. & Talma, S. (1998), 'Antagonistic Patterns and New Technologies', in C. Disco & B.J.R. van der Meulen (eds), *Getting New Technologies Together*, Berlin: de Gruyter, pp. 285-306.

Rohracher, H. (1998), *Technology Policy: Would Social Studies of Technology Make a Difference?*, Heft 27, Graz: IFZ.

Rosen, P. (1993), 'The Social Construction of Mountain Bikes: Technology and Postmodernity in the Cycle Industry', *Social Studies of Science*, **28**, (3), pp. 479-513.

Rosen, P. (1996), 'On Artifacts, Analysis and Alliteration: Theory and Politics in Constructivist Technology Studies', *Social Studies of Science*, **26**, pp. 705-711.

Rosenberg, N. (1976), *Perspectives on Technology*, Cambridge: Cambridge University Press.

Rossel, P. (1999), 'Le Courant de Pensée "Social Shaping of Technology" (SST) et son Évolution', in P. Rossel *et al.* (eds), *Les Systèmes d'Innovation*, Institut de Recherches Économiques et Régionales, Neuchâtel: Université de Neuchâtel.

Rothwell, R. & Gardiner, P. (1984), 'The Role of Design in Product and Process Change', *Design Studies*, **4**, (3), pp. 161-180.

Rothwell, R. & Gardiner, P. (1985), 'Invention, Innovation, Re-Innovation and the Role of the User: a Case Study of British Hovercraft Development', *Technovation*, **3**, pp. 167-186.

Rüdig, W. (1989), *Towards a 'New' Political Science of Technology*, Strathclyde Papers on Government and Politics no. 63, Glasgow: University of Strathclyde.

Russell, S. (1986), 'The Social Construction of Artefacts: a Response to Pinch and Bijker', *Social Studies of Science*, **16**, (2), pp. 331-346.

Russell, S. (1991), *Interests and the Shaping of Technology: an Unresolved Debate Reappears*, STA Research Programme Working Paper no. 4, Wollongong: University of Wollongong.

Russell, S. & Bunting, A. (1998), 'Privatisation, Electricity Markets and New Energy Technologies', paper to conference *Technology in the Making*, RSSS, ANU, Canberra, 2-3 Jul 1998.

Russell, S. & Williams, R. (1987), *Opening the Black Box and Closing it Behind You: On Micro-Sociology in the Social Analysis of Technology*, Edinburgh PICT Centre Working Paper no. 3, RCSS, Edinburgh: Edinburgh University.

Russell, S. & Williams, R. (2001), 'Concepts, Spaces and Tools for Action? Exploring the Policy Potential of the Social Shaping Perspective', Chapter 4 in this volume.

Sahal, D. (1985), 'Technology Guide-Posts and Innovation Avenues', *Research Policy*, **14**, (2), pp. 61-82.

Saviotti, P. & Metcalfe, S. (eds) (1991), *Evolutionary Theories of Economic and Technological Change: Present State and Future Prospects*, London: Harvard.

Schmidt, S. & Werle, R. (1992), 'The Development of Compatibility Standards in Telecommunications: Conceptual Framework and Theoretical Perspective', in M. Dierkes & U. Hoffmann (eds), *New Technology at the Outset*, Boulder CO: Westview, pp. 301-326.

Schmidt, S. & Werle, R. (1998), *Coordinating Technology: Studies in the International Standardisation of Telecommunications*, Cambridge MA: MIT Press.

Schneider, V. (2000), 'Evolution in Cyberspace: the Adaptation of National Videotext Systems to the Internet', *The Information Society*, **16**, (4).

Schot, J. (1991), *Technology Dynamics: an Inventory of Policy Implications for Constructive Technology Assessment*, paper W35, Den Haag: NOTA.

Schot, J. (1992), 'Constructive Technology Assessment and Technology Dynamics: the Case of Clean Technologies', *Science, Technology and Human Values*, **17**, pp. 36-56.

Schot, J. (1998), 'Constructive Technology Assessment Comes of Age', paper to *EASST'98: Cultures of Science and Technology: Europe and the Global Context*, Lisbon, 30 Sep - 3 Oct 1998.

Schot, J., Hoogma, R. & Elzen, B. (1994), 'Strategies for Shifting Technological Systems: the Case of the Automobile System', *Futures*, **26**, (10), pp. 1060-1076.

Schot, J. & Rip, A. (1996), 'The Past and Future of Constructive Technology Assessment', *Technological Forecasting and Social Change*, **54**, pp. 251-268.

Schumm, W. & Kocyba, H. (1997), 'Recontextualisation and Opportunities for Participation: the Social Shaping of Implementation', in C. Clausen & R. Williams (eds), *The Social Shaping of Computer-Aided Production Management and Computer-Integrated Manufacture*, COST A4, Luxembourg: Office for Official Publications of the European Communities, pp. 49-62.

Schwarz, M. & Thompson, M. (1990), *Divided we Stand: Redefining Politics, Technology and Social Choice*, London: Harvester Wheatsheaf.

Schwarz, M. (1993), 'The Technological Culture: Challenges for Technology Assessment and Policy', *Science and Public Policy*, **20**, (6), pp. 381-388.

Senker, J. (1995), 'Tacit Knowledge and Models of Innovation', *Industrial and Corporate Change*, **4**, (2), pp. 425-447.

Shackley, S. & Wynne, B. (1996), 'Representing Uncertainty in Global Climate Change Science and Policy: Boundary-Ordering Devices and Authority', *Science, Technology and Human Values*, **21**, pp. 275-302.

Sharp, M. & Pavitt, K. (1993), 'Technology Policy in the 1990s: Old Trends and New Realities', *Journal of Common Market Studies*, **31**, (2), pp. 129-151.

Shaw, B. (1985), 'The Role of Interaction between the User and Manufacturer in Medical Equipment Innovation', *R&D Management*, **15**, (4), pp. 283-292.

Silverstone, R. & Hirsch, E. (eds) (1992), *Consuming Technologies: Media and Information in Domestic Spaces*, London: Routledge.

Singleton, V. & Michael, M. (1993), 'Actor-Networks and Ambivalence: General Practitioners in the UK Cervical Screening Programme', *Social Studies of Science*, **23**, pp. 227-264.

Slaughter, S. (1993), 'Innovation and Learning During Implementation: a Comparison of User and Manufacturer Innovations', *Research Policy*, **22**, pp. 81-94.

Smit, W.A., Elzen, B. & Enserinck, B. (1998), 'Coordination in Military Socio-Technical Networks: Military Needs, Requirements, and Guiding Principles', in C. Disco & B. van der Meulen (eds), *Getting New Technologies Together: Studies in Making Sociotechnical Order*, Berlin: Walter de Gruyter, pp. 71-106.

Smits, R., Leyten, J. & den Hertog, P. (1995), 'Technology Assessment and Technology Policy in Europe: New Concepts, New Goals, New Infrastructure', *Policy Sciences*, **28**, pp. 271-299.

Sonneborn, C. & Russell, S. (1999), *Green Power in Context*, Australian Cooperative Research Centre for Renewable Energy, Occasional Paper.

Sorge, A. *et al.* (1983), *Microelectronics and Manpower in Manufacturing*, Aldershot: Gower.

Sørensen, K.H. (1992), 'Towards a Feminized Technology? Gendered Values in the Construction of Technology', *Social Studies of Science*, **22**, pp. 5-31.

Sørensen, K.H. (1996), 'Learning Technology, Constructing Culture: Socio-technical Change as Social Learning', STS Working Paper no. 18/96, Centre for Technology and Society, Trondheim: University of Trondheim.
Sørensen, K.H. (1998a), 'Infrastructure, Regulation and Innovation: Dimensions of an Interactive Technology Policy', unpublished paper.
Sørensen, K.H. (1998b), 'A Sociotechnical Legacy? a Note on Trajectories and Traditions in Norwegian Research on Technology and Work', in K.H. Sørensen (ed), *The Spectre of Participation: Technology and Work in a Welfare State*, Oslo: Scandinavian University Press, pp. 1-23.
Sørensen, K.H. (ed.) (1999), *Similar Concerns, Different Styles? Technology Studies in Europe, vol. II*, COST A4, Luxembourg: Office for Official Publications of the European Communities.
Sørensen, K.H. (2001), 'Social Shaping on the Move? On the Policy Relevance of the Social Shaping of Technology Perspective', Chapter 2 in this volume.
Sørensen, K.H. & Levold, N. (1992), 'Tacit Networks, Heterogeneous Engineers and Embodied Technology', *Science, Technology and Human Values*, **17**, (1), pp. 13-35.
Sørensen, O. (1999), 'NOAH, Getting Everyone on Board: Dominating Development Arenas by Networks and Standards', paper to conference *Bringing Materiality (Back) into Management*, Gilleleje, 21-23 Oct 1999.
Spilker, H. & Sørensen, K.H. (2001), 'Feminism for Profit? Gender Politics in the Norwegian Multimedia Industry', Chapter 9 in this volume.
Steward, F. (1995), 'Risk Analysis and Rival Technical Trajectories: Consumer Safety in Bread and Butter', in A. Rip, T. Misa & J. Schot (eds), *Managing Technology in Society: the Approach of Constructive Technology Assessment*, London: Pinter, pp. 111-136.
Steward, F., Conway, S. & Overton, M. (1995), 'Actors and Interactions in the Management of Innovation: Analysing Network Nodes and Links', in D. MacLean, P. Saviotti & D. Vinck (eds), *Designs, Networks and Strategies, vol. 2*, Brussels: European Commission DGXIII, pp. 133-151.
Summerton, J. (1992), *District Heating Comes to Town: the Social Shaping of an Energy System*, Linköping Studies in Arts and Sciences no. 80, Linköping: Linköping University.
Summerton, J. (ed) (1994), *Changing Large Technical Systems*, Boulder CO: Westview.
Summerton, J. (1997), 'Heroes, Giants and Critics: on Building Bridges between Systems Approaches, ANT, and STS', paper to workshop *Actor Network and After*, Keele University, 10-11 Jul 1997.
Tait, J. & Williams, R. (1999), 'Policy Approaches to Research and Development: Foresight, Framework and Competitiveness', *Science and Public Policy*, **26**, (2), pp. 101-.
Teece, D.J. (1988), 'Technological Change and the Nature of the Firm', in G. Dosi *et al.* (eds), *Technical Change and Economic Theory*, London: Pinter, pp. 256-281.
Thomas, P. (1996), 'The Devil is in the Detail: Revealing the Social and Political Processes of Technology Management', *Technology Analysis and Strategic Management*, **8**.
Tierney, M. & Williams, R. (1990), *Issues in the Black-Boxing of Technologies: What Happens When the Black Box Meets Forty Shades of Grey?*, Edinburgh PICT Working Paper no. 22, RCSS, Edinburgh: Edinburgh University.

Timmermans, S. & Berg, M. (1997), 'Standardization in Action: Achieving Local Universality through Medical Protocols', *Social Studies of Science*, **27**, pp. 273-305.

van den Belt, H., and Rip, A. (1987), The Nelson-Winter/Dosi model and synthetic dye chemistry. In W. Bijker, T. P. Hughes and T. Pinch (eds), *The Social Construction of Technological Systems,* Cambridge, MA: MIT Press, pp. 135-158.

van Lente, H. (1993), *Promising Technology: the Dynamics of Expectations in Technological Developments*, Delft: Eburon.

van Lente, H. & Rip, A. (1998a), 'Expectations in Technological Developments: an Example of Prospective Structures to be Filled in by Agency', in C. Disco & B. van der Meulen (eds), *Getting New Technologies Together: Studies in Making Sociotechnical Order*, Berlin: de Gruyter, pp. 195-220.

van Lente, H. & Rip, A. (1998b), 'The Rise of Membrane Technology: from Rhetorics to Social Reality', *Social Studies of Science*, **28**, pp. 221-254.

van Oost, E. (1998), 'Aligning Gender and New Technology: the Case of Early Administrative Automation', in C. Disco & B. van der Meulen (eds), *Getting New Technologies Together: Studies in Making Sociotechnical Order*, Berlin: Walter de Gruyter, pp. 179-202.

Vergragt, P. (1988), 'The Social Shaping of Industrial Innovations', *Social Studies of Science*, **18**, pp. 483-513.

Verheul, H. & Vergragt, P. (1995), 'Social Experiments in the Development of Environmental Technology: a Bottom-Up Perspective', *Technology Analysis and Strategic Management*, **7**, (3), pp. 315-326.

Vincenti, W.G. (1994), 'The Retractable Airplane Landing Gear and the Northrop "Anomaly": Variation-Selection and the Shaping of Technology', *Technology and Culture*, **35**, pp. 1-33.

Vincenti, W.G. (1995), 'The Technical Shaping of Technology: Real-World Constraints and Technical Logic in Edison's Electrical Lighting System', *Social Studies of Science*, **25**, (3), pp. 553-574.

Vogel, D. (1986), *National Styles of Regulation: Environmental Policy in Great Britain and the United States*, Ithaca NY: Cornell University Press.

von Hippel, E. (1976), 'The Dominant Role of Users in the Scientific Instrument Innovation Process', *Research Policy*, **5**, pp. 212-239.

Wajcman, J. (1991), *Feminism Confronts Technology*, North Sydney: Allen & Unwin.

Wajcman, J. (1994), 'Technological A/Genders: Technology, Culture and Class', in L. Green & R. Guinery (eds), *Framing Technology: Society, Choice and Change*, St Leonards NSW: Allen & Unwin.

Weber, K.M. (1998), 'Innovation Diffusion and Political Control in the Energy Sector: a Conceptual Approach and Empirical Evidence from the Case of Cogeneration in the UK', in R. Coombs *et al.* (eds), *Technological Change and Organisation*, Cheltenham: Edward Elgar, pp. 210-238.

Weber, K.M. (1999), *Innovation Diffusion and Political Control of Energy Technologies. a Comparison of Combined Heat and Power Generation in the UK and Germany*, Heidelberg: Springer/Physica.

Weber, K.M. (2001), 'The Political Control of Large Socio-Technical Systems: New Concepts and Empirical Applications from a Multi-Disciplinary Perspective', Chapter 12 in this volume.

Weber, K.M. & Hoogma, R. (1998), 'Beyond National and Technological Styles of Innovation Diffusion: a Dynamic Perspective on Cases from the Energy and Transport Sectors', *Technology Analysis and Strategic Management*, **10**, (4), pp. 545-565.

Weber, K.M., Hoogma, R., Lane, B. & Schot, J. (1999), *Experimenting with Sustainable Transport Innovations: a Workbook for Strategic Niche Management*, Sevilla/Enschede: Institute for Prospective Technological Studies.

Weber, K.M. & Paul, S. (1999), *Political Forces Shaping the Innovation and Diffusion of Technologies: an Overview*, SEIN Working Paper no. 4.

Webster, A. (1991), *Science, Technology and Society*, Basingstoke: Macmillan.

Webster, J. (1990), *Office Automation: the Labour Process and Women's Work in Britain*, Hemel Hempstead: Harvester Wheatsheaf.

Webster, J. (1991), 'Advanced Manufacturing Technologies: Work Organisation and Social Relations Crystallised', in J. Law (ed.), *A Sociology of Monsters: Essays on Power, Technology and Domination*, London: RKP, pp. 192-222.

Webster, J. & Williams, R. (1993), 'Mismatch and Tension: Standard Packages and Non-Standard Users', in P. Quintas (ed.), *Social Dimensions of Systems Engineering: People, Processes, Policies and Software Development*, New York: Ellis Horwood, pp. 179-196.

Weingart, P. (1984), 'The Structure of Technological Change: Reflections on a Sociological Analysis of Technology', in R. Laudan (ed.), *The Nature of Technological Knowledge: Are Models of Scientific Change Relevant?*, Dordrecht: Reidel, pp. 115-142.

Wijnberg, N.M. (1994), 'National Systems of Innovation: Selection Environments and Selection Processes', *Technology in Society*, **16**, (3), p. 313.

Williams, R. (1987), 'Democratising Systems Development: Technological and Organisational Constraints and Opportunities', in G. Bjerknes *et al.* (eds), *Computers and Democracy: a Scandinavian Challenge*, Aldershot: Avebury, pp. 77-96.

Williams, R. (1997a), 'Globalisation and Contingency: Tensions and Contradictions in the Mutual Shaping of Technology and Work Organisation', in I. McLoughlin & M. Harris (eds), *Innovation, Organisational Change and Technology*, London: International Thomson Business Press, pp. 170-185.

Williams, R. (1997b), 'Social Learning in Multimedia', paper to COST A4/TSER conference *The Promise of Technology*, Copenhagen, 2-3 Oct 1997.

Williams, R. (1997c), 'The Social Shaping of a Failed Technology? Mismatch and Tension between the Supply and Use of Computer-Aided Production Management', in C. Clausen & R. Williams (eds), *The Social Shaping of Computer-Aided Production Management and Computer-Integrated Manufacture*, COST A4, Luxembourg: Office for Official Publications of the European Communities, pp. 109-132.

Williams, R. (1997d), 'The Social Shaping of Information and Communication Technologies', in H. Kubicek *et al.* (eds). *The Social Shaping of Information Superhighways: European and American Roads to the Information Society*, Frankfurt/New York: Campus/St Martin's, pp. 299-338.

Williams, R. (1999), 'ICT Standard Setting from an Innovation Studies Perspective', Proceedings of *First IEEE Conference on Standardisation and Innovation in Information Technology (SIIT '99)*, Aachen, pp. 251-262.

Williams, R. & Edge, D. (199), 'The Social Shaping of Technology', *Research Policy*, **25**, pp. 865-899.
Williams, R., Faulkner, W., & Fleck, J. (eds) (1998), *Exploring Expertise*, Basingstoke: Macmillan.
Williams, R., Stewart J. & Slack, R. (2000), *Social Learning in Multimedia*, Final Report to European Commission, DGXII TSER, Research Centre for Social Sciences, Edinburgh: University of Edinburgh.
Windrum, P. (2000), 'Knowledge Pools and Innovation Networks in E-Commerce: the Integrating Role of Knowledge-Intensive Services', paper to SEIN workshop *Innovation Networks: Theory and Policy*, London, Jan 2000.
Winner, L. (1993), 'Upon Opening the Black Box and Finding it Empty: Social Constructivism and the Philosophy of Technology', *Science, Technology and Human Values*, **18**, (3), pp. 362-378.
Winskel, M. (1998), *Privatisation and Technological Change: the Case of the British Electricity Supply Industry*, PhD thesis, Edinburgh: University of Edinburgh.
Woolgar, S. (1991a), 'Configuring the User: the Case of Usability Trials', in J. Law (ed.), *A Sociology of Monsters: Essays on Power, Technology and Domination*, London: RKP, pp. 57-102.
Woolgar, S. (1991b), 'The Turn to Technology in Social Studies of Science', *Science, Technology and Human Values*, **16**, (1), pp. 20-50.
Wyatt, S. (1998), *Technology's Arrow: Developing Information Networks for Public Administration in Britain and the United States*, Maastricht: Universitaire Pers Maastricht.
Wynne, B. (1983), 'Redefining the Issues of Risk and Public Acceptance: The Social Viability of Technology', *Futures*, **15**, (1), pp. 13-32.
Wynne, B. (1988a), 'Technology as Cultural Process', in E. Baark & U. Suedin (eds), *Man, Nature and Technology: Essays on the Role of Ideological Perceptions*, Basingstoke: Macmillan, pp. 80-104.
Wynne, B. (1988b), 'Unruly Technology: Practical Rules, Impractical Discourses and Public Understanding', *Social Studies of Science*, **18**, pp. 147-167.
Wynne, B. (1995), 'Technology Assessment and Reflexive Social Learning: Observations from the Risks Field', in A. Rip, T. Misa & J. Schot (eds), *Managing Technology in Society: the Approach of Constructive Technology Assessment*, London: Pinter, pp. 19-36.

Glossary: Some key Social Shaping concepts

At the beginning of Chapter 3 we noted the daunting proliferation of concepts in SST, which points both to the richness of this intellectual terrain, and to a tendency of scholars in a rapidly evolving field to coin new terms to address their particular concerns and contexts more precisely. This profusion however creates a barrier to those outside the field, which may not be an undue problem for scholars but which can be unhelpful if SST wishes to build bridges with practitioners and encourage them to explore, discuss with us and use our insights. Given our goal of making the achievements of SST more widely available, we have provided a glossary of a selection of key concepts.

For each term we have indicated representative authors using it, and provided a definition. Where possible we have given or adapted the proponents' own definitions; otherwise the explanation represents our interpretation of the general use of the concept.

In the last column we list related concepts we have either included in the glossary (in italics) or introduced and discussed elsewhere in Chapter 3. Terms grouped in this way are not to be read as synonyms, and the list as a whole is not to be taken as a consistent and coherent scheme of concepts that can all be used together.

This is a glossary, not an encyclopaedia. We have had to be selective and to group related terms. We include those we judge have gained, or seem to be gaining, significant currency in SST studies. We have had to leave out many other terms whose use so far remains largely limited to specific authors. We have also omitted many concepts now widely used in SST which are familiar from and common to other frameworks of social theory. The authors listed with each concept are representative rather than exclusive, and not necessarily the originators. Our selection and omission will not please everyone.

While we hope our attempted definitions are useful, we stress that the significance of a novel concept is unlikely to be apparent in the definition alone, but rather in the use and effect of the term:

- the way it foregrounds different aspects of institutions and processes;
- the attention it draws to new patterns, generalisations, differences or commonalities;
- the way it may elevate insights to a general level;
- its connotations – the narratives or images it invokes; and
- the extent it helps reorientate discourse on an issue.

The terms are inescapably metaphorical (McLoughlin 1999, pp. 7-8). A choice between comparable concepts is a judgement about how each allows us to make sense of empirical material in a way consistent with our other principles of explanation.

Term	Representative authors	Concept and comments	See also
Actor network	Callon, Law, Latour	An actor, sociotechnical entity or technology, conceptualised as an emerging and increasingly stabilised network of associations between diverse material and non-material elements – artefacts, humans, texts, symbols, concepts, etc. ANT usually follows the network-building strategies of a central actor. It stresses the mutual constitution and transformation of elements in the process, and the generation of social phenomena – agency, knowledges, institutions, power – as *effects* of network-building.	see section 3
Alignment	Molina	Processes in which the views and interests of innovation players – members of a sociotechnical constituency, potential customers – are aligned in successful innovation.	*closure, coordination, networks* see section 7
Appropriation	various	Processes in which technologies are adopted and incorporated by users into particular social and technical settings, including local practices and culture.	*domestication, embedding, innofusion, social learning, contextual-isation* see section 9

Term	**Representative authors**	**Concept and comments**	**See also**
Closure	Pinch & Bijker	By analogy with the emergence of scientific consensus, a process by which, or the point at which, interpretations of an artefact (or an institution or process) by different social groups - conceptions of design and use, value and significance, or in particular, its 'self-evident' essential character - are brought into agreement or one interpretation becomes dominant.	*interpretative flexibility*, *stabilisation* see sections 3 & 7
Codes / encoding	Feenberg, Mackay & Gillespie, Mollinga & Mooij	System of rules (/process) through which broader social structures pattern local technological practice, dominant social objectives and criteria are incorporated, or (Feenberg 1991) 'the construction and interpretation of technical systems [is brought] into conformity with the requirements of a system of domination'.	orientation complex, structural filters see section 8
Co-evolution, co-production, mutual shaping	various	Process in which technologies and social organisation or culture, or specific elements of each, are jointly created – in contrast to depictions of two separate spheres interacting or of one determining the other; e.g. industrial automation products are condensations of past work organisation as much as tools for transforming organisation.	see section 6

Term	Representative authors	Concept and comments	See also
Configurations, configurational technologies	Fleck	'[M]ore or less unique assemblies of [material, organisational, procedural] components, some standardly available, others specially developed, built up to meet the particular requirements of user organisations' (Fleck 1988b). Stresses that complex technologies, especially IT systems, are often not acquired as integrated systems but assembled to suit specific applications and contexts. Highlights the scope for innovation during implementation and diffusion, and different possible technology supply strategies.	see section 6
Constructive technology assessment	Rip, Schot	Strategy for steering and managing technological change which seeks to integrate continuous or iterative anticipatory assessment of sociotechnical outcomes and effects into design and development as well as implementation. CTA aims to overcome the traditional institutional and temporal separation of policies for innovation and control.	reflexive technology development see section 10
Critical events	Vergragt,	Crucial points in innovation processes at which directions are decided, changed or consolidated, and which profoundly affect subsequent possibilities.	moments of transformation, *innovation journey*, decision points, occasions, circle of uncertainty see sections 7 & 10

Term	Representative authors	Concept and comments	See also
Delegation	Latour	Strategy, process or act of allocating a social control function to a material artefact – in contrast to organisational or cultural mechanisms.	see section 6
Development arena	Jørgensen & Sørensen	'[C]ognitive space that holds together the settings and relations that comprise the context for product and process development that includes: • a number of elements such as actors, artefacts and standards that populate the arena; • a variety of locations for action, knowledge and visions that define the changes of this space; and • a set of translations that has shaped and played out the stabilisation and destabilisation of relations and artefacts' (Jørgensen & Sørensen, in this volume). Accommodates analysis of the development and interaction of multiple networks.	arenas of negotiation see section 10
Distributed innovation	various	Structure of innovation process with constitutive activities dispersed in space, connected in diverse and asynchronous ways and involving heterogeneous actors (Rammert, in this volume). Contrasts with depictions of predictable sequence, stability and functional organisation in innovation.	*innovation journey* see sections 7 & 10

Term	Representative authors	Concept and comments	See also
Domestication	Lie, Sørensen, Silverstone	Process by which technologies are made to work and are given meaning by individuals, collectives, institutions – acquired, placed, interpreted and integrated into practices. Emphasises the active role of users in defining the use and significance of technologies in everyday life.	*appropriation*, trivialisation see section 9
Embedding/ disembedding/ re-embedding	Rip, Schot	Process of integrating technologies into/extracting from/ reintegrating into local contexts of use.	(re)contextual-isation, *appropriation*, *entrenchment* see section 9
Enrolment	Latour, Callon	Process in network-building in which actors' support is gained for development of a sociotechnical entity, their role defined and their interests and identities orientated to suit.	see sections 3 & 7
Entrenchment	Collingridge	Process in which a technology becomes firmly embedded in its use context, particularly through the establishment of supporting institutions or infrastructure, systemic interdependence, economic advantage, shared evaluations, and their mutual reinforcement; or the stable state of such a sociotechnical arrangement. Highlights and accounts for the difficulty of removing, changing or controlling the technology.	*embedding* see section 9

Term	**Representative authors**	**Concept and comments**	**See also**
Guiding principle	Elzen, Enzerink, van de Poel	A rule or a set of fundamental rules – those at the top of the hierarchical set which forms a regime – which patterns innovative activity.	see section 8
Heterogeneous engineering	Law	Practice or process in which engineers and other technical workers grapple with a variety of social relations and actors as well as technical artefacts in producing technological change. Stresses the requirement for diverse forms of knowledge and practice.	see section 7
Innofusion	Fleck	Innovation in the process of diffusion – the way in which artefacts, rather than being fixed at the design stage, are transformed in their implementation and use, in the course of the 'struggle to get the technology to work in useful ways at the point of application' (Fleck 1988a). An important mechanism through which user needs and requirements are discovered and incorporated. Contrasts with conventional notions of diffusion of stable artefacts or techniques.	see section 9
Innovation journey (or biography)	Rip, Rammert	The course of development of an innovation. Stresses non-linearities and branchings, and acknowledges that artefacts may be transformed radically from the original concept.	technological drama see section 7
Interpretative flexibility	Pinch, Bijker	Scope for the attribution by different groups of different meanings to an artefact, according to their different backgrounds, purposes and commitments.	boundary object, trading zone, *closure* see sections 3 & 7

Term	Representative authors	Concept and comments	See also
Network externalities	Arthur, David	Increased benefits of a network technology (e.g. telephone) to an individual as the number of other users increases. May explain both the difficulties in establishing new inter-operating technologies and the stability of entrenched technologies and standards.	*path dependence*, *entrenchment* see section 7
Niche	Vergragt, Kemp, Schot, Hoogma & others	A protected space for the early development of innovations, in which the effects of selection pressures are lessened or suspended.	protected space see sections 5 & 8
Obduracy	Law	The difficulty for a technology developer of overcoming technical constraints and problems or of changing an existing entity, often seen as an emerging property of a sociotechnical network and therefore attributable to material and social factors and their interweaving.	malleability, irreversibility see section 7
Paradigm	Dosi	Shared outlook or mindset within a community of technical practitioners which guides and constrains processes and directions of innovation. Consists of exemplars and search heuristics – assumptions about appropriate directions to pursue and problems to work on; knowledge, techniques and materials to draw on; model solutions and explanations; criteria of progress. Indicates means by which wider economic and political objectives and commitments are mediated into technical practice.	*frame*, style, *selection environment*, *trajectory* see section 8

Term	Representative authors	Concept and comments	See also
Path dependence	Arthur, David	Emergence of constraints on change of direction through cumulative investment or progressive entrenchment – in effect, advantages one technological option, often by virtue of its earlier appearance and contributes to the exclusion of alternatives.	lock-in, irreversibility see section 8
Product space	Cawson, Haddon & Miles	Conceptual space representing general agreement on the functionality required for a new product but in which competing designs with overlapping functionality are put forward.	*visions* see section 7
Promise-requirement cycle	van Lente	Interactive and iterative process in which visions of the form, functionality and significance of a technology are articulated and matched to requirements and expectations of potential users and wider constituencies.	see section 7
Regime	various	Multi-layered set of rules and grammar operating in and derived from the complex of 'scientific knowledges, engineering practices, production process technologies, product characteristics, skills and procedures, and institutions and infrastructures that make up a technology' (Geels & Schot 1998). A regime '[guides] technological development and its embedding in society ... [and] orients actions and perceptions' (Rip & Schot, in this volume). The concept captures structural aspects of institutions – beyond the sociocognitive and technical community focus of *paradigm* – which pattern innovation and adoption.	*paradigm, selection environment* see section 8

Term	Representative authors	Concept and comments	See also
Script/ inscription	Akrich, Callon, Latour	Set of rules or meanings embedded in an artefact, procedure or representation which attempts to prescribe the behaviours of users and conditions of use so as to enable the realisation at a distance of the function of the entity as intended by the developer or controller.	see sections 6 & 9
Selection environment	Nelson & Winter	Set of conditions favouring the development and adoption of particular kinds of innovation. Associated with an evolutionary metaphor for technological innovation: technological activity produces variations from which are selected those best suited to the context.	*regime*, variation see section 5
Social learning	various	Protracted process entailed in creating and appropriating new technologies, in which developers, implementers and users learn from experience and interaction. The process is seen not just in individual and cognitive terms but as necessarily social and political and entailing institutional change: the concept stresses negotiation and interaction among a wide range of actors, subject to conflicts and differences of interest and power. It flags the active and reflexive role of actors involved in a 'combined act of discovery and analysis, of understanding and giving meaning, and of tinkering and the development of routines' (Sørensen 1996). The concept serves to alert participants, managers and policy makers to the necessity of the process and what is required to facilitate it.	learning economy, *appropriation*, *domestication*, supplier-user coupling, intermediaries, reflexive technology development see sections 9 & 10

Term	Representative authors	Concept and comments	See also
Sociotechnical	various	Character of processes, knowledges and entities in which social and technical elements are closely interwoven. The adjective is used to stress the pervasive technological mediation of social relations, the inherently social nature of all technological entities, and indeed the arbitrary and misleading nature of distinctions between 'social' and 'technical' elements, institutions or spheres of activity.	seamless web, *co-evolution*, *co-production* see section 6
Sociotechnical constituencies	Molina	'[D]ynamic ensembles of technical constituents (expertise, tools, machines, etc.) and social constituents (people and their values, interest groups, etc.) which interact and shape each other in the course of the creation, production and diffusion of specific technologies' (Molina 1992a). Analysis focuses on how the resources needed to build new technologies are assembled, and how this process shapes innovation.	see section 7
Sociotechnical ensembles	Bijker	Continually changing associations of social, technical, semiotic, etc. elements. Stresses the heterogeneity and contingency of assemblages and may avoid the connotations of 'system'.	see section 6
Sociotechnical landscape	Rip, Schot	Context of existing sociotechnical institutions which pattern specific processes of technological change.	see section 8

Term	Representative authors	Concept and comments	See also
Sociotechnical systems	Hughes	Assemblages of technical, organisational and other elements. Concept tends to stress interdependence of elements, importance of functional adjustment of elements and linkages for overall performance, and a system-level dynamic determining development and constraining change. Arguably inappropriate for assemblages where the system metaphor would overemphasise stability and internal coherence.	see section 6
Stabilisation	various	Process by which a technological form becomes settled from a period of conflict, negotiation or indeterminacy, particularly as visions for the technology take social and material form, or, for network theorists, by virtue of the stabilisation of the heterogeneous relations of which it is a product or a part.	*closure* see section 7
Strategic niche management	Hoogma, Kemp, Schot, Weber	Set of principles and practices for creating and managing the protected spaces in which a set of related innovations can be developed and introduced. Entails both pursuing appropriate changes in regimes influencing those developments and in turn using the emerging cluster of developments to further that shift.	*niche* see section 5 & 8

Term	Representative authors	Concept and comments	See also
Technological frame/framing	Bijker	Structure of rules and practices (or process of applying such) which enable, guide and constrain technological development in specific areas. Contains heterogeneous elements – social and cognitive; includes exemplary artefacts, scientific theories, values, goals, test protocols, tacit knowledge, central problems and related strategies. The analysis depicts the frame being drawn on, and in part constituted, in the process of stabilisation, and both needing sustaining and undergoing continual change.	see section 8
Trajectory	Dosi, Nelson & Winter	Direction of advance of a technology within a generic form, often expressed in terms of change in a specific measure of performance or characteristic accepted as representing improvement. Represents 'the pattern of normal problem-solving activity' (Dosi) – in part the result of constraints exerted by the associated paradigm.	envelope, avenue, corridor, *paradigm* see section 8
Translation	Latour, Callon	One of a sequence of transformations undergone by a developing sociotechnical entity. In contrast to the metaphor of physical diffusion in which artefacts and ideas are transported unchanged from one context to another, translation indicates that they are (and must be) transformed in the process by the actors involved. Or, in the process of building an actor-network, or of alignment and stabilisation more generally, the allocation or (re)definition of attributes or roles of actors, or a change (as in *delegation*) in the mode of achievement of a function.	

Term	**Representative authors**	**Concept and comments**	**See also**
Translation (or transformation) terrains	Bijker	Local settings which may pattern (parts of) an innovation process. Concept draws attention to features of those spaces – and differences between them – in which crucial changes in an emerging sociotechnical entity take place.	*sociotechnical landscape*, transformation spaces see sections 8 & 10
Vision	various	Developers' conception of the form and features of a new technology, its functions and benefits, and the required new sociotechnical order in its domain of application. Developers deploy a vision to mobilise support and resources; the vision sharpens and shifts as the constituency is developed.	poles of attraction, agenda formation, *product space*, framing object see section 7

4. Concepts, Spaces and Tools for Action? Exploring the Policy Potential of the Social Shaping Perspective

Stewart Russell and Robin Williams

1 CHANGING CONTEXTS AND CHALLENGES FOR TECHNOLOGY POLICY

Introduction

This is a time of profound and promising change for technology policy. Traditional policy frameworks, centred around promoting technology research and development, have been called into question. A search is under way for policies which are more comprehensive, addressing the consumption and use of technologies, and their socio-economic outcomes, as well as promoting supply. The European Commission's Fifth Framework Programme (FP5) marks an important development in these respects: it seeks to integrate Research and Technological Development (RTD) with measures to support technology uptake, and is geared towards European policy goals – of sustainability and improvements to the quality of life as well as traditional concerns with economic competitiveness and growth.

National industrial and other infrastructure needs, and military goals, provided the rationale for some technology policies and national programmes up to the 1970s. These typically assumed support for selected areas of basic science, and state sponsorship and organisation of certain major projects and support services, as the appropriate mechanisms. Then from the 1980s on, programmes, policies and instruments in technology policy proliferated as national governments, and increasingly European agencies, acknowledged the importance of innovation for economic goals in times of fundamental economic restructuring and growing international competition. These policies typically sought to stimulate innovative activity in general and to provide supportive conditions, and to support the development of key industries or initiatives in producing generic technologies (Caracostas & Muldur 1998).

A third generation of technology policies is now evident at least in the general orientation of European Union programmes – though the fundamentally new perspective it represents is still permeating at different rates through national policies and programmes.

This new orientation stresses:

a) the complexity of the innovation process

- the changed environment and character of innovation, production and consumption, particularly in the speed and unpredictability of change and the uncertainties surrounding the innovation process and its outcomes,
- the globalisation of industries and markets, the diverse forms and distributed loci of knowledge and innovation, the involvement of a wider range of social groups, and the need for greater cooperation in national and European innovation systems;

b) the complex goals of policy intervention

- the diverse goals, specified in social and environmental as well as market terms, which need to be integrated with the dynamics of innovation,
- integrating the goals of promotion and control over technological innovation, in technology steering which also recognises the difficulties of achieving direct control,
- organisation of initiatives, support mechanisms and policy around socio-economic and other goals or problem areas, rather than around technologies;

c) new sites and spaces for intervention

- measures to encourage technology application and consumption as well as supply,
- an orientation towards user requirements, a need for greater inclusivity and involvement of user constituencies, and the importance of social learning processes;

d) complex strategies and new tools for intervention

- the need for integrated or complementary policies for technological innovation, based on a systemic analysis of technology,
- at the same time, the need for greater specificity of policies for particular technical domains, sectors and industries,
- new criteria for selecting projects for public support and new means of monitoring and evaluating them.

Various factors have stimulated this shift in the thinking underlying policies for technology. The context of science and technology policy is becoming more challenging. The pace and dynamism of technology development continue to increase. This brings greater uncertainties and an increasing array of choices to technological decision-making. At the same time it seems that technologies are becoming a more intimate part of everyday culture and practices. Whereas the characteristic technologies of previous decades such as nuclear power or weapons technologies were large centralised systems, often built and operated for the state and presented as largely the preserve of a narrow policy and technical elite, information and communication technologies today are penetrating every aspect of our lives at work, in civil society and in the home. Technology is a more immediate part of our lives, and the concerns and needs of the many kinds of user are more critical to successful innovation. At the same time many citizens are expressing a wish to be more closely involved in choices over technology.

The new style of policy is thus encouragingly consistent with the broad approach of research and theorising in the social shaping of technology (SST). SST has contributed to this rethink. However the communication between technology studies and technology policymakers has not been particularly effective. The critique of 'linear' supply-driven models, for example, started with Freeman's classic work more than a quarter of a century ago (Freeman 1974). Though explicit linear models of technology policy are firmly out of fashion today we should note that linear assumptions about the process of innovation have shown remarkable resilience (Tait & Williams 1999).[1]

This book seeks to improve the communication between technology policy and technology studies, and in particular work in an SST perspective. SST has particular relevance for those involved in and concerned about technology decision-making. However policy makers and academic researchers to some extent live in different worlds and cultures. SST can be criticised for having largely failed so far to present its findings in a way that is readily understood and seen as relevant and useful for policymakers grappling with urgent and difficult decisions. This is in part a question of the languages deployed by academics – and one objective of this book is to provide an explanation of the terms and concepts that have emerged from SST. However it also indicates that SST researchers need a better understanding of the exigencies facing decision makers and special skills and expertise they deploy.[2]

In this paper we start by examining the movements and influences which have led to a rejection of orthodox models of technology dynamics and technology policy. Policies which relied on a science and technology 'push', failed to take into account the inherent uncertainties of innovation, and they neglected socio-economic and environmental outcomes at the same time that broader constituencies were becoming more concerned with unanticipated and undesirable impacts and were seeking involvement in technological decision-making. We then trace the quest for a model which acknowledges coupling between stages in innovation processes, and in particular interactions between development and use via markets and other mediations. We argue that the resultant models, while a step in the right direction, are still inadequate for the complexities of socio-technical change. A number of currents of change, including the research and theorising in SST surveyed in this book, point to a much more fundamental reorientation of thinking about policies for technology. It is one which paradoxically both acknowledges much greater complexity in the dynamics of socio-technical change and the influences on it, and hence implies a more subtle view of possible interventions, but at the same time greatly opens up the scope and terrain of intervention. We examine the contribution that SST work has made and can make to this new approach, and suggest the forms of interaction between the academic and policy spheres that can be most fruitful.

The Failure of the Orthodox Linear Model of Technology Policy

The success of science-intensive industries in the decades after the second world war, and the enormous advances made possible by 'Big Science' during the war provided the rationale for the adoption across most developed states of a technology policy framework that looked to the promotion of advances in core sciences and technology to drive technical advance and economic growth.

This policy framework was consistent with a modernist perspective in which technical progress was synonymous with social progress. Technological change was taken to be the major driver of social change. The view entailed a further technological determinist assumption that new technologies required particular sorts of social arrangements. The dominant concern in social policy was thus a very narrow one: how quickly and smoothly could a society adjust to meet the exigencies of new technology? It was argued that the biggest risks would arise from a failure to ensure uptake of technology and rapid adaptation. The role of social science in this view was limited to monitoring and predicting future technological trajectories, so that it could assess the future 'socio-economic impacts' of new technologies.[3]

Over the last two decades, the marked shortcomings of this traditional perspective have been progressively revealed. Both of its key elements – the emphasis on technology push and its inadequate model of the socio-economic and other implications of technological change – have been called into question.

The failure of science/technology push policies – uncertainties in technical innovation

Traditional technology policies in the European Union and most OECD countries pursued economic growth and competitiveness through support for research and technological development (RTD). They assumed a linear model of innovation, involving a one-way flow of information, ideas and solutions from basic science, through applied RTD, to industrial production and the diffusion of finished artefacts through the market to consumers. The basic driver of technological innovation was seen as advances in underlying scientific and technological knowledge, arising perhaps through public funded research and development, and the resulting creation of new technological artefacts.

However this 'science and technology push' model proved ineffective in delivering successful technological advances, let alone the economic and social benefits that were expected to accrue. There was an increasing number of examples in which technologies developed in the laboratory were not taken up on a commercial basis. In contrast to the confident promises of engineers and promoters, it turned out that new technologies often failed and, even when implemented, did not yield the predicted improvements in wealth creation and quality of life. Technological innovation proved to be a fundamentally uncertain endeavour. As well as the narrowly technical difficulties encountered, it often proved difficult to fit new technological offerings to social need and to build new markets.

As a result, public technology policies based on such linear models were increasingly seen as unhelpful – not least because of their privileging of technological supply and their artificial division of innovation into separate phases. It was increasingly recognised that more sophisticated technology policy approaches were needed which spanned the whole life-cycle of a technology, from research and development, to its commercialisation, consumption and use.

Lack of attention to unanticipated and undesirable outcomes of technology

The traditional model, with its emphatically positive attitude towards new technologies, tended to downgrade consideration of socio-economic outcomes of technological change. To the extent that negative outcomes were acknowledged, they were largely presumed to be modest, and were seen as a necessary but acceptable corollary of the presumed benefits of technology. Since the 1970s, in particular, there has been increasing recognition that technologies could often bring *unanticipated* and *undesired* consequences, and on a significant scale. The starting point for this was growing awareness of environmental and other risks.[4]

Regulation of the hazards of technological activities was, of course, one of the first areas of public intervention in technology development. However, regulatory apparatuses emerged in response to individual environmental or health crises and have been fragmented across different domains of technological activity and across different kinds of hazard, divided for example between those affecting consumers, workers and the physical environment. Regulation has thus typically been reactive – adjusting *ad hoc* to the emergence of new technological hazards. It has been largely separated from RTD policies.

There has been growing dissatisfaction then with existing arrangements for dealing with such risks. The criticism has targeted both the institutional separation between bodies dealing with the promotion of technology and those grappling with the unwanted consequences of the technologies thus emerging, and the shortcomings in practice of a fragmented regulatory structure. The system of responding to the impacts of technological change on health, the environment, workplaces and wider social patterns appears to react too late and offer limited genuine opportunities for change.

The emphasis of traditional technology policy on technology promotion, rather than control of effects, reflected the modernist assumption that the unwanted costs of a technology were a necessary evil resulting from a trajectory of technological advance that was taken for granted – as if technology developed according to some kind of internal techno-economic logic. In contrast, a growing body of critical thought suggested that the particular socio-economic outcomes of a technology, its costs and benefits, were not an inevitable, technical matter but arose from the ways in which a technology was designed and implemented. Environmental concerns, for example – culminating in the recognition of global climate change – raised questions about the extent to which existing technologies had been designed without taking into account social goals such as pollution avoidance, and suggested the possibility of creating more environmentally-sustainable technologies and industrial systems. This kind of thinking has broadened the technology policy agenda beyond a narrow concern solely with technology promotion to consider choices in the design and application of technologies. It has raised questions of how to assess in advance the prospects and implications of new technologies and how to select appropriate technological options accordingly.

A broadening of the range of concerns and actors in technology decision-making

These factors suggest that publics and governments are becoming more demanding about technology: they seek more benefits and fewer risks, and are consequently placing more extensive and stringent requirements on what technology delivers. At the same time, people are no longer satisfied to leave technological decision-making to experts. In particular there is widespread distrust of committed experts who have promoted technologies that did not live up to expectations and failed to alert the wider population to the potential risks. Many individuals and groups want to become more closely involved in technological decision-making. For example, trade unions and movements for workplace democracy and participation sought involvement in decision-making over technological change at work. Women's groups and religious and moral bodies have become engaged in debates about the new reproductive technologies and the new genetics. Probably the largest movement comprises the variety of interests concerned with environmental issues and sustainability. It is increasingly accepted that technology decision-making should no longer be the preserve of a narrow coterie of technical specialists and policymakers. Various publics are also wanting to contribute to debates – and are arguing that their knowledges, commitments and concerns are as relevant to decision-making as those of experts.

Alongside this has developed a broader view of the spaces in which influence may be exerted over technological development. Policy is no longer limited to the formal realm of technology policy-making – in particular the allocation of RTD budgets – but includes other areas of public policy – for example in the social domains in which technologies are applied and used. There have been numerous attempts to influence system design and development through direct engagement with engineers – in such initiatives as participatory design at work. Finally, with distributed technologies such as ICTs, the consumer is emerging as an important actor: there is increasing recognition of their active role in selecting from supplier offerings, in deploying them for particular purposes, and in struggling to get technologies to work and be useful.

The Search for an Interactive Technology Policy Framework

The search for new, more effective and more sophisticated approaches to technology policy only partly stems from this growing recognition of the failings of orthodox approaches to technology policy in highly complex, rapidly changing high-technology fields. It also reflects the emergence of more complex and stringent demands from policy-makers and the public. We see two related demands emerging from politicians and senior policy-makers: first for evidence-based policy, with a stronger basis in science and social science knowledge; and second for more integrated policy responses. The BSE crisis, for example – and a series of other health-related controversies around food – has drawn attention to the shortcomings of the separation between on the one hand policies for agriculture and food production, which largely emphasise application of technologies to cut costs and intensify food production, and on the other hand emerging health policies and a new commercial dynamic in retailing, both of which stress the quality of food and are underpinned by much more traditional criteria of quality.

Coupling technology supply to markets

The need for more integrated approaches to technology policy – both public policy and commercial strategy – that link technology to markets was first signalled almost 30 years ago by Chris Freeman. Freeman (1974) stressed the importance of coupling between technology supply and its user markets – linkages which are today analysed by evolutionary economists as the basis for the learning economy.

However, linking new technologies to the market of potential users has posed considerable challenges. This has been particularly notable in information and communications technology (ICT) in which an accelerating rate of change in core technologies has been associated with enormous turbulence in product markets. The falling costs of computing and communications has opened up the possibility of new kinds of products and services. But what will these look like? There has been considerable uncertainty about which products will prove sufficiently attractive to consumers to be commercially viable – particularly given expectations that radical technological change would both allow very different ways of meeting user needs and create new markets. In a context where there is no existing customer base, knowledge of earlier products and markets may no longer be a reliable guide. Meeting 'user need' became a rallying call for industry and for government alike – as in the later stages of the ESPRIT (European Strategic Programme of Research in Information Technology) Programme in the European Commission's Fourth Framework Programme (4FP). These uncertainties were given greater salience by the potentially enormous costs of research and development of new knowledge-intensive technological systems. In industries like ICT these uncertainties about future markets and competitive challenges have combined. Technological changes have tended to cut across existing product markets and industrial sectors. The brief history of this sector has been characterised by the displacement of many established players by new market entrants. Correctly anticipating – and shaping – evolving markets and user needs are key to the commercial survival of firms in these sectors (Myrhvold 1999).

These uncertainties regarding user acceptance of new technologies have become more significant in recent years because technologies such as ICT are being taken up by an ever growing share of the population. Greater scale increases the potential development costs and the commercial benefits of successful systems. It also increases the difficulties faced by suppliers in knowing their user market. Furthermore these systems are becoming part of everyday life – at work, in the community and in the home – in a much more direct way than before. Over the first four decades of the application of computers in administration, for example, most workers had only indirect experience of computing. As the technologies moved out from the domain of technical specialists, they became subject to a range of new and potentially differing demands, from managers and end-users at work, and from people using them for entertainment in the home. User demands have become more critical. It is no longer sufficient to generalise about 'the user'. Although some user needs are generic, many are more specific. ICT applications are increasingly bound up with *social practice*. Industry-specific applications must support the competitive strategy and coordination processes of the organisation – they are 'organisational technologies', replete with information about the specific organisation and its sectoral context. ICT applications in everyday life are subject to perhaps more diverse and challenging user expectations – not just functional requirements of productivity and cost-effectiveness, but also complex personal goals relating to individual and group identity and satisfaction.

These questions about how a technology is domesticated and socialised are not, of course, restricted to ICTs. A similar array of issues is coming onto the horizon, for example, in the application of biotechnology in new medical treatments and in agriculture (Green 1991).

From the linear model to 'linear plus'

The European Commission's Fourth Framework Programme FP4 and comparable national programmes, such as the 'foresight' initiatives launched in the 1990s by a range of developed economies, gave greater attention to questions of dissemination and application as well as the generation of new technological knowledge. Various measures were proposed to assist the transfer of technologies from public sector research laboratories and their commercial exploitation. These developments, though important, reflect only a partial rethinking of the traditional model. Rather than abandon the linear model of innovation altogether, they represent attempts to counteract the reasons for its ineffectiveness. For example, measures were proposed to assist the dissemination of new technological knowledge from the research and education sector to industry, ignoring the fact that much industrial innovation is rooted in knowledge generated within industry itself. This limited modification has been described as a 'linear plus' model (Tait & Williams 1999): it still seeks to get the innovation process to operate in an effectively linear manner.

It is not necessarily the case that policies based on linear models can never be effective. Linear models, or at least improved linear models that at least acknowledge some interaction and feedback between stages, can apply as reasonable approximations in some areas. In certain circumstances, where a technological paradigm and infrastructure of institutions and complementary technologies have already been established, stages of innovation may proceed effectively sequentially, driven by the supply of new technology offerings. The development of new drugs is perhaps a case in point, where, providing well-established criteria have been followed for demonstrating efficacy and safety, pharmaceutical firms can be fairly confident about the market for new products. Where the linear model is a reasonable representation, it can be a rather efficient way of organising innovation (Tait & Williams 1999).

It is necessary however to identify which technology areas or industries such models fit, and not assume that they apply generally. Where concepts of products and services are novel and diverge from existing social and technical templates, effective innovation requires other kinds of adaptation and learning, and knowledge and other inputs from a broader range of players. The linear and 'linear plus' models both tend to downplay the importance of the array of non-technical actors – players downstream in the innovation process and others involved indirectly – and of their knowledge, experience and concerns.

The challenge is thus to develop innovation models that deal adequately with the full range of relevant players and configurations of innovation activities. This calls for more attention to demand-side interests and the process of appropriation of new technological capabilities. This is not, however, to suggest a shift to the obverse of the technology-push model: an equally simplistic demand-pull picture. Instead it is necessary to tackle the complex interactions between demand-side and supply-side actors, and with others who may not be directly involved but who may influence technology choice as regulators, as policymakers, and informally by shaping opinions. In the search for such an integrated model, linear plus is a step in the right direction but one that does not go far enough.[5]

2 REORIENTATING THINKING ON POLICIES FOR TECHNOLOGY

Technology policies are being reassessed across the developed world, and we see encouraging developments both in the terms of general debate and in programmes and practices. In particular, the European Community Fifth Framework Programme (5FP) represents a challenging rethink of previous technology policy frameworks. It signals a break from the technology-push models of the past and instead proposes an integrated approach, encompassing technology promotion and exploitation and at the same time geared towards broad social and policy goals such as competitive and sustainable growth. The emphasis is upon interdisciplinary research to support this effort. It will bring together social sciences as well as natural science and engineering disciplines, in a targeted effort, guiding individual RTD and uptake projects and linking them into clusters around certain strategic technological targets.[6]

The integrated model proposed by 5FP builds on and complements developments in thinking about the process of technological innovation: the critique of linear models and the espousal of an interactive model of innovation, as well as a recognition of the need for interdisciplinary approaches which bring together insights from social and policy sciences, business studies and economics, and the disciplines of science and engineering. Indeed in many respects, the challenges posed by 5FP may have run ahead of theoretical understanding and our ability to make useful generalisations about policy approaches and solutions.

Alongside major stimuli in the policy world itself – the experience of governments' attempts to promote and control technologies, and the problems analysts and indeed practitioners have identified with existing policy frameworks – SST and a range of other academic work have at least pointed to a major reorientation of conceptions of technology policy. The broad implications of this shift are:

- a much wider scope for the objects of policy, even if it is narrowly defined as government intervention to influence sociotechnical change;
- a more integrated approach to the range of policy areas that impinge on technological change;
- a broader view of possible means and points of intervention for steering technology;
- a wider notion of how technology steering is achieved and by whom;
- a different conception of the objectives and the modes of intervention.

The effect of SST work in general, described in Chapter 3, has been to open up innovation processes to scrutiny, to unpack their structure in time and space, and to stress the thoroughly social nature of what may have been assumed to be technical choices, logics or practices. The implication of much recent SST work has been to extend that focus of investigation and hence possible intervention downstream from research, design and development, to use and appropriation, and across a wider range of relevant actors. It indicates a wider range of points of intervention on an innovation journey – though differences in how effective and easy it may be at each – and a wider range of appropriate objectives, instruments and criteria for evaluation. It points to the inclusion of technology policy and specific interventions alongside other actions in the analysis of shaping – to treat them as 'endogenous', arising in the process of innovation and affected recursively by them, rather than somehow being external and tacked onto them – and therefore a more reflexive understanding of interveners' roles in that process.

This new view is reflected in a range of writings on policy issues which seek to move away from a linear understanding of the innovation process. However, the endeavour throws up new difficulties. If the relationship between technology and society is not unidirectional and determinate, but involves a variety of more or less loose linkages at different levels, the setting for technology policy becomes extremely complex. An interactive model which starts by highlighting linkages between technology supply and user markets rapidly encounters questions surrounding the different structures of markets, the diversity of intermediate and final users, and the negotiated and unstable character of demand as needs evolve in the face of social and technical change. The picture of innovation is one of much greater complexity and unpredictability, and one which poses a range of new problems for technology policy formation and for the contribution of SST to this.

The new view also forces an examination of the relationship between technology policy and wider policy domains. Alongside state interventions intended to promote innovation we find a wide range of government policies and actions on technical and knowledge infrastructures, industry and market restructuring, and environmental and health regulation, which set fundamental conditions affecting the adoption of technologies. Thus numerous other areas of policy have direct or indirect effects on technological innovation and adoption. Yet typically the connections between these interventions have been neglected and little has been done to coordinate them.

The analysis makes clear that no one point or level of intervention will be adequate – particularly an exclusive focus on the design and development phases of innovation. It opens up a wider range of points of influence, and draws attention both to tensions between different means of intervention and to opportunities for synergy and reinforcement.

Another major shift is toward more modest ambitions in policy. This follows from the emphasis in SST on the complexity and unpredictability of innovation directions and outcomes, and the complex internal interdependencies in technologies and industries. But it is also consistent with a number of other influences:

- the discouraging experience of many initiatives to restructure organisations using ITs, of interventions to control major technological projects, and of initiatives to introduce wider participation in design processes;
- general critiques of forecasting and synoptic rationality;
- the failure of social engineering or technological fixes;
- analyses of the limitations of government intervention and the widespread actual retreat of government from planning, infrastructure development and other strategic intervention.

Thus technology steering will look much less like the traditional picture of omnipotent and omniscient central direction. It will be much more like modulation and orchestration of the existing dynamics of innovation or technology management. It will entail providing conditions for social learning processes: first, a more detailed understanding of innovation processes, especially the internal workings of firms and technical communities, and second, sensitive and effective feedback to enable their redirection. It may entail inducement or encouragement or information to enhance existing processes, or new elements with the leverage to redirect those processes. It will be a continuing process of anticipation and intervention. It will be attempting to influence an innovation journey, 'managing contingencies and uncertainties' (Deuten, Rip & Jelsma 1997), rather than executing a grand plan as if that can be done with finality.

The shift faces the serious criticism that intervention in technological change will become unintentionally conservative; it will resign itself to minor tinkering with agendas and directions of development set by powerful interests. It will be deemed impossible to pursue strategic social objectives that conflict with them. The complexity and limitations of intervention may become excuses for inaction.

Clearly strategic political goals for technologies can be set and pursued. As Weingart (1984) observes, alliances of government agencies, engineers, and corporations acting as quasi-public agencies with heavy subsidies, have on occasions pushed through major projects without market discipline or rigorous economic appraisal and with little regard for public acceptability. It can be argued that some were barely performing adequately in technical terms let alone viable in market conditions. Moreover the theoretical shift we have identified comes at a time when clearly major changes can still be made when there is political will – such as the dramatic restructuring of several countries' electricity industries, with abrupt changes in what had been taken as deeply entrenched technological systems – and when the redirections envisaged as needed for environmentally sustainable technological systems are ambitious indeed. At the same time there is a marked move in government, alongside liberalisation and privatisation, away from active planning of industrial and technological change in favour of markets, and increasing delegation of policy work to consultants.

It is thus crucial, in this move to a more modest conception of possible intervention, to rethink the ways in which strategic social objectives can be formulated, pursued and maintained. Paradoxically while the analysis implies a more limited idea of the control which can be exerted directly on innovation processes by government and public, it does open up a much greater range of possible forms and points of intervention, and promises more effective oversight in the long run.

For strategic moves like that towards an objective of sustainability at industry or sectoral level, we believe SST work promises a better understanding of possible routes, and points and means of intervention in existing regimes. Many formulations of sustainable development have identified current unsustainable practices and their impacts, and envisaged scenarios of new technologies and in some cases social correlates. The major weakness of many calls for sustainability however, is in failing to spell out a path towards the desirable future, to identify means of intervening in current patterns and institutions, and to diagnose possible obstacles and sources of resistance. There is a much greater chance of producing the desired outcomes if we take as a starting point the dynamics of current development, select points and means of leverage carefully, and are prepared for continuing and flexible intervention as the changes unfold. An SST understanding thus means we should be in a better position to advise on interventions that can realistically contribute to such long-term publicly articulated goals.

3 THE CONTRIBUTION OF SST TO A NEW TECHNOLOGY POLICY

We contend that SST analyses can make a substantial contribution to the current rethinking of approaches to technology policy: in general through a reconceptualising of its key problems and concerns, and specifically formulating or improving particular forms of policy analysis and practice.[7] It is clear indeed that some general ideas from SST already have had significant effect on the new policy thinking we have described.[8] However, the difference from traditional perspectives on technology, in which the dynamics of technological change tended to be neglected and its nature and roles oversimplified, means that the process under way represents a substantial reorientation in understanding and approach.

The review of social shaping research in Chapter 3, and the other contributions to this book, are based on the assumption that examining possible changes in intervention in sociotechnical development requires an improved understanding

- of the dynamics of sociotechnical development, so as better to identify possible points of intervention;
- of the dynamics of policy intervention – examining the characteristics of current policy instruments and initiatives; analysing the effects of existing interventions into processes of innovation; and developing ways of understanding the likely effects of new forms of intervention (Weber, Chapter 12 in this volume).[9]

This new understanding, however, does not feed into policy-making in a single and simple way, and we are concerned to set out carefully both how it can be useful and what cannot reasonably be expected of this or for that matter any other socio-economic research.

It is highly unlikely that its use will be direct and instrumental in the manner depicted by technocratic policy models – in which policy is somehow deduced rationally from research (Rayman-Bacchus, Williams & Bechhofer 1998). The understanding will seldom translate into general or detailed prescriptions for intervention – though of course research groups with a deep knowledge of a particular area may be justified in making specific recommendations. Nor can SST work provide generally applicable *methods* for formulating policy.

Finally, the analysis indicates it will seldom be possible to make valid generalisations across broad technology areas and contexts. The growing body of empirical evidence attests to the complexity of the process of technological innovation; the many actors involved; and the operation of a range of factors which may be in tension. It highlights the contradictory pressures on innovation processes and the uneven and contingent nature of their outcomes. The picture that SST reveals is far removed from ideas of steady, cumulative, global technological progress. Various writers have stressed, for example, the local and differentiated cultures of technology supply (Vincenti 1994, Hård 1994). Similar points emerge about the local character of user communities and markets and the relationship between local markets and contexts of use and how they may – or may not – be transformed into more global markets.[10]

Comparative work in SST points to the diversity of developments and their outcomes between different domains of life, technical fields, industries and regions. This diversity is often not sufficiently recognised by practitioners and decision-makers who may take for granted many aspects of their national or sectoral context.

Correspondingly, then, the whole question of intervention becomes more complex. Given the importance of local contexts and contingencies, analysts have cautioned against making strong generalisations for public policy or business strategy, for example about the role of particular institutional and policy settings in supporting successful innovation.

Though we argued at length in Chapter 3 that it is possible to produce valid and useful general claims about technology dynamics and intervention, it is certainly the case that the pitfalls of over-generalised approaches to technology policy have been the major reason for a shift away from a search for 'cookbook' recipes for success and towards a processual approach (Clark & Staunton 1989, Williams & Edge 1996).

Processual Approaches to Innovation – Tools for Reflexivity and Social Learning

At first sight, then, SST approaches might seem to offer little assistance for hard-pushed decision-makers in government and industry. Their challenge to this work is, understandably, to go beyond exercising a cautionary and critical voice and to offer some stronger general lessons. We argue that several sets of developments in the work represent promising and productive ways forward.

First is the growing body of empirical research findings in SST. These are steadily offering a more adequate and more comprehensive account of the process of technological innovation and how this varies across settings – differing technical domains, differing socio-economic contexts of use, differing national, cultural, economic and policy contexts.

Thus the case study basis of much SST research, and the dangers of invalid generalisation, do not rule out robust general lessons: on the contrary, with appropriate qualifications and sensitivity to the dangers of translation, SST work can help interveners understand how insights can be usefully transported between areas. That is to say, the value of analyses of specific technologies or applications, with a better understanding of the range of conditions of which they are representative, is that practitioners meeting a new situation will have an improved set of templates or paradigmatic accounts with which to recognise its features and know how to approach its treatment. At a different level, a better understanding of the broad sociotechnical landscape, which forms the context for particular industries or technologies, can help policy makers identify changes needed in fundamental conditions affecting innovation.

Second is the development of tools for analysing innovation and concepts for making sense of the features and patterns. Though the diversity revealed by these SST studies belies any simple correlation between context, process and outcomes of innovation, these studies have allowed the development of a rich conceptual framework which helps to map out points of regularity and similarity. SST has been extremely productive in its theoretical and conceptual contribution, and many of its concepts are geared, more or less explicitly, to guiding intervention.[11]

Some authors have sought to make these insights available to non-academic audiences and more useful – for example through a 'tool-box' approach, whereby a range of frequently encountered processes and issues are identified. These are offered not as one-size-fits-all generalisations, but as *simplifications* about innovation processes together with indications of the contexts in which they are relevant. Molina's 'socio-technical constituencies' model (1997), for example, identifies issues concerning the internal *alignment* of groups of diverse players involved in technology development. Other instances of readily applicable frameworks emerging from SST are *constructive technology assessment* and *strategic niche management* (see Rip & Schot, Chapter 5 in this volume; Geels, Chapter 13 in this volume).

The third development concerns the activities and reflexive understanding of the players involved in the innovation process itself: it focuses on the adaptive behaviours and learning opportunities available to them. With its emphasis on processes, on their complex and shifting contexts, and on the limitations of foresight and planning, this work brings to centre stage the question of how well individuals, groups or organisations – or, for that matter, an economy as a whole – can react creatively to unpredictable change and to particular exigencies. The concept of *social learning* suggests exploring the activities of various actors in solving problems in particular contexts (*learning by doing*) and then in deploying that knowledge more broadly (*learning by interacting*) (Sørensen 1996). It is seen not as a process of dissemination of knowledge, but of selection, adaptation and transformation (Williams, Slack & Stewart 2000). This concept parallels a wider growing concern with processes such as learning and capability building[12] and with the importance of ideas of the 'knowledge economy' and the 'learning organisation' (Gibbons *et al.* 1994)

A further basic premise of the application of SST work therefore is that enhanced understanding is crucial in structuring and facilitating that learning: for developing consciousness of capabilities and skills, and understandings of contexts and possibilities, and for establishing appropriate organisations and networks. The analyses will similarly prove relevant for devising and conducting evaluations of existing and future interventions such as support programmes and regulations.

New Modes of Using and Interacting with SST Work

We emphasise then that the value of SST research is not as a set of finished solutions for decision-making – of the sort implied by the 'recipe book' approach. It will be much more like a tool box for selective and reflective application, knowledge which policy makers can make use of in different ways according to their custom, training, purpose and circumstance.

We envisage a wide range of often indirect ways in which the work should feed into policy practice:

- providing background information and a deeper understanding;
- as a provocation or challenge, or to reorientate thinking;
- raising awareness of a wider range of possible directions and outcomes;
- providing new more appropriate narratives, models, categories or metaphors;
- clarifying a problem and structuring discourse around the problem and possible solutions;
- sensitising to the identities and roles of a wider range of players and interests;
- increasing awareness of uncertainties and dynamism in innovation and possible barriers to rapid change;
- sensitising to the range of accounts of an area from different perspectives.[13]

Accordingly the role of SST analysts is, as we have stressed, much less likely to be in trying to set out detailed prescriptions, and more in mapping the terrain of innovation, clarifying and monitoring the processes, and informing choices in the light of strategic objectives.

The whole shift we have outlined in the conception of appropriate intervention points to a need for a continuing interaction between policy makers and SST analysts in developing not only general guidance of the sort represented in this report, but also advice on specific projects, industries and sectors.[14]

NOTES

1 Perhaps an important factor underlying the repeated return of policymakers to 'linear', supply-driven models is that they appear to hold out the enormously enticing promise of a world in which the development, uptake and outcomes of new technologies can be planned and controlled, together with rather straightforward mechanisms for intervening – through support for technological development (Tait and Williams 1999).

2 This points to a two-fold weakness in technology policy research. There has been a tendency towards an unhelpful separation between short-term technology policy studies and evaluations, which have often been descriptive and not well theorised, and the conceptually richer analyses emerging from SST. This gulf needs to be overcome. Further, policy itself thus needs to be more firmly integrated as an object of SST analysis, so that policy instruments and objectives are treated as interacting with other influences on sociotechnical development.

3 We can detect the intellectual heritage of this kind of perspective in the requirements in parts of the European 3rd and 4th Framework Programmes to conduct assessments of the 'economic and social impacts' of technologies developed under their auspices.

4 An important turning point in attitudes to new petrochemical technologies was associated with Rachel Carson's book, *Silent Spring*, which documented the damage to wildlife associated with the post-war use of synthetic pesticides and other biologically active materials. Further challenges arose in the 1970s as a succession of health and environmental hazards testified to the possible dangers associated with new technological activities (asbestos, PVC, radiation) – dangers which science and technology had failed to predict (Faulkner, Fleck & Williams 1998).

5 Although focusing in this account upon the shortcomings of frameworks which emphasised 'science and technology push', we should also note their mirror image in those policy approaches which gave primary emphasis to the role of markets. Such approaches became prevalent in the 1980s, notably in the UK. The high-profile failure of a number of large-scale state-sponsored technologies (e.g. the Anglo-French Concorde supersonic airliner) informed arguments that the state was not effective in 'picking winners' in the future technology race, and that such choices should be left to the market. However this resort to market-based systems of selection had its own difficulties. There was a growing awareness that markets could fail in certain circumstances – for example in relation to the needs of economically marginal groups, costs that were non-monetarised or 'externalised', and failures to achieve critical mass for market viability of unproven technologies. Moreover the resort to short-term market exigency left unanswered problems of technology policy. Today this dichotomy between state and market provision seems unhelpful. Instead the key issues concern how best to integrate public and private provision. The broader implication is that neither a 'science push' nor a 'market pull' model is adequate. A realistic and useful policy framework needs instead to address the interaction between them.

6 This is not to say that 5FP has overcome all the failings of previous approaches. The legacy of earlier linear policy frameworks is represented, for example, in the very structure of the programme in the distinction it makes between research and take-up activities. This separation overlooks the important innovative opportunities which may arise as new technologies are implemented and used.

7 In seeking a closer engagement with policymaking, we do not wish to erode the distinction between SST researcher and decision-maker – or intervener more generally. This would not only downplay the considerable expertise that many decision-makers and others

deploy, and the specific responsibilities and exigencies to which they are subject, which are far removed from the world of research. The shift to being an advocate of particular decisions would also conflict with one of the core tenets of constructivist approaches – about the need for symmetry in analysis. Thus, following Bloor's sociology of science, SST has from the outset insisted on the need for 'methodological relativism', involving a sceptical and even handed treatment of various competing technological alternatives, precisely to avoid analysis becoming blinkered by commitment to particular technological paths and claims (Pinch & Bijker 1984). Though there is a good argument which many researchers would accept for a stronger orientation of SST work towards questions of intervention, clearly the community of researchers must maintain a balance between explicitly policy or politically driven work, and work aimed at enhancing its analysis. The two depend on each other in a variety of ways (Webster 1991, p. 42).

8 It is difficult to track the diffusion of particular ideas from the research domain and their 'impacts' upon technology policy and related discourses in academic or policy circles. Much depends upon the receptiveness of policy makers to particular messages and analyses. However SST forms an important part of the dialogue that has helped move discussion forward.
In turn, it is clear that the growth of technology studies has been in large part a response to practical issues of planning and managing technology and the dramatic social change associated with it, and to dealing with perceived opportunities and problems. As Sørensen (Chapter 2 in this volume) argues, 'the emergence of technology studies in Europe may be seen to indicate that the concerns of policy makers may play a rather constructive role in the development of academic research.'

9 This involves increased attention by SST researchers to policy processes, drawing insights from parallel advances in policy analysis with compatible emphases on process and on network and boundary formation, etc. (Sabatier & Jenkins-Smith 1993).

10 It is not just that the substantive analysis depicts much more complex processes. It also raises questions about what might otherwise have been taken as relatively simple matters – by challenging and problematising accepted concepts, distinctions, boundaries and models of causality. The effect is not comforting but it is necessary.

11 The variety of new concepts which SST work has proposed can be daunting, and a key objective of the review and glossary in Chapter 3 is to provide some structure and order to these. They are obviously crucial to the improved understanding of innovation processes and contexts. Each in effect attempts to provide a better form of simplification, condensation, metaphor, narrative, or bracketing of complexity – more analytically rigorous and/or pragmatically useful than the assumptions and generalisations in conventional accounts. Explicit discussion of these concepts and their advantages should help sensitise interveners to the assumptions behind the various frameworks and the explanatory principles on which accounts are constructed. It should also indicate the consequences or dangers of taking one rather than another, and what form is best for what purpose.

12 See e.g. the papers in *Technology Analysis and Strategic Management*, Dec 1999.

13 The diversity of SST work itself has important implications for its application in policy. In the review in Chapter 3, we comment on the various concepts and frameworks, and identify the extent to which they overlap. We have not thought it appropriate, however, to try and judge which are 'best'. Rather, we would expect those applying these ideas to assess which are most appropriate, singly or in combination, in particular problem domains. It may be instructive to look at an area from several viewpoints, to get partial insights, and perhaps to move between them when the limits of one become apparent. Those evaluating the models more pragmatically may be more inclined to mix or adapt approaches in a potentially productive way than the theorists propounding them.
Since there is no self-evident translation of analyses into policy prescriptions, it is to be expected that arguments about the significance of particular phenomena will sometimes conflict. If, as we advocate, there is to be a continuing dialogue between policy communities and SST researchers, then it may be better to be aware of the range of models and to use advice from different quarters. We have noted in Chapter 3 the different development of SST in different countries: interveners must recognise there will be national peculiarities of technological concerns, systems, social and policy environments, etc. that make translation of insights across national boundaries questionable.

14 We stress that SST is not the only body of work which should provide the basis for this reshaped technology policy approach, and that a number of the policy implications drawn

out in this report have been pointed out in other currents in technology studies. In particular, much technology policy is obviously concerned with the structural and systemic character of a nation's or region's innovation activity. Developments in innovation theory based in evolutionary economics tackle the issues much more directly at this level, and have already offered a greatly improved understanding over conventional economics.

BIBLIOGRAPHY

Caracostas, P. & Muldur, U. (1998), *Society, the Endless Frontier: a European Vision of Research and Innovation Policies for the 21st Century*, European Commission DGXII Science Research Development Studies EUR 17665, Luxembourg: Office for Official Publications of the European Communities.

Clark, P. & Staunton, N. (1989), *Innovation in Technology and Organisation*, London: Routledge.

Deuten, J.J., Rip, A. & Jelsma, J. (1997), 'Societal Embedment and Product Creation Management', *Technology Analysis and Strategic Management*, **9**, (2), pp. 219-236.

Faulkner, W., Fleck, J. & Williams, R. (1998), 'Exploring Expertise: Issues and Perspectives', in R. Williams *et al.* (eds), *Exploring Expertise*, Basingstoke: Macmillan, pp. 1-27.

Freeman, C. (1974), *The Economics of Industrial Innovation*, 1st edition, Harmondsworth: Penguin.

Gibbons, M. *et al.* (1994), *The New Production of Knowledge: the Dynamics of Science and Research in Contemporary Societies*, London: Sage.

Green, K. (1991), 'Shaping Technologies and Shaping Markets: Creating Demand for Biotechnology', *Technology Analysis and Strategic Management*, **3**, (1), pp. 57-76.

Hård, M. (1994), 'Technology as Practice: Local and Global Closure Processes in Diesel Engine Design', *Social Studies of Science*, **24**, (3), pp. 549-585.

Myrhvold, N. (1999), transcript of interview, *In Business*, BBC Radio 4, 11 Nov 1999.

Molina, A. (1997), 'Insights into the Nature of Technology Diffusion and Implementation: the Perspective of Sociotechnical Alignment', *Technovation*, **17**, pp. 601-626.

Pinch, T. & Bijker, W. (1984), 'The Social Construction of Facts and Artefacts: or How the Sociology of Science and the Sociology of Technology Might Benefit Each Other', *Social Studies of Science*, **14**, pp. 399-444.

Rayman-Bacchus, L., Williams, R. & Bechhofer, F. (1998), *The Dynamics of Exploiting Social Science Research: Towards Modelling the Social Process*, Science and Technology Studies Working Papers, no. 3, Edinburgh: University of Edinburgh.

Sabatier, P.A. & Jenkins-Smith, H.C. (1993), *Policy Change and Learning: an Advocacy Coalition Approach*, Boulder CO: Westview Press.

Sørensen, K.H. (1996), *Learning Technology, Constructing Culture: Socio-Technical Change as Social Learning*, STS Working Paper no. 18/96, Centre for Technology and Society, Trondheim: University of Trondheim.

Tait, J. & Williams, R. (1999), 'Policy Approaches to Research and Development: Foresight, Framework and Competitiveness', *Science and Public Policy*, **26**, (2), pp. 101-.

Vincenti, W.G. (1994), 'The Retractable Airplane Landing Gear and the Northrop "Anomaly": Variation-Selection and the Shaping of Technology', *Technology and Culture*, **35**, pp. 1-33.

Webster, A. (1991), *Science, Technology and Society*, Basingstoke: Macmillan.

Weingart, P. (1984), 'The Structure of Technological Change: Reflections on a Sociological Analysis of Technology', in R. Laudan (ed.), *The Nature of*

Technological Knowledge: Are Models of Scientific Change Relevant?, Dordrecht: Reidel, pp. 115-142.

Williams, R. & Edge, D. (1996), 'The Social Shaping of Technology', *Research Policy*, **25**, pp. 865-899.

Williams, R., Slack, R. & Stewart, J. (2000), *Social Learning in Multimedia*, final report of EC TSER project, RCSS, Edinburgh: University of Edinburgh.

5. Identifying *Loci* for Influencing the Dynamics of Technological Development

Arie Rip and Johan W. Schot

INTRODUCTION

New technologies eventually get embedded in society, and their impacts depend on the processes of contextualisation. In this overall co-evolution of technology and society, a variety of actors are interested in influencing technological change in terms of their own goals, be it market success, strategic advantages, quality of life, sustainability. It is through the actions and interactions of these actors, guided by their assessments and occasional, more or less systematic anticipations that technologies evolve and are adopted and adapted.

A challenge, not just for technology developers, but increasingly also for policy makers and critical societal groups, is to influence technological change at an early stage, when irreversibilities have not yet set in and one can hope to sway the balance between desirable and undesirable impacts. The dilemma involved has been articulated by Collingridge (1980): when control of technological change is still possible, knowledge of eventual impacts (and how these will arise) is so limited that the direction to go is unclear. His knowledge and control dilemma becomes less stark if one looks at the whole developmental process and tries to understand the dynamics and evolving patterns. Constructive Technology Assessment (CTA) addresses this specific argument: understanding of heterogeneous and often contested developments allows constructive interaction and gradual and cumulative shifts in desirable directions (Daey Ouwens *et al.* 1987; Rip, Misa & Schot 1995). At the time, this was a programmatic claim. In a later state-of-the-art paper, ongoing anticipation and open-ended learning were emphasised, and a number of CTA strategies were identified and discussed (Schot & Rip 1997).

One could argue that Collingridge's dilemma is still with us: the emphasis on learning does not imply that the right directions will be found in time (also because attempts to alter directions will be contested). Phrased in this general way, it is part of the *condition humaine*. A closer look at how technological change occurs in our societies, combined with a shift in perspective (and thus ambition) on how to influence technological change, allows us to improve on Collingridge and on earlier CTA strategies. The key notion is that there are preferred *loci* for influencing; windows of opportunity as it were. In this paper, we develop this notion for the specific context of technological developments initiated in firms (or other technology-promoting organisations) – a context which is dominant in the present era. The approach is generally applicable, however.

To set the scene, we first introduce further dilemmas (or paradoxes), and note that such dilemmas are somehow resolved in ongoing practices. Attempts to do better have to link up with such ongoing practices, and recognise the dynamics involved.

A first tension (or dilemma, or paradox) is the difference between functionalities originally envisaged for a technology and the eventually dominant ones. The functions the telephone now fulfils are very different from the ones envisaged originally: communicating between operations in the centre and in the periphery of the town, and piping concerts from the concert hall in the city centre to the suburbs (Fischer 1992). In other words, intervention strategies cannot simply be based on actors' intentions at the time, and their predictions of eventual achievements.

Especially in R&D-based innovations, the development trajectory optimises the new process or product as such, but its eventual success requires re-contextualisation, a process which cannot be anticipated fully, let alone determined, at the earlier stages of the trajectory (Verbund sozialwissenschaftliche Technikforschung 1997, p. 20). A striking example is the negative reception, by the deaf community, of cochlea implants (Garud and Ahlstrom 1997). This example also shows that the problem is not just a cognitive one (how to anticipate the unpredictable), but also a socio-political one (technological development is a matter of insiders, but will be exposed to outsiders). This is a dilemma, and not just a matter of blinkers, because there are costs involved in taking wider contexts into account at an early stage (Deuten *et al.* 1997).

The second tension (and paradox) strikes at the heart of the claim that understanding the dynamics of technological change and its embedding in society will allow intelligent intervention. Recent sociology of technology, together with the empirical part of economics of technology, provide a rich (if sometimes patchy) understanding of the dynamics of technological development and its embedding in society (Rip & Kemp, 1998). But this understanding is essentially retrospective, based on historical case studies and surveys. The patterns and regularities found in this way may be extrapolated into the future, but at a risk: circumstances may be different. In fact, they will be different already because of preceding technological developments and their dynamics, and their being recognised and understood for what they are.

Since actors will act strategically, including action based on the understanding of earlier dynamics, this may then shift the dynamics into a new pattern – which may even undermine the basis for their action. Self-negating prophecies, as in the case of warnings that are heeded, are an example of such a shift.

An intriguing further example is provided by Moore's Law, which has held for more than three decades now because actors direct their efforts and co-ordinate their actions with the continuation of Moore's Law as the frame of reference (Van Lente & Rip 1998). As a news report in *Science* (1996) phrases it: 'Researchers around the globe are working furiously to extend the life of Moore's Law by coming up with alternative chip-patterning techniques for use when current lithographic tools hit the wall.' It is exactly because of such efforts, driven by innovation competition, that advances in chip technology remain predictable. As soon as firms decide to adopt another strategy, and go for alternatives, Moore's Law would lose its hold, and thus its validity.

Does this imply that the insights of sociology and economics of technology are to no avail? Indeed, sociologists, economists and political science scholars of technology should be modest about their contribution. But there **is** a contribution: first the negative one, the message that the ambition of guaranteeing the achievement of a desired goal is futile; and second, it specifies ways to modulate ongoing dynamics in the hope of getting closer to one's goals. This is like Charles Lindblom's productive 'muddling through' approach (cf. Lindblom & Woodhouse 1993), but with an added point: understanding the dynamics of development allows one to identify opportunities for intervention, and specify how such interventions can be productive. This will not resolve all complexities of anticipation and intervention, but will go some way to mitigate them.

As a first step, therefore, we introduce insights from technology dynamics. Concretely, we will use the metaphor of an innovation journey – which actually refers to the underlying phenomenon of emerging path-dependencies – to analyse dynamics (contingencies as well as recurrent patterns), and show that one can reconstruct the innovation journey in terms of actors' perspectives and actions, as well as interactions and their outcomes.

THE FIRST STEP: USING OUR UNDERSTANDING OF TECHNOLOGY DYNAMICS TO CREATE A MAPPING TOOL

Recent sociology and economics of technological change offer important insights (Bijker *et al* 1987; Dosi *et al* 1988; Tushman & Anderson 1997). For our purpose, we select two main points.

First, the recognition of contingencies and the tension with the need to anticipate, somehow. In Van de Ven *et al.* (1989), the studies of product creation processes within a firm are presented as innovation journeys with their setbacks and shifts. There is no path given in advance, the actors create a path by walking (sometimes stumbling along). Such innovation journeys include tentative anticipations on embodiment in society, and this is increasingly important, e.g. for biotechnology firms (Rip & Van de Velde 1997). Actors, e.g. the focal firm, will reduce the complexities by creating a concentric picture of the firm in increasingly wider environments (Deuten *et al.* 1997). If one takes technology, rather than a firm, as the focus, a variety of organisations (firms as well as public agencies and NGOs) are important. Non-linearities, branching, and path-dependencies then become even more striking (Rip 1993). In other words, anticipations by firms and other technology-developing organisations should take these complexities on board.

Second, and in spite of the *prima facie* contingencies, patterns emerge through linkages, alignment and networks. This holds for innovations, as in the well-known contrast between regular (or incremental) innovations and 'architectural' (or radical) innovations, where competencies as well as market relations are disrupted and have to be built up again (Abernathy & Clark 1985). And it holds for sectors and regimes, as in Garud (1994)'s analysis of 'fluid' and 'specific' structures, and Callon's (1998) closely related distinction between 'hot' and 'cold' situations. The importance of linkages and alignments extends beyond innovations and industry structures to the embedding of technology in society, which includes mutual adaptation with other products and actors, and articulation of acceptability (Rip 1995).

Generally speaking, there is co-evolution of innovations and industry structures, and more broadly, of technology and society, and there are definite patterns in the co-evolution (Rip & Kemp 1998). While there will always be contingencies, there is also linkage creation, increasing alignment, and thus a certain amount of predictability, or at least, reasonable foresight.

Our question about intelligent intervention can now be reformulated, in a first round, as one of anticipation on eventual contextualisation of a novel technology (as part of the innovation journey), and identification of possibilities for intervention. A mapping and diagnostic tool is necessary to do this, and this implies that the complexities of the real world have to be simplified into stylised facts, which must be sufficiently rich to capture complexity, but also simple enough to allow application across a variety of cases.

To develop such a tool, the first simplification is to focus on typical activities in innovation journeys. These activities (like invention, development, prototyping, introduction, diffusion) may occur sequentially, but always with feedback and feed-forward loops. Also, the identity of the 'travellers' (the actors as well as the technological options being developed) may change in the course of the journey. The case of the telephone is a clear example, but it is visible in almost all innovations. Anticipation on selection environments, and more generally, on context is important, and this may well be internalised and institutionalised. Test laboratories and consumer panels being, by now, obvious examples (Van den Belt & Rip 1987).

Actors involved with developing and introducing a new technological option will immediately recognise this way of capturing the dynamics of development. But it has a definite 'concentric' bias, in the way it starts with novelty creation and follows its journey over time and across space. As an analyst, one could, instead, focus on sectors and regimes from the beginning, or even more broadly, on evolving socio-technical landscapes, or techno-economic paradigms as Freeman calls them (Rip & Kemp 1998). Such a non-concentric view is not biased toward success, and this is important because many new options die at an early stage. For the question of influencing overall developments, attention to reviving options left by the wayside, or shifting the success criteria more generally, is important (even if it may well be difficult to nurture such alternatives so that they become real options).

This argument about actors and analysts is mirrored in practice by the contrast between insiders (who will take a concentric view) and outsiders (for whom the new technological option is only one item, and possibly an unwanted intruder) – cf. Garud and Ahlstrom (1997). Monitoring of the environment is done by the insiders. While this monitoring has an increasingly broader scope as the innovation journey progresses, there will also be surprises – when the environment strikes back, and upsets incipient paths and path dependencies.

These considerations set the scene for the second simplification, which addresses the problem of keeping the overall picture in mind, in spite of a concentric bias. The so-called 'techno-economic network' mapping approach of Callon *et al.* (1992) is a tool to map linkages and networks as they occur, and reminds the analyst (and then the actor) of the whole breadth of aspects involved in innovation. The simplification, to keep the mapping exercise manageable, is to reduce these aspects to four 'poles' to characterise activities: science, technology, market, and regulation, each of them operationalised in terms of dominant intermediaries and interactions (say, the four Ps: publications, patents, profits, and performances) (De Laat 1996). These 'poles' reflect the historically evolved, more or less institutionalised situation in present-day societies. They cannot cover all the details and complexities, but are a useful first approximation.

The two simplifications together: the journey metaphor and alignments categorised in terms of four 'poles', allow a useful visualisation of actual innovation and application/adoption in a two-dimensional scheme (Figure 5.1), with progress along the path of the innovation journey on the vertical axis, and the Science, Technology and Market/Society poles of techno-economic networks on the horizontal axis. (Figure 5.1, shows only three poles, but adds 'regulation' and 'society' to the market pole to indicate that there are more aspects.)

This visualisation covers a dynamic process, and allows progressive snapshots of an evolving situation. In the beginning, only the top of the map is in place, and the remainder is still in the future. Over time, as the innovation journey progresses, further parts of the map materialise. In the early stages, the market/society pole only figures in terms of market studies, early promises, and other expectations. Such expectations guide actions, but precariously: they may construct a 'house of cards' that breaks down when the effort to maintain it becomes too heavy (Van Lente 1993).

The advantage of introducing this mapping tool (in spite of its simplifications) is that it visualises the present as well as the future. Actors tend to project a linear future, defined by their intentions, and use this projection as a road map – only to be corrected by circumstances. The present mapping tool forces actors to consider the non-linearity of evolution, accept the complexity, and thus become more effective.

Both insiders and outsiders can profit from this mapping tool (and the understanding of the dynamics of technological change which informs it). Insiders can, and should, forget their linear projections, and accept that their attempts to make a difference cannot always be successful. Outsiders are invited to consider opportunities for intervention, defined by the dynamics of the innovation journey.

We think three types or clusters of activities can be distinguished where the innovation journey enters into a new phase: build-up of a protected space, stepping out into the wider world, and sector-level changes, and we will arrange our discussion accordingly.

Figure 5.1 Innovation Journey in Context

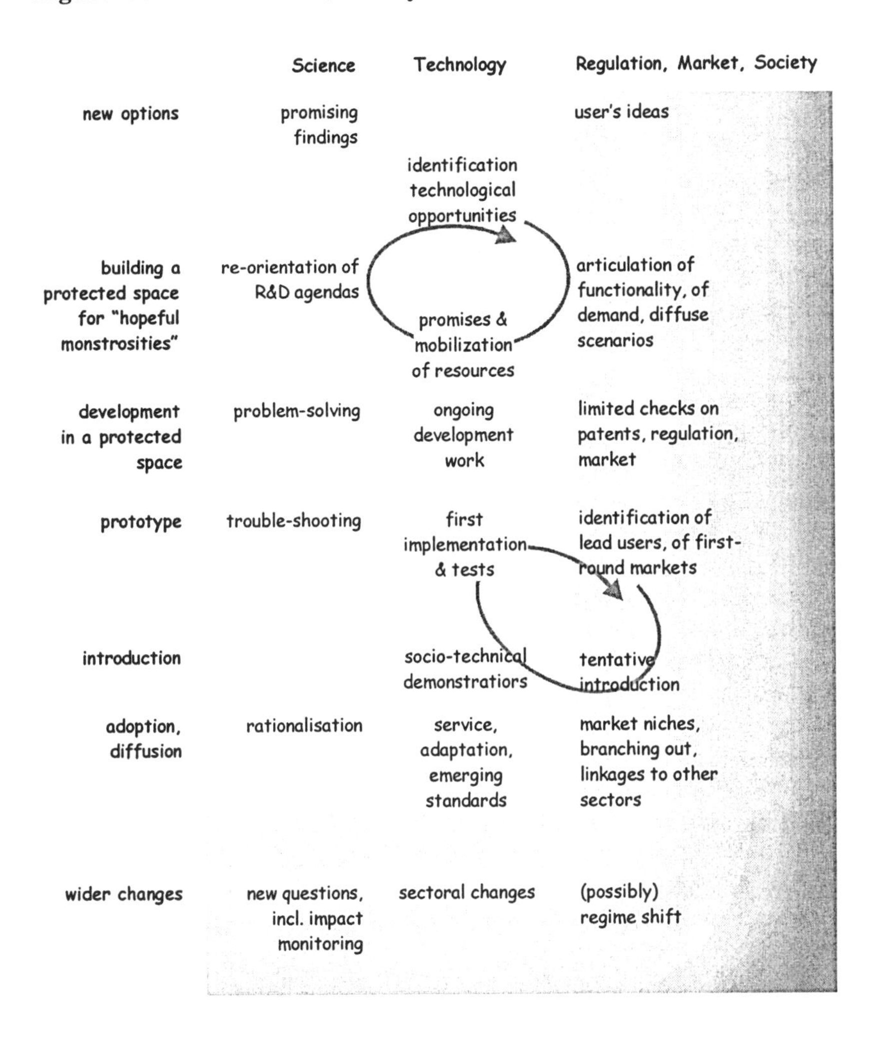

BUILDING-UP A PROTECTED SPACE FOR 'HOPEFUL MONSTROSITIES'

The simplified story-line of the innovation journey-in-context starts with the identification of a technological opportunity: a new option, or the pressure to find a solution for a problem. Such options may derive from R&D findings or scientific advance in general, but other sources remain important. The story of the zipper is a case in point. 'New combinations' (to use Schumpeter's suggestive phrase) are important. Suggestions or questions from (professional/knowledgeable) users can also identify opportunities, as Von Hippel (1976) has made clear.

The role of science varies, but has often to do with the discovery or modification in the laboratory of an effect which is linked with potential application. An example is the discovery of high-temperature superconductivity, which led to speculations about more efficient magnetic trains (eventually, other applications of this new laboratory phenomenon turned out to be more realistic, e.g. detection systems for very weak magnetic signals).

In the pharmaceutical industry, the search for 'leads' is a recognised activity, and this has to do with the level of articulation of functions to be fulfilled: (re)searchers have a good idea of what they should be looking for. In other sectors, the articulation of functions, and thus search for opportunities, is more *ad hoc*. From the world of science, there is a continuous stream of ideas and promises, but with reference to broad and diffuse functionalities only (for membrane science and technology, Van Lente & Rip (1998) provide a detailed study of the dynamics).

For our analysis, the key point is that such technological opportunities start out as 'hopeful monstrosities' (Mokyr 1990, Stoelhorst 1997): full of promise, but not able to perform very well. Actors will make more specific promises (to sponsors) to mobilise resources to be able to work on the technological opportunity, and nurture it into a semblance of functionality – what is called 'proof of principle'.

Such promises anticipate, and thus further articulate functions and possible societal demand. Because they also specify what the material, device or artefact should be able to do ('performance'), this identifies and stabilises an R&D agenda. As Van Lente (1993) phrased it, a promise-requirement cycle is started up and shapes the trajectory.

There will be a lot of hype and hand-waving, and actors will take-up different positions (exaggerate, or underplay the promise for fear of creating disappointments and a backlash). Quality control of the rhetoric is important (for all actors), but cannot, by definition of the situation, be definitive. Certain patterns have become recognisable, though (the 'gee whizz' factor; goldrush/bandwagon, and decline when the 'nomads' have travelled on). We note in passing that quality control is also necessary on the part of resource providers, within a firm or in the public domain, who often have a portfolio to fill with promising projects, and are eager to find them.

The net effect of the networking and resource mobilisation is the emergence of a protected space for promising R&D, for developing the technological opportunity. Part of the protection stems from a (precarious) agreement over a diffuse scenario about functions to be fulfilled and their societal usage. The nature of the protected space, its boundary agreements, the rules and heuristics derived from the promises that were made, together determine choices and directions. Work within the protected space thus proceeds according to its own dynamics, with only occasional checks with the scenario of usage (if at all).

The advantages are clear; they are recognised and consciously applied, in an extreme version, as skunk works, as when IBM constituted a separate group, with its own resources and outside regular management control, to develop its PC. (A practice pioneered as a conscious management tool by Lockheed aircraft company in the 1940s (Valéry 1999).)

The risks derive from the fact that the diffuse scenario which legitimated the creation of the protected space remains diffuse. This may create problems later on, which require repair work and/or (unexpected) shifts in direction. And sometimes, the promise will turn out to be empty after all.

STEPPING OUT INTO A WIDER WORLD

At some moment of time, a decision is taken (or emerges) to go for prototypes or other attempts at demonstrating a working technology. Activities include prototyping, exploring/optimising production, trouble shooting and rationalisation through further research, implementation and learning about usage, and preliminary market testing. In the case of biotechnology, regulation and acceptability also become real issues, and might direct efforts in particular directions.

These activities are much less self-contained than the earlier research and development, and fall prey to intra- and inter-organisational tensions. Different interactions and management styles are called for. Garud and Ahlstrom (1997) contrast the 'enactment style' of the insiders with the 'selection cycles' in which outsiders are involved as well. The complexity is all the more troublesome because time pressures are often very large at this stage. As Deuten *et al.* (1997) emphasise, the need to learn (in order to introduce the innovation successfully) may be great, but there may not be space for learning.

A useful way of assessing the complexity is to map the sociotechnical couplings that occur, for example with other actors in the chain, with government bodies, with third parties (for example, consumer groups), and check for balance and possible path-dependencies. Some of these couplings are already present for other reasons (e.g. alliances), or have been made at an earlier stage (Deuten *et al.* 1997).

With the tentative introduction of the new product or process, with a few customers ('lead users'), or in a 'societal' experiment (often in collaboration with public authorities, as in the case of electric vehicles), the complexity increases, but also the opportunities for real-world learning and subsequent modification of the product. Because specific sociotechnical couplings are introduced, path dependencies may occur: certain niches are created for learning, so the kind of learning depends on the niches, and this may not be adequate to the demands and selections in the wider world. In addition, the visibility of the project/product increases which will have repercussions.

Sociotechnical demonstrators are important, but not always possible in a direct way. For some sociotechnical systems, social experiments are done: real-life, but experimental use, to learn about the system, about use, and about articulation of demand (see Schot *et al.* 1994, Weber *et al.* 1999 on electric vehicles). These are full-blown sociotechnical demonstrators, in a situation where the devices are available, but the system is still uncertain.

When the devices are still uncertain, there can be trials with prototypes. Often, this is only possible in collaboration with an intermediate or professional user, who as it were provides the system, at least, the system as s/he is utilising up till then. So the 'real life' aspect becomes more dubious.

Market introduction, an important concern for marketers and higher levels in a firm, is thus a gradual process rather than a point decision. The non-linear nature of adoption of new products and technologies, while recognised retrospectively, requires lateral thinking, away from the present functionalities and prospects of the product. An interesting example is the use of industrial enzymes not for the projected improvement of main industrial processes but to stonewash denim for jeans - a niche market and an occasion to get production and utilisation experience with such enzymes relevant for broader industrial usage. An earlier example (instructive in many other ways as well) is how poly-ethylene became a successful polymer – because of non-critical demand provided by the unexpected Hula Hoop craze in the middle 1950's (McMillan 1979).

The 'market' is neither one-dimensional nor homogeneous, and demand is only gradually articulated in response to supply: in the late 19th century, there was no articulated demand for automobiles, and producers gradually learned to distinguish relevant product attributes in interaction between new technical possibilities and customer responses (Abernathy *et al* 1983). The point continues to be important, in the large as with modern biotechnology and multi-media technology, and in the small (Green 1992, Bower and Christensen 1995).

There must be something like a protected space for the new product, so that it will survive an otherwise too harsh selection environment. At the same time, limitation to the particular protected space creates a product that survives only within that space. This may well be the final outcome: a product existing in one market niche; fuel cells, in spite of their recognised general promise, really only functioned in space applications (Schaeffer 1998). But introducers of new products often want more, and explore further market niches. In general, a strategy of ever widening 'niche management' is in order (Weber *et al.* 1999). In the case of electric vehicles, the differential success of various introduction strategies support this claim (Hoogma 2000).

SECTOR-LEVEL CHANGES

The new product and/or new technology branches out in various ways. There have been some retrospective studies of these processes: of the telephone, developed with a view to other uses (Fischer 1992, de Sola Pool 1983); electric lighting between 1880 and 1930 is another interesting example (Marvin 1988; Nye 1990). Such historical studies are very important to broaden the perspective and create a sensibility, with actors keen on immediate success for their projects, for cross-linkages and shifts.

Branching of niches leads to niche 'piling' (Schot 1998): heaped on top of each other, the niches add up to something more than their simple sum. There is a cumulative effect of further varieties of application: suppliers orient themselves to the new technology, economies of scale and scope are exploited, and recognition, by users, of further possibilities (think of telephone and electric lighting again) creates new sociotechnical linkages. The sector starts to change, and its relations with other sectors change. The latter can become so important that the technology driving such changes by being taken up widely is called a pervasive technology, and characteristic of a new techno-economic paradigm (Freeman 1992).

Path dependencies occur for a variety of reasons: increasing returns (Arthur 1996), *de facto* standard setting (David 1985), sunk investments in competencies and culture (Burgelman 1994, North 1990). The outcome is the advent of a technological regime: an emerging and then stabilised set of rules guiding technological development and its embedding in society (Rip and Kemp 1998, Kemp *et al.* 2000).

The development of computers in the 1940s and 1950s (and beginning of 1960s) provides an interesting example. Computers started out being part of existing computing and automation regimes, and only gradually developed their own dynamics (linked to software, and in particular programming with the help of programming languages, and an infrastructure including compilers, rather than by creating physical linkages). By 1965, the tables were turned, and the computer plus software led developments rather than having to adapt. A computer regime was established (Van den Ende & Kemp 1999).

As with emerging rules and institutions in general, a reversal occurs in which a precarious product of actions and interactions of actors, requiring care and repair all the time, turns into a stable regime which orients actions and perceptions. This is the way to understand how design hierarchies become established (Clark 1985), and the regime concept can be used to broaden the notion of design hierarchy (Van de Poel 1998).

Cumulative effects may thus lead to the emergence of new regimes and/or shifts in existing regimes. This is a multi-actor, multi-level process, in which no single actor can sway the balance intentionally. Actors will attempt to do so, of course, jockeying for position in the newly emerging games and regimes, and involving themselves in strategic alliances. In standard setting in information and communication technology and in consumer electronics such processes are very visible. While the actors involved, as well as the media reporting on the struggles, may think in terms of heroic stories in which power and cleverness of the actors determines the outcome, the cumulative process of increasing interdependencies and sunk investments is the major explanation.

HOW TO INFLUENCE DEVELOPMENTS

If actors want to exert influence and change an emerging path in another direction, they face the momentum that has been built up and the loss of malleability because of increasing alignment. Internal actors are constrained by their inclusion in the dynamics, and external actors have to overcome the distance between inside and outside (cf. Garud & Ahlstrom 1997). A contest of forces may ensue, and one which may become larger than the original issue. This is how radio-active waste became an issue which may well subvert the further diffusion of nuclear energy production (Albert De La Bruhèze 1992), and the present controversies about genetically modified products - Frankenstein food - may lead to similar impasses.

The alternative approach is to see path dependency as an opportunity. If one can help shape the path and its ensuing dependencies at an early stage, there is no need to interfere later on: the irreversibilities along the path will take care of maintaining direction.

To do so in practice, the first requirement is to understand the dynamics of such developments in context. In general, it is important to enrich the innovation journey, as it occurs anyway, by anticipations and feedback. Anticipations of outcomes (including impacts of the technology on society) must be a ongoing concern, rather than *ad hoc* efforts to persuade a sponsor or regulator that the journey can continue. The learning made possible through scenarios (especially important at an early stage), through socio-technical demonstrators, and through the recognition of niches, must feed back into the development work.

Particularly interesting is the identification of preferred *loci* for intervention: just before 'gelling', it is still possible to exert influence, while there is some assurance that a real difference will be made because it will become part of the trajectory. Three such *loci* are visible in our present analysis: when a protected space for early development emerges; at the first introduction into the wider world; and when niches start to branch and pile.

The second requirement is intelligent intervention: it is not brute force but playing with the dynamics which will make a difference. This applies to attempts at intervention by outsiders as well. Modulation (with some orchestration) of the dynamics appears to be the right approach. With the many actors involved, and the heterogeneity of their interests and strategies, there is no guarantee that a coherent direction will evolve. A 'shadow of authority' may be necessary to break through impasses. Credibility pressures, for example in relation to environmentally-friendly products and processes, can play such a role.

Authority and credibility pressure are also routes through which public interest considerations about desirable directions can be brought to bear on the dynamics of development. While this indeed happens, for example in technology forcing regulation by governments and in public debates and consensus conferences about new technologies and their eventual impacts, there are also limitations deriving from the outsider position.

Intelligent intervention works out differently for the three *loci* we identified, because the dynamics and the extent of embedding in society are different. Insider and outsider interventions will be shaped differently, of course, but one can formulate general approaches.

Locus i

Modulate promise-requirement cycles, and the attendant resource mobilisation activities, so as to build a forceful agenda (for work in the protected space) on which general interests appear in addition to (short-term) actors' interests. Assess the balance of sociotechnical couplings, also from the point of view of desired societal aims. It is important to force some articulation of the diffuse scenario, even if this cannot be done with the help of a 'sociotechnical demonstrator' yet. There is nothing to demonstrate at this stage. But one can present ideas and expectations, and learn from responses by prospective users, opinion leaders. Socio-technical scenarios could be built as a kind of prospective sociotechnical demonstrators.

Locus ii

Modulate the introduction of a new process or product, seeing it as an experiment (in society rather than in a laboratory) through which one can learn about the technology and its impacts. As Weber *et al.* (1999, p. 11) note: 'Demonstration projects often aim at convincing others of the usefulness of a certain innovation, while [one should] aim to explore and learn in a quasi-controlled manner about the practicalities outside of the R&D setting.' In the cases they have studied in the domain of transport (including electric vehicles), such learning does not always occur because of short-termism and political exigencies.

Locus iii

Modulate the cumulative processes which may lead to regime changes. This is the most uncertain and precarious of the three *loci*, because of the involvement of many actors in many places, and the limited influence each individual actor can have on the outcome. Hoogma (2000) has developed a metric for niche creation and expansion. He shows, for the introduction of electric vehicles in a number of countries, that conforming to existing patterns of use and/or existing socio-technical configurations is necessary as a first step. Van de Poel (1998), after analysing eight cases of regime change, emphasises the twin requirements of creating openings for transformation by undermining the legitimacy of the existing regime, and building a technical agenda and developing technical alternatives to fill up the space so created.

In this chapter, we have not addressed the full scope of the challenge how to influence technological development at an early stage. We were able to identify three *loci* where anticipatory intervention can make a difference, but there are other such *loci*. Test labs, for example, are an accepted way of anticipating on later contexts, and can be seen as another instance of 'gelled' alignment, now as an institution rather than a stage in a process (in quasi-evolutionary terminology: an [institutionalised] nexus between variation and selection, cf. Van den Belt & Rip 1987, Schot 1992).

Our focus has been on industrial innovation, and in particular on new materials, devices and other products. In the development of large systems, on the other hand, other features may be prominent because of the larger need for co-ordination. The overall approach will be the same, however: instead of a rational planning and management approach, the vicissitudes of development and emerging alignments will be the starting point. (Interestingly, within RAND Corporation, the home of systems analysis, this same point was made and supported through case studies by a small group of analysts including Kenneth Arrow and Richard Nelson; see Hounshell 2000.)

Our analysis indicates what productive requirements will be on anticipation (e.g. articulate diffuse scenarios) and on intervention (modulation rather than force). There is, however, no guarantee that enlightened intervention will lead to desirable outcomes. In other words, we have not resolved the assessment-and-control dilemma. What we have done, instead, is to reformulate it, from a stark contrast between knowledge and control at the beginning and at the end, to ongoing processes of anticipation and intervention. The further step was to emphasise the importance of learning and modulation, as two sides of one reflective process.

BIBLIOGRAPHY

Abernathy, William J., and K. B. Clark (1985), Innovation: Mapping the winds of creative destruction. *Research Policy* **14,** pp. 3-22.

Abernathy, W.J., K. B. Clark and A.M. Kantrow (1983), *Industrial Renaissance -- Producing a Competitive Future for America*, New York: Basic Books, pp. 25-26.

Albert De La Bruhèze A.A. (1992), *Political Construction of Technology: Nuclear Waste in the United States, 1945-1972.* Delft: Eburon Press.

Arthur, W. B. (1996), Increasing Returns and the New World of Business. *Harvard Business Review*, July-August, pp. 100-109.

Bijker, W., Hughes, T.P. and Pinch, T. (eds.)(1987), The Social Construction of Technological Systems. New Directions n the Sociology and History of Technology, Cambridge (MA): MIT Press.

Bower, J. L., and C. M. Christensen, Disruptive Technologies: Catching the Wave. *Harvard Business Review*, Jan-Feb 1995, pp. 43-53.

Burgelman, R. A. (1994), Fading Memories: A Process Theory of Strategic Business Exit in Dynamic Environments, *Administrative Science Quarterly* **39**, pp. 24-56.

Callon, M. (ed.) (1998), *The Laws of the Market*, Oxford: Blackwell.

Callon, M., P. Larédo and V. Rabeharisoa (1992), The Management and Evaluation of Technological Programs and the Dynamics of Techno-economic Networks: The case of the AFME. *Research Policy* **21,** pp. 215-236.

Clark, K. B. (1985), The Interaction of Design Hierarchies and Market Concepts in Technological Evolution. *Research Policy* **14,** 235-251.

Collingridge, David (1980), *The Social Control of Technology*, London: Frances Pinter.

Daey Ouwens, C., P. van Hoogstraten, J. Jelsma, F. Prakke and A. Rip, (1987), *Constructief Technologisch Aspectenonderzoek. Een Verkenning* Den Haag: Staatsuitgeverij, NOTA Voorstudie 4.

David, P. A. (May 1985), Clio and the Economics of QWERTY. *American Economic Review* **75** pp. 332-337.

De Laat, B. (1996), *Scripts for the Future. Technology Foresight, Strategic Evaluation and Socio-technical Networks: the Confrontation of Script-based Scenarios*, Amsterdam: University of Amsterdam, PhD Thesis.

Deuten, J. J., and A. Rip (2000), Narrative Infrastructure in Product Creation Processes, *Organization*, **7**, (1) 67-91.

Deuten, J. J., A. Rip and J. Jelsma (1997), Societal Embedment and Product Creation Management, *Technology Analysis & Strategic Management*, **9**, (2) pp. 219-236

Dosi, G. & Orsenigo, L. (1988), 'Coordination and Transformation: an Overview of Structures, Behaviour and Change in Evolutionary Environments', in G. Dosi *et al.* (eds), *Technical Change and Economic Theory*, London: Pinter, pp. 13-37.

Fischer, C.S. (1992), *America Calling. A Social History of the Telephone to 1940*, Berkeley, Cal.: University of California Press.

Freeman, C. (1992), *The Economics of Hope. Essays on Technical Change, Economic Growth and the Environment*, London: Pinter Publishers.

Garud, R. (1994), Cooperative and Competitive Behaviors During the Process of Creative Destruction. *Research Policy* **23**, 385-394.

Garud, R., and D. Ahlstrom (1997), Technology Assessment: a Socio-cognitive Perspective. *Journal of Engineering and Technology Management* **14**, pp. 25-48.

Green, K. (1992), Creating Demand for Biotechnology; Shaping Technologies and Markets. In R. Coombs, P. Saviotti, and V. Walsh (eds), *Technological Change and Company Strategies. Economic and Sociological Perspectives*, London, New York etc: Harcourt, Brace & Jovanovitch Publishers pp. 164-184.

Hoogma, R., (2000), *Exploiting Technological Niches. Strategies for Experimental Introduction of Electric Vehicles.* Enschede: Twente University Press.

Hounshell, D. A. (2000), The Medium is the Message, or How Context Matters: The RAND Corporation Builds an Economics of Innovation, 1946-1962, in A. C. Hughes and T. P. Hughes (eds), *Systems, Experts, and Computers. The Systems Approach in Management and Engineering, World War II and After*, Cambridge, Mass.: MIT Press, pp. 255-310.

Kemp, R., A. Rip, and J. Schot (2000), 'Constructing Transition Paths Through the Management of Niches,' to be published in R. Garud & P. Karnoe, *Path Dependence and Creation,* Mahwah, N.J.: Lawrence Erlbaum Associates.

Lindblom, C. E. and E. J. Woodhouse (1993), *The Policy-Making Process, Third Edition*, Englewood Cliffs, N.J.: Prentice Hall.

Marvin, C. (1988), *When Old Technologies Were New. Thinking About Electric Communication in the Late Nineteenth Century*, New York: Oxford University Press.

McMillan, F. M. (1979), *The Chain Straighteners*, London and Basingstoke: MacMillan Press.

Mokyr, J. (1990), *The Lever of Riches*, New York: Oxford University Press.

North, D. C. (1990), *Institutions, Institutional Change and Economic Performance*, Cambridge: Cambridge University Press.

Nye, D. E. (1990), *Electrifying America. Social Meanings of a New Technology, 1880-1940*, Cambridge, Mass.: MIT Press.

Pool, I. de Sola (1983), *Forecasting the Telephone. A Retrospective Technology Assessment*, Norwood, N.J.: Ablex Publishing Corporation.

Rip, A. (1993), Cognitive Approaches To Technology Policy. In S. Okamura, F. Sakauchi, I. Nonaka (eds), *New Perspectives on Global Science and Technology Policy*, Tokyo: MITA Press, pp. 81-96. The Proceedings of NISTEP Third International Conference on Science and Technology Research.

Rip, A. (1995), Introduction of New Technology: Making Use of Recent Insights from Sociology and Economics of Technology. *Technology Analysis & Strategic Management* **7**, (4) 417-431.

Rip, A., and R. Kemp (1998), Technological Change. In S. Rayner and E.L. Malone (eds), *Human Choice and Climate Change*, Columbus, Ohio: Battelle Press, **2**, Ch. 6, pp. 327-399.

Rip, A., T. J. Misa, and J. W. Schot (eds) (1995), *Managing Technology in Society. The Approach of Constructive Technology Assessment*, London: Pinter Publishers.

Rip, A., and R. te Velde (1997), *The Dynamics of Innovation in Bio-Engineering Catalysis. Cases and Analysis*, Seville: European Commission Joint Research Centre, Institute for Prospective Technology Studies, Technical Report Series EUR 17341 EN.

Schaeffer, G. J. (1998), *Fuel Cells for the Future. A Contribution to Technology Forecasting from a Technology Dynamics Perspective*, Enschede: Twente University Press.

Schot, J.W. (1992), Constructive Technology Assessment and Technology Dynamics, The Case of Clean Technologies. *Science, Technology & Human Values* **17** 36-56.

Schot, J.W. (1998), The Usefulness of Evolutionary Models for Explaining Innovation. The Case of the Netherlands in the Nineteenth Century. *History and Technology* **14** pp. 173-200.

Schot, J., R. Hoogma and B. Elzen (1994), Strategies for Shifting Technological Systems. The Case of the Automobile System. *Futures* **26**, pp. 1060-1076.

Schot, J., and A. Rip (1997), The Past and Future of Constructive Technology Assessment. *Technological Forecasting and Social Change* **54** pp. 251-268.

Science: Can Chip Devices Keep Shrinking? 13 December 1996, p. 1834.

Stoelhorst, J-W. (1997), *In Search of a Dynamic Theory of the Firm. An Evolutionary Perspective on Competition Under Conditions of Technological Change with an Application to the Semiconductor Industry*. Enschede: University of Twente, PhD thesis.

Tushman, M. & Anderson, P. (1997), *Managing Strategic Innovation and Change*, New York/Oxford: Oxford University Press.

Valéry, N., Innovation in Industry. *The Economist*, 20 Feb 1999, pp. 1-128, at p. 18.

Van den Belt, H., and A. Rip (1987), The Nelson-Winter/Dosi model and synthetic dye chemistry. In W. E. Bijker, T. P. Hughes and T. J. Pinch (eds), *The Social Construction of Technological Systems. New Directions in the Sociology and History of Technology*, Cambridge, MA: MIT Press, pp. 135-158.

Van den Ende, J., and R. Kemp (1999), Technological Transformations in History: How the Computer Regime Grew out of Existing Computing Regimes. *Research Policy* **28**(8) 833-851.

Van de Poel, I. (1998), *Changing Technologies. A Comparative Study of Eight Processes of Transformation of Technological Regimes*, Enschede: Twente University Press, PhD thesis.

Van de Ven, A. H. *et al.* (1989), *Research on the Management of Innovation: The Minnesota Studies*, New York: Harper & Row.

Van Lente, H. (1993), *Promising Technology. The Dynamics of Expectations in Technological Developments*, Delft: Eburon.

Van Lente, H., and A. Rip (1998), The Rise of Membrane Technology: From Rhetorics to Social Reality, *Social Studies of Science* **28** pp. 221-254.

Van Lente, H., and A. Rip (1998), Expectations in Technological Developments: An Example of Prospective Structures to be Filled in by Agency. In C. Disco and B.J.R. van der Meulen (eds), *Getting New Technologies Together*, Berlin: Walter de Gruyter, pp. 195-220.

Verbund sozialwissenschaftliche Technikforschung: *Mitteilungen*, Heft 19, München 1997.

Von Hippel, E. (1976), The Dominant Role of Users in the Scientific Instrument Innovation Process, *Research Policy* **5** pp. 212-239.

Weber, M., R. Hoogma, B. Lane and J. Schot (1999), *Experimenting with Sustainable Transport Innovations. A Workbook for Strategic Niche Management*, Seville/Enschede.

6. The Cultural Shaping of Technologies and the Politics of Technodiversity

Werner Rammert

THE SOCIAL SHAPING PERSPECTIVE SPECIFIED: THE MEANING OF CULTURE IN TECHNOLOGICAL DEVELOPMENT AND ITS IMPLICATIONS FOR TECHNOLOGY POLICY

Traditional discourses on technology gave little attention to culture and the specificity of social institutions: the long-standing and prevalent technological determinist view as a case in point. This was associated with a technocratic type of technology policy. Similarly, later economic analyses of technological development focused on rational choices between different technologies, on the return of investments in R&D and on the impact of technological change on economic growth and the wealth of nations (see e.g. Freeman & Soete 1993). Universal criteria of efficiency were imputed. This led to the *politics of techno-globalisation*: all firms follow the best practice of the champion in a technological field, and all nations reorganise their innovation system in line with the winner in the international technological competition.

However differences of cultures and social institutions have significant consequences, too. The cultural diversity of technical practices, the cross-breeding of different engineering traditions and the mix of regional techno-regimes may become the unexpected source of start-ups of innovative firms, emerging technological fields and new networks of innovation. The cultural shaping view which is presented in this paper emphasizes the role of different symbolic orientations, patterns of practices and institutional regimes in the process of technological development. It opens the space and offers the tools for a new kind of technology policy that I call the *politics of technodiversity*. This involves a variety of projects and programmes; it includes a diversity of collective actors; and it takes advantage of different local and institutional conditions.

The cultural shaping approach extends and specifies the broad body of knowledge that is collected under the label of social shaping or social construction of technology. It is one aim of this paper to extend this focus towards institutional uncertainty and towards the processes of cultivating new technologies (see Russell & Williams, Chapter 3 in this volume). The *social shaping* of technology is now a widely accepted perspective (see Cronberg & Sørensen 1995). It is a well-established research strand in technology studies to analyse how social relations affect the precise design characteristics of particular technologies, how they influence the choice between alternative paths of technical development, and how they are fostering or inhibiting particular technologies (MacKenzie & Wajcman 1985: 24).

The shaping of technologies, however, cannot be conceived as a single closure process and a local deliberate decision that takes place during the design phase and outside institutionalised spheres (see Pinch & Bijker 1987; Kline & Pinch 1994; Bijker 1995). Social constructivism has been criticised for neglecting the multi-stage character of technical development and overlooking the constraints of established social structures (see Russell 1986; Vergragt 1988; Rosen 1993; Hård 1993). The development of technologies should better be conceived as a continuous process of creative variation, taking place in and between various technology projects, enacted by different social actors, closed and re-opened by negotiations in multiple arenas of conflict and selected by some institutional filters (Rammert 1997).

We summarise the debate about the shortcomings of social constructivism and say more precisely what the term *'shaping'* in our concept of cultural shaping means:

- There are multiple influences on the technological development, not a single one.
- Shaping involves two-way interaction, not a one-way process of determination.
- It is a kind of 'heterogeneous engineering' (Law 1987) or 'extended translation' (Callon 1993; 1995), not a homogeneous representation of a social form in the technical medium.
- It is situated in particular 'time-space zones' (Giddens 1987: 148); it is not a universal act.
- There are plenty of critical events (Van den Ven and Garud 1994) during the whole technological development process, from inception to use; there is not a single point of decision or a limited time for closure.

If the shaping process is defined in this way, it opens up the door for a combined cultural and institutional approach of shaping technology. The *notion of culture* itself suffers from too much interpretative flexibility. One can define culture as a special sphere of values and norms or even as a particular system besides economy, politics, or technology, as Talcott Parsons did and many sociological followers are still doing. One can define culture as the ensemble of all ideas, artefacts and practices which are highly valued in a society and draw a distinction towards profane culture or simple civilisation. But in this context of shaping technologies I shall follow the broad anthropological perspective and define *culture* as a special 'frame' (Geertz 1983: 21) or mode,

- how things are viewed differently,
- how things are done differently, and
- how these activities are institutionally arranged differently.

Consequently, culture is opposed not to economy or politics, but to certain abstract and formalised concepts of them. This concept of culture contains a particular comparative perspective which emphasizes empirical practices and which deconstructs the myths of universality. Thus, culture does not imply highbrow and idealism, but is rooted in everyday life and pragmatism. This notion of culture encompasses three elements:

- configurations of valued signs and symbols (semantic aspect),
- patterns of practices (pragmatic aspect), and
- regimes and styles: how something is instituted (syntactic or institutional aspect).

Cultural orientations are rooted in processes of signification and semantic framing of action and perception. They are condensed into visions, paradigms, cultural models or directing metaphors. Patterns of practice rise in the processes of pragmatic structuring of action. They crystallise into dominant habits and different styles of activities. Institutional regimes grow out of processes of syntactic regulation of action. They gain durability as national traditions and established arenas of negotiation. In a certain way, this conception of culture is paralleled by the Cultural Studies approach. This approach is well-known for its studies of the production and consumption of cultural artefacts in the field of media and mass communication. But it also concentrates on processes of 'meaning-making' and on 'social practices' (du Gay *et al.* 1997: 11).

If culture is defined in this way, it can be demonstrated that cultural patterns are shaping the development of technologies more intensely and more frequently than we usually assume. I shall argue in the first part that the cultural shaping of technology plays a significant role in the design of artefacts, in the direction of technological development and in the diversity of engineering traditions, user cultures and innovation regimes.

The subject of the cultural shaping approach is the distributed process of innovative actions encompassing all kinds of actors and practices: technological, economic, political, and cultural. When we examine the political implications of this framework, it is evident that the scope and scale of conventional technology policy must be changed. I shall show below that interventionist types of technology policy geared towards particular technology programmes or issues need to be complemented or even replaced by an interactive and network-oriented type of innovation policy that caters for technological and institutional diversity across the whole innovative process and in the long run.

Both parts of the paper are connected by the central argument that the cultural shaping view recommends the particular politics of technodiversity.

HOW TO COPE WITH THE CIRCLE OF UNCERTAINTY: THE SHIFT FROM A TECHNO-ECONOMIC TO AN INSTITUTIONALIST-CULTURAL VIEW

The idea of economic shaping of technologies reaches far back in the tradition of the Social Sciences. Karl Marx and Max Weber unanimously believed that economic criteria finally determine the rate and direction of technical change. However examining the two main strands of economic theorising about technological development indicates a slight shift from a techno-economic to an institutionalist-cultural view: the neo-classical approach of rational choice and the evolutionary approach of routines and technological regimes.

The adherents of the neo-classical style of reasoning dissolve technological development into many technological choices. These choices are conceived as rational decisions between substitutable goods. This type of techno-economic reasoning has been widely criticised for its empirical improbability and theoretical inconsistency. Herbert Simon demonstrated that only 'bounded rationality' can be achieved and that strategies of 'satisficing' can be realistically expected. Kenneth Arrow (1962) emphasised the critical role of 'uncertainty' and argued that an economic calculus cannot be applied to processes of decision-making in the cases of invention, research and development. The neo-classical approach to Economics fails to explain technical change, because it has developed an 'over-rationalised' conception of economic action.

The proponents of an evolutionary approach to the economic shaping of technologies have learnt from the shortcomings of the techno-economic view. They emphasise local variations from routine behaviour and the selective retention of technological paradigms and regimes (see Nelson and Winter 1982). They argue that organisational 'routines of decisions' and technological 'rules of thumb' characterise the firms' innovative behaviour more than 'rational choices'. The normative concept of a market equilibrium is replaced by a more empirical and broader perspective of 'selective environments' that includes market relations, state regulations, and socio-cultural preferences. The emergence of the new is explained by the unconscious modification of search routines and the unplanned confluence of different R&D strands. The success of a new technology is derived from its adaptability to the selective institutional environments, not from an universal technological superiority. The new technologies and relevant social institutions influence one another in a kind of co-evolution. In this way the techno-institutional interdependency stabilises into a new technological regime (see Dosi 1982; Nelson 1994).

This Nelson-Winter-Dosi model of technical change can be criticised for the deterministic overtones in the concept of technological trajectories and technological regimes (see Rip & van den Belt 1987), for the neglect of the actors' capacities to intervene, and, moreover, for the neglect of their capabilities for creative and reflective action. The evolutionary approach to the economics of technical change shows some weaknesses, because it is based on an 'under-rationalised' concept of innovative action. But this approach introduces the notion of institutional environments and non-rational orientations into economic reasoning and thereby paves the way for an institutionalist-cultural view. A conceptual change from economic determination and rational decision to institutional shaping and routine selection has taken place.

Both concepts of economic shaping have difficulties in coping with the central problematics of innovation. Any really new technology is fundamentally uncertain in many aspects. Any firm which decides to invest in, and any government agency which wants to foster the development of, a new technology can neither rely on sound economic calculation nor upon stable technological prospects. This eminently uncertain character of new technologies at the outset constitutes a sequence of problems that I call *the circle of uncertainties* (Rammert 1999). This circle consists of a wide range of uncertainties which are mutually intertwined in many ways. If an actor wants to develop a new technology or if s/he decides to follow one path of innovation, s/he is confronted with a great number of uncertainties:

- whether s/he gets access to the information about this technology,
- whether s/he is able to select the relevant information out of the flood of information,
- whether s/he has the capacity to process and to convert it into useful knowledge,
- whether the innovation comes up with a feasible technical product,
- whether this product can be produced economically,
- whether a new market can be established,
- whether the users accept the product and tolerate its unintended consequences,
- whether the developer gets a fair return to his/her investments and risks,
- whether his/her property rights are sufficiently protected, and
- whether the product meets the compatibility requirements of technical standards and legal norms.

As we conceive technological development as a heterogeneous and distributed process, where decisions, actors, aspects and artefacts are dispersed over time and space, we have to take into account a multiplication of uncertainties. If the complexity of the innovative situation is increased in such a way that economic techniques of calculation cannot be applied any longer, then cultural patterns of orientation and experience-based practices are substituted for economic accounting.

Three kinds of cultural patterns can be distinguished in this techno-economic context. First, visions and technological paradigms can be seen as cultural orientation complexes: sets of ideas that help orient and co-ordinate technological development, if its complexity and its openness cannot be reduced by other 'rational' means. Second, building routines out of experimental inquiry, learning by doing, and recursive learning can be described as another cultural pattern of practice to handle the insecurity of unknown situations. Third, the cultural orientation complexes and patterns of practice are embedded in institutional regimes with longer duration and wider expansion. Paradigms of technological solutions pave the way towards stable paths of technological development; the mutual adaptation of technological construction and institution-building crystallise into particular regimes of innovation. Practices that are experimentally produced and that because of successful experiences are often reproduced turn into traditions and particular patterns of institutional arrangements.

Cultural orientation schemata, cultural patterns of practice, and institutional regimes together shape the development of technologies at times and places where markets fail and firms cannot calculate. The territories of high technology development and the periods of early technology generation constitute these particular uncertain and complex zones. A sensitive and sustainable technology policy should concentrate on these early times of the technological development when the conception and design of new technologies are pre-shaped, because the variety of technological alternatives is rich, and the paths of development are still open.

TECHNOLOGY PROJECTS AND INNOVATION REGIMES: THE CULTURAL CONSTRUCTION OF THE DESIGN AND THE INSTITUTIONAL SHAPING OF THE TECHNOLOGICAL DEVELOPMENT

New technologies do not arise as hard artefacts and finished systems, although they are often presented to us in this way. They are part of a broad stream of conceptual variation and experimental design – as many historical studies about the shaping of technologies have demonstrated. Controversies about technological superiority and quarrels about the safety of a technical design indicate the inherent openness and contingency of technical development. They can only be limited by institutional closure and cultural consensus mechanisms. Economic criteria or technological parameters are not effective at this early stage of development, because they are dependent variables rather than fixed points of reference. Any technical construction must to this extent be seen as a cultural construction, because the choice, the evaluation and the configuration of technical elements are shaped by cultural patterns.

Particular *technology projects* are the concrete places where the general interaction between these cultural patterns and technological potentialities are organised. These technology projects restrict the openness and the contingency of technical development, because promoters follow particular visions of use, because engineers choose certain concepts, and because firms establish different traditions of design.

According to traditional concepts, technologies are supposed to be shaped foremost by parameters of technological perfection and criteria of economic efficiency. But with new technologies, who knows what it means to be 'better'? Better under which aspects? Better for how long and in which place? And better in comparison to which alternative? With our two-way interaction view of shaping in mind, technological parameters can be seen as cultural artefacts. They are the products of cultural evaluation and social negotiation processes. The features of a new technology are, besides other factors, mainly dependent on three cultural forces:

- first, on the *visions of function and use*, including an orientation by paradigms, an interpretation of the function and an image of the prospected user, which align the further technical development,
- second, on the *concepts and styles of engineering*, which grow out of different academic, professional and organisational cultures and which are inscribed in the technical design, and
- third, on *traditions and regimes*, which reflect the particular attitudes and established relations between the actors in the field and which stabilise the way the development of technologies is institutionalised.

The cultural shaping of technology takes place in processes of symbolic interpretation, material inscription and social institutionalisation.

Visions of function and use open up new options and paths of technological development and at the same time limit the wide range of other possibilities. Visions bundle heterogeneous elements within an integrating perspective (see Dierkes, Hoffmann and Marz 1999). The closer we move to the inception and generation stage of new technologies, the circle of uncertainty increases and the variety of possible projects increases. Visions of how to use a device and for what purpose give a first orientation to technical development. For instance, the idea of substituting a programmable machine for the disciplined operations of human computers (Turing 1937; Heintz 1995) pointed the way to hardware and software development in computer technology. The vision of computers as 'augmented knowledge workshops' (Engelbart 1962) and as 'personal dynamic media' (see Kay & Goldberg 1988; Mambrey, Paetan & Tepper 1995) later diverted computer development from big calculating machines to the new trajectory of small personal computers. Examples indicate that *cultural interpretations can have technological and economic consequences.*

Concepts and styles of engineering are parts of design cultures. They are based on engineering routines and standard solutions. They give more concrete orientation than visions. They determine the particular design of a technical system, for instance whether you can interact with the computer via keyboard, natural language or screen contact. The design depends on the user model that designers have in their heads. The conceptualisation of machines, programs or networks shapes the frames of social relations. For instance, once a design concept is chosen, it is inscribed into the techno-structure. This fixes who does the work, the bank or the client; who may intervene, the manager or the employee; who gets access to the relevant data, the administration or the citizen. *This process of inscribing concepts of use, user models, and supplier-user relations into technical systems can be seen as a cultural construction of technology.*

What seems to be the result of an explicit rational technological or economic decision, can sometimes be revealed as routine-based and an inherent feature of engineering culture, such as implicit models, hidden curricula, or *traditions* of engineering. Engineers often stick to their proven concepts, to their learnt styles of engineering and to the established state of the art, even if they are designing new technologies. In following routines and tested traditions they want to minimise uncertainties and to keep away the risk of non-functioning. Besides the 'technological momentum' of large technological systems (Hughes 1983) and the 'irreversibility' of complex networks (Callon 1993), this structural conservatism of the engineering culture (Knie 1989) is an important factor, which can explain why new visions quickly lose their variety-enforcing power and why different conceptual options are constricted to traditional lines of development.

The telephone, for instance, was in the beginning seen as a one-way media like the established telegraph. Following this one-way transport concept German telecommunication engineers designed the telephone in the first years as a local device for the prolongation of the telegraphy system. Hungarian engineers experimented with a one-to-many communication concept and pre-structured the star-like pattern of radio mass communication. The vision of a two-way media only became dominant some years after the patent application and led to the well-known network-like architecture of the telephone system (Rammert 1993).

The development of computer technology offers another example. Although the visions of a 'universal machine', of a 'personal tool', and of a 'media of communication' existed – remember Turing's vision of an intelligent machinery (Turing 1950) and other early inventors like Vannevar Bush, Douglas Englebart, Carl Adam Petri and Allan McKay – for a long time computers were built and improved according to a single technological tradition. It was dominated by the paradigm of the calculating machine. The development aimed at the design of supercomputers for central and faster calculations (MacKenzie 1991). The computer engineers continued to improve the speed, the reliability centralising of technical system; but they did not search for alternative uses and different designs. Military influence and the conservative engineering style were responsible for this 'natural trajectory' in computer development from the forties up to the sixties of the last century.

Technologies cannot be separated from the techno-structure of which they are a part. A machine, for instance, is coupled with other machines to function as a production system; a machine is integrated in a whole system of technical infrastructure which provides for the necessary energy, material, and information; the whole technical system is associated with many social institutions, like organisations and legal norms, which constitute the socio-technical complex to fulfil a social function. But functions are defined in a cultural context, and the corresponding institutions grow along historically chosen pathways. These socio-technical complexes or large technological systems are shaped by national traditions and particular institutional frameworks (see Hughes 1983; Mayntz and Hughes 1988; Mayntz and Schneider 1988; Kubicek, Dutton & Williams 1997). They are not the same all over the world and they are not organised like functionally specialised systems. The features and the arrangement of institutions constitute different *technological regimes.*

The American Fordist regime of car mass-production differs from the Japanese Toyota regime of intelligent and flexible production. In Germany, one can observe a high quality incremental innovation regime in the machine tool and chemical industries which is suited to developing complex goods and which needs long term trust relations between the developers and the users of new technologies. In the USA and Great Britain, the institutional framework favours radical innovation in biotechnology and microelectronics (compared to the co-operation and regulation culture as in Germany. Here, risk-taking and strong competition are rewarded even in the higher education system (see Soskice 1996; Hage & Hollingsworth 2000).

We have seen that the cultural construction of technologies can sometimes be more crucial to economic success and social diffusion than technical improvement alone. Real new-to-the-world-technologies bring up a lot of uncertainties, concerning the choice of technical parameters, the test of adequate functioning, envisaged user groups, and the commercial pay off. This circle of uncertainty is interrupted by cultural concepts which guide the technical development by implicit paradigms and explicit visions. It is reduced by routines and rules that build the institutional framework and embed technical development in society (see the contributions in Dierkes & Hoffmann 1992). What at a later point of time appears to be determined by parameters of technological perfection and by criteria of economic calculation, can at an earlier point of time be disguised as the contingent product of cultural construction and can be analysed as an institutional shaping of technological development by routines and regimes.

The policy implications of this insight are evident. It is at these early times of construction and development that a vivid debate about possible different visions should be initiated. The greater the variety of ideas and concepts of design and use that can be mobilised in the early technology generation stages, the greater are the opportunities to find a creative and socially acceptable path of technological development in the long run. An evolution of technology whose characteristics are not made explicit and whose consequences are not debated easily runs into irreversibility (see Callon 1993) and risks getting locked-in (see Arthur 1988) in later times. Further, it is not wise to apply established technological parameters and economic criteria to radically new technologies. Innovations need different criteria of testing and evaluation. This paradoxical relation requires a politically mediated process of experimental projects and experience-seeking in protected places. The development of solar technologies, windmills or electric cars should therefore be developed in such 'niches' of evolution (see Rip & Schot, chapter 5 in this volume; Hård and Knie 1994). The institutions that are built up in close cooperation with established technologies must also be reshaped to give room for a new co-evolution of technologies-in-action and institutional regime-building. New technologies need new markets and new forms of work and organisation. And they need new tools, and a new kind of technology policy, that caters for its uncertain, open and distributed character.

MULTIPLE POLICY ARENAS AND LOCAL USER CULTURES: THE PUBLIC AND PRIVATE CONSTRUCTION OF TECHNICAL PRACTICES

The development of new technologies does not end with their production, but continues during the diffusion and use of new technical devices, too. And so does the cultural shaping of technology via recursive learning processes between developers and users.

How do new technologies disseminate into private households? An engineer of the development department would answer that new devices convinced people because of their technical efficiency. A business economist would emphasise that they were selling goods with a reasonable price. These conditions of diffusion are not sufficient. The users also must make out a meaning and recognise the usefulness of a device. They must be able to appropriate it and to fit it into their everyday life routines without problems. Even more, new technologies must offer playgrounds to experiment and invent new uses. The Sony Walkman, for instance, was designed to be used by young people or professional musicians only listening to the music of their choice when they are not at home. But after a short time, people of all ages and professions invented a diversity of uses indifferent places which later on influenced the design that became more and more distinctive (du Gay et al. 1997).

The meaning and usefulness of a new technology are not self-evident. The visions of the inventors, the concepts of the engineers, and the images of the salesmen are only proposals for a sensible use of the new technical artefacts, though they have pre-structured the production of prototypes. From the outset on, there is experimentation on possible uses and negotiation about the reasonable use of a new technology. Within the firm, since no-one knows what the user really wants, the different departments are in competition to configure the user (see Woolgar 1992 and Akrich 1995).

Many collective actors participate in this process wherein the meaning and reasonable use of a new technology is defined and negotiated. No single actor can impose his/her concept of computer use to the others. For instance, the quick diffusion of micro computers in American households was mainly caused by the self-augmenting and self-organising cultural movement of hackers and computer clubs (Allerbeck & Hoag 1989). Neither the persuasive marketing strategies of the producers nor the political decision to introduce computer education at schools were responsible for that success. *The non-military and non-commercial use of computers can be seen as a cultural invention, a creative process of re-defining and domesticating (see for this term Sørensen 1995) the big calculation machine and its use.*

The cultural shaping view encompasses two interdependent processes: the public construction of the sensible uses and rules how and for which sake to use the new technology, and the private construction of the meanings and styles of a new technical practice. We have observed the *public construction* of the micro computer as a useful private gadget in Germany after 1970 in three arenas (see Rammert 1996). Producers and sellers of computers on the one side and user clubs and consumer organisations on the other side constituted a *techno-economic arena*: They discussed the problems of a fair price and negotiated about service and quality standards. Struggles between the police, government institutions, and data-bank owners on the one side and hackers and political groups fighting for the freedom of information or the protection of personal data on the other side established a *political arena*. Those engaged negotiated about the definition what is a legitimate, what legal, and what is criminal use of computers. In the *cultural arena*, quite different groups, such as computer artists and marketing people, scientists and computer clubs, were competing together in creating new applications and cultivating new styles of computing.

This public construction of a new technology's meaning and usefulness functions as a backstage for the *private construction* of a new technical practice. At first, we have to realise the difference of *places* where computer use is practised. It makes a big difference, whether an information system is used at work or at home or in public places. In the work setting, one can expect that the employee has the professional competence to use it and will follow the use instructions. But particular professional styles and organisational cultures will also shape the style of computing. In the private setting at home, we will meet a great variety of skills and interests: systems should therefore be open for creative and experimental uses and give enough room for emotional needs and amusement (Dholakia 1994). In public places, the system designer has to imagine the 'occasional user' (Kubicek & Taube 1994) with limited skills and a low frustration tolerance. If we look at the new communication technologies, like the Walkman, the mobile phone and the internet, we can observe that even the limits between the places are dissolved. Music listening, a private amusement at home or a collective event in a concert hall, is turned into a very individual experience and private listening in the public domain when one uses the Walkman (du Gay *et al* 1997: 114). If we focus on the new information technologies and the different places where they are practised, it can be seen that the cultural concepts of the 'standardised user' and of the conventional distinction between private and public worlds are fading now.

Second, the differences between *cultural milieu and life styles* have to be considered. Beyond traditional class divisions sociologists observe an 'individualisation' and 'multiplication' of life styles (Beck 1986). Private users cannot any longer be conceived as 'passive consumers', buying goods driven by nearly the same motives and using the products in nearly the same manner. In particular the new information technologies which are open to flexible and creative use require the 'active user', who shapes and appropriates the technical system according to his individual needs and personal style.

The micro computer, for instance, does not move into the household as a fact and finished good, but it is constructed and configured along sub-cultural images and tastes. As the computer is eminently suitable as a projection surface, like a Rorschach-test (see Turkle 1982), it would be quite unreasonable to expect its adoption to invoke a unitary, one-dimensional computer user; whether the freak with a pale face, or the man with a mechanical character (Weizenbaum 1976; Pflüger & Schurz 1987). As our deep empirical study of about fifty computer users, (men and women of all social classes and aged between 18 and 60) demonstrated, one will find different ways of defining the computer, different styles of using the computer, and different social and cultural impacts conditioned by these differences of domestic computer cultivation (Rammert 1996; see also Turkle 1984; Eckert *et al.* 1991).

CONSEQUENCES FOR AN EXTENDED TECHNOLOGY POLICY: CARING FOR TECHNODIVERSITY AND CREATING INTERACTIVE NETWORKS OF INNOVATION

The cultural shaping view broadens and intensifies the perspective on technological development. It focuses on the whole process of innovation. It includes both the early stages of the inception and generation of new technologies and the later stages of diffusion and use. It analyses the complex interrelations between all aspects of the innovative process. That is why more conventional views of technological or economic shaping of technologies are too crude and restricted to lay the foundations for an extended and evidence-based innovation policy. Their utility remains limited; insofar as the technological view succeeds in turning its method from mere technological forecasting into multi-optional scenario building, and insofar as the economic view is prepared to put more emphasis on the social embeddedness of technological choices and on the co-evolutionary character of technical change. Additionally, the cultural shaping view stresses the role of actors, their practices, and their relations with one another in negotiating arenas. Innovation is seen as a distributed process of various activities, enacted or mediated by heterogeneous elements and happening at dispersed places and at different times (see Rammert 1998).

These insights should have consequences for the formulation and practice of technology policy. An innovation policy based on evidence rather than disciplinary myths needs a high awareness of these complex interrelationships and a deep understanding of how technological projects are turned into dominant designs and what is needed to assure and to shape the innovative process. It can no longer rely on the mere monitoring of technological trends and policies to strengthen promising paths of development by pushing big projects and entrenching long-term programmes (see for new concepts Rip, Misa & Shot 1995). As social shaping research has shown, this technocratic kind of technology policy remains blind to both social implications and excluded alternatives. It runs the risk of subsiding big companies; excluding other social actors; and getting locked-in in an unwanted technological trajectory. An innovation policy that is informed by cultural shaping research cannot restrict itself to the economic problems of looking for the one best technological solution, of searching for the most efficient allocation of resources and for bench-marking the highest performance. Such a rationalistic view is not based on sound empirical work, but on formal and normative models. It narrows the political perspective to aspects of commercial exploitation and best management practices. At the same time, the conditions of long-term innovativeness may be undermined when the range of possible technology projects is reduced by strong economic calculations, when the variety of practices and institutional settings is stream-lined towards a globalised standard model of innovation management.

The shift from a techno-economic view to an institutionalist-cultural view implies a shift in technology policy. The fundamental problem of an innovation policy that refers to an enlarged space of shaping has to cope with the *paradoxical problem of reducing uncertainty without strangling creative diversity*. Creative diversity refers to three main sources of innovation:

- 'technological diversity' – the knowledge bases of various scientific and technological research fields;.
- 'actor diversity' – the heterogeneity of innovative actors and the distributedness of the innovative activities;
- 'institutional diversity' – the variety of institutional settings, national regimes and forms of social embeddedness.

Technology studies have opened the black box of technology. Its students have analysed the implicit cultural models of rational man, gender, social inequality or exploitable nature. By this work of making explicit the tacit dimension of design and the hidden curriculum of development they are multiplying the paradigms and defining perspectives. These insights need to be transferred into the cultural arenas where public debates about visions of use and the value of concepts should be initiated. At this early stage of inception and technology generation, it is the main task of a sensitive technology policy to open minds to the multiplicity of perspectives and the variety of possible paths of technological development. It is not sufficient to do this by the conventional tool of scientific policy consulting; a democratic and sustainable innovation policy requires the organisation of public discourses in different arenas in order to mobilise a broad spectrum of ideas and views. After this period of variety enforcement, a publicly controlled closure process needs to be induced. Different visions, concepts, and expectations have to be melted into the amalgam of an enriched cultural model, a kind of leitmotif, that is stable enough to give orientation to further development and flexible enough to give room for reorientation, if necessary.

This idea is far removed from traditional technology policies to find and fix a putative 'one best way' and to fit a ready-made technology definitely in the existing technical infrastructure. The new innovation policy acknowledges the paradoxon of opening and limiting the range of innovation paths. It augments the diversity of knowledge bases and visions. It initiates discourses and mediates conflicts in public arenas. Thereby a variety of ideas and distinctive views are linked with one another. As the outcome of these negotiations, a leitmotif can be finally composed that weaves the different and dispersed threads of technological development together. Under this semantic aspect, *the politics of technodiversity involves creating a space for multiple options and a variety of ideas and views and connecting them loosely by presenting leitmotifs.*

A second aspect of cultural shaping concerns the process of pragmatic structuring: Technological development is seen as a distributed process of innovative activities. Different social actors are involved who follow particular standards of rationality and who are guided by often opposing interests. Different practices of research scientists, inventors, entrepreneurs, lawyers, users and political system builders have to be matched with one another. The key problem is to balance the driving and inhibiting forces in order to smooth the critical passage points between the different social worlds without sweeping away productive frictions. Scientific research practices, technological development practices and economic innovation pratices should be orchestrated in a better way to enhance new technologies and products, but early patent claims should not inhibit the research process (as feared in the field of biotechnology). The one-way relation between developer, producer and user of new technologies should be turned into an interface of recursive innovation. Time, money, and trust should not be gambled away, as it is done with the traditional strategy, initially developing a technology under isolated efficiency and profitability criteria, and afterwards improving usability, applying attractive design, and caring about the range of sensible uses. One can imagine a temporal coupling of the technological engineering and the cultural construction processes, e.g. iterative, evolutionary software development where time is reserved for learning with users (see Floyd 1987), or the integration of technology assessment in the early development process, while spaces for experimentation still exist (Rip & van den Belt 1988; Rammert *et al.* 1993; Wynne 1995).

Recursive innovation means that the practices of the other social worlds, e.g. of the appliers or users, should be represented and integrated in the early design and developmental process. Reflexive innovation calls for a competence to understand the rationality of others and to take the role of the other participants in the distributed innovation process. Standardisation, contracts and market competition are also strong co-ordination forces, but they have the tendency to impose one 'best' technology too rigidly upon the other participants. The politics of technodiversity requires a reflexive type of innovation policy that caters for the plurality of perspectives and practices. It prefers connecting diverse actors by interactive patterns of practices and by trust relations instead of strong standards. Thereby it keeps the space for different paths of design and development open as long as possible and it arranges platforms for joint innovative actions and recursive learning between the actors of the innovation network.

The innovation process is split into many events that take place in different institutional spheres. In modern societies, the production and distribution of new knowledge has been mainly organised in the academic and public research system. The utilisation of new ideas has been concentrated in particular public agencies and in the R&D-departments of the industrial system. The setting of priorities, the allocation of resources, and the regulation of the whole 'national system of innovation' (see Lundvall 1992 and Nelson 1993) have been the legitimate tasks of the political system. But actually, the established national systems of innovation that were mainly based on the functional autonomy of institutional spheres and on the coordination via market or via hierarchy are dissolving. The standard biography of an innovation that runs from inception to diffusion, stage by stage, is breaking up. Innovations are shaped at the same time in many different 'functional arenas' which are embedded in national institutional regimes, and loosely connected on a global level by 'idea-innovation-networks' (see Hage & Hollingsworth 2000; Hollingsworth 2000).

On the one side, markets are an efficient mechanism to coordinate heterogeneous and dispersed activities like those of the innovation process. But they require a certain accountability. Markets fail when uncertainties grow too high and time horizons expand. That is why liberal technology policies that favour commercialisation and de-regulation will not be successful in the long run. Though legal and bureaucratic impediments will be abolished, the gaps between heterogeneous fields and the divide between winners and losers are widened. On the other side, state and hierarchy are efficient means of coordination because they can establish accountability and certainty within their own limits. But bureaucratic organisations fail, when differences should be maintained and the time horizon should be kept open, as is required for the innovation process. Therefore neo-corporatist strategies for technology policy will only show a limited success. It subordinates heterogeneous forces and fields under common projects, programs, and priorities. This strong alignment of different visions and innovation cultures risks dampening scientific creativity and deadening entrepreneurial innovative capacities.

Networks of organisation are a third co-ordination mechanism that avoids some disadvantages and combines the advantages of the other two. They are based on four principles: complementarity of resources, mutual adaptation, reciprocity of exchange, and trust-relations (Powell 1990: 296). Negotiation maintains the flexibility of markets without exhibiting the latter's indifference towards goods and actors. Trust relations reduce uncertainties without eroding creative differences as radically as organisations do. A reflexive and sustainable type of innovation policy should be based on a post-Schumpeterian mode of innovation (Rammert 1998) going beyond focusing only on markets or hierarchy. It centres around a particular kind of innovation networks, which can be defined as heterogeneous networks of collective interactive learning. They differ from the many other networks, e.g. the closed personal networks of clans and rotary clubs, the strategic networks of interorganisational management, or the policy-networks of neo-corporatist governance (see Freeman 1991 and Powell & Smith-Doerr 1994).

A new innovation policy has to shape an institutional regime that encourages the creation of such interactive innovation networks in various technological fields and counteracts strategies whereby one principal agent tries to control and exploit the network or exclude others from the network. The politics of technodiversity requires a balanced process of institutional learning between different actors. This process emerges and succeeds, if the state and its agencies participate themselves as one actor amongst others in the process of shaping technologies; if interactive networking between the heterogeneous innovative actors is facilitated, if reciprocal awareness and trust are raised thereby, and if a translation process between the scientific, economic, and political culture is organised that offers shared leitmotifs and makes sense to all.

At the end, we can see that the shift of theoretical approaches in science and technology studies suggests a change of technology policy. We learned from the cultural shaping approach that technologies are shaped more intensely by the hidden dimensions of practices and the patterns connecting them together than by explicit technological criteria and articulated economic choices. This insight should have consequences for the governance and management of technological development. That does not mean that the established strategies and tools are always inefficient and should be completely replaced by new ones. Certainly, it remains important for some kind of traditional technologies and for the later stages of technological development to formulate explicit goals and to make clear decisions about priorities and programmes. But it will be more important in the long run to support spaces where diverse technological development projects and different practices of using technology may emerge. It will be more favourable for the sake of innovativeness to create particular arenas of negotiation which give easy access to non-official groups and non-governmental organisations. These multiple arenas should be turned into public places where networks of innovation may develop and where different collective actors may learn and benefit from one another. It will be a prudent policy, if the politicians and administrators are aware of the productive effects of differing institutional designs and of different institutional regimes. They should limit their strategies of simplification and standardisation. If they decide to maintain a certain degree of institutional diversity, the restrictive effects of benchmarking can be avoided. Institutional diversity can be used to enforce creative comparisons and to encourage experimental cross-breeding.

Under this perspective of cultivating diversity, the globalisation of economic and technological development and the unification of the European states can be seen as a great opportunity to benefit from all kinds of diversity. The politics of technodiversity proposed reaps benefits from the fact that Europe is rich in cultures, traditions and institutions. It encourages producing a great variety of ideas, projects and perspectives and thereby raising technological diversity. It controls the conditions of exchange and bargaining between scientific, economic and political actors and gives excluded groups access to the public arenas. Thereby it raises actor diversity. Finally, it connects ideas, institutions and people by promoting the building of interactive innovation networks and by stimulating institutional learning whereby institutional diversity is raised.

BIBLIOGRAPHY

Akrich, M. (1995), 'User Representations: Practices, Methods and Sociology', *Managing Technology in Society*. A. Rip, T. Misa and J. Shot (eds), London: Pinter: 167-184.

Allerbeck, K. and Hoag, W. (1989), 'Utopia is Around the Corner', Computerdiffusion in den USA als soziale Bewegung. In: *Zeitschrift für Soziologie*, 18, 1: 35-53.

Arthur, B. (1988), 'Competing Technologies, Increasing Returns and Lock-in by Historical Events', In: *Economic Journal*, **99**: 590-607.

Arrow, K. (1962), 'Economic Welfare and the Allocation of Resources for Invention', *The Rate and Direction of Inventive Activity*. National Bureau of Economic Research (ed.), Princeton University Press.

Beck, U. (1986), *Risikogesellschaft*. Frankfurt/M: Suhrkamp.

Bijker, W. E. (1995), *Of Bicycles, Bakelites, and Bulbs. Towards a Theory of Sociotechnical Change*. Cambridge, MA: MIT Press.

Callon, M. (1993), 'Variety and Irreversibility in Networks of Technique Conception and Adoption', in D. Foray and C. Freeman (eds), *Technology and the Wealth of Nations*, London: Pinter, 232-268.

Callon, M. (1995), 'Technological Conception and Adoption Network: Lessons for the CTA Practitioner', in A. Rip, T. Misa and J. Shot (eds), *Managing Technology in Society*, London: Pinter, 307-330.

Cronberg, T. & Sørensen K. H. (eds) (1995) *Similar Concerns, Different Styles? Technology Studies in Western Europe*, Proceedings of the COST A4 workshop, Ruvaslahti, Finland, 13 - 14 January 1994, COST A4, **4**, Luxembourg: Office for Official Publications of the European Communities..

Dierkes, M., Hoffmann, U. and Marz, L. (1996), *Visions of Technology*. New York: St. Martins Press.

Dierkes, M. and Hoffmann, U. (eds) (1992), *Technology at the Outset. Social Forces in the Shaping of Technological Innovations*. Boulder, CO: Westview Press.

Dholakia, R. R. (1994), 'The Plugged-in Home: Marketing of Information Technology to U.S. Households', in P. Zoche (ed.), *Herausforderungen für die Informationstechnik*. Heidelberg: Physica im Springer-Verlag, 86-100.

Dosi, G. (1982), 'Technological Paradigms and Technological Trajectories', *Research Policy* **11**, 147-166.

du Gay, Hall, S., James, L., Mackay, H. and Negus, K. (1997), *Doing Cultural Studies - The Story of the Sony Walkman*, London: SAGE/Open University

Eckert, R. *et al.* (1991), *Auf digitalen Pfaden. Die Kulturen von Hackern, Programmierern, Crackern und Spielern*. Opladen, Westdeutscher Verlag.

Elster, J. (1983), *Explaining Technical Change*. Cambridge: Cambridge University Press.

Engelbart, D. (1962), Augmenting human intellect - A conceptual framework, AFOSR-3223, Stanford Research Institute, Menlo Park; reprinted under the title 'The Augmented Knowledge Workshop'. In, A. Goldberg (ed.), *A History of Personal Work Stations*. New York: Addison Wesley 1988.

Floyd, C. (1987), 'Outline of a Paradigm Change in Software Engineering', in G. Bjerknes *et al.* (eds), *Computers and Democracy*. Avebury: Aldershot, 186-203.

Foray, D. and Freeman, C. (eds) (1993), *Technology and the Wealth of Nations*, London: OECD.

Freeman, C. and Soete, L. (1993), 'Conclusions', in D. Foray and C. Freeman (eds), *Technology and the Wealth of Nations,* London: OECD: 389-400.

Freeman, C. (1991), 'Networks of Innovators: a Synthesis of Research Issues', *Research Policy* **20**, 5: 499-514.

Geertz, C. (1983), *Dichte Beschreibung. (Thick Description)*, Frankfurt: Suhrkamp.

Giddens, A. (1987), *Social Theory and Modern Sociology*, Cambridge: Polity Press.

Hage, J. and Hollingsworth, R. (2000), *A Strategy for Analysis of Idea Innovation Networks and Institutions*, Working Paper TUTS-WP-5-2000, Technical University Berlin.

Hård, M. (1993), 'Beyond Harmony and Consensus: A Social Conflict Approach to Technology', *Science, Technology and Human Values*, **18**, (4), 408-432.

Hård, M. and Knie, A. (1994), 'The Ruler of the Game: The Defining Power of the Standard Automobile', in K. Sørensen (ed.), *The Car and its Environment. The Past, Present and Future of the Motorcar in Europe,* COST A4, **2**, Luxembourg: Office for Official Publications of the European Commission.

Heintz, B. (1995), ''Papiermaschinen'. Die sozialen Voraussetzungen maschineller Intelligenz', in W. Rammert (ed.), *Soziologie und künstliche Intelligenz. Produkte und Probleme einer Hochtechnologie*, Frankfurt/M: Campus, 37-64.

Hollingsworth, R. (2000), *Doing Institutional Analysis - Implications for the Study of Innovations* (unpublished paper), Madison: University of Wisconsin.

Hughes, T. P. (1983), *Networks of Power - Electrification in Western Society, 1880 – 1930*, Baltimore: John Hopkins University Press.

Hughes, T.P. (1987), 'The Evolution of Technological Systems', in Bijker, W. *et al.* (eds), *The social construction of technological systems*, Cambridge: MIT, 51-82.

Kay, A. and Goldberg, A. (1988), 'Personal Dynamic Media', in A. Goldberg (ed.), *A History of Personal Workstations,* New York: Addison Wesley.

Kline, R. and Pinch, T. (1994), 'Taking the Black Box off its Wheels: The Social Construction of the American Rural Car', in K. H. Sørensen (ed.), *The Car and its Environment. The Past, Present and Future of the Motorcar in Europe*, Luxembourg, COST A4, **2**: Office for Official Publications of the European Commission, 69-92.

Knie, A. (1989), *Das Konservative des technischen Fortschritts*, Berlin: WZB-paper FS II 89-11,

Kubicek, H. and Taube, W. (1994), 'Die gelegentlichen Nutzer als Herausforderungen für die Systementwicklung', *Informatik-Spektrum*, 17, 347-356.

Kubicek, H., Dutton, W. and Williams, R. (eds) (1997), *The Social Shaping of the Information Highway: European and American Roads to the Information Society*, Frankfurt: Campus.

Law, J. (1987), 'Technology and Heterogeneous Engineering: The Case of Portuguese Expansion', in W. Bijker, T. Hughes and T. Pinch (eds), *The Social Construction of Technological Systems.* Cambridge, MA: MIT Press, 111-134.

Lundvall, B.-A. (ed.) (1992), *National Systems of Innovation: Towards a Theory of Innovation and Interactive Learning*, London: Pinter.

MacKenzie, D. (1991), 'Notes Towards a Sociology of Supercomputing', in T. de La Porte (ed.), *Social Responses to Large Technical Systems,* Dordrecht, 159-175.

MacKenzie, D. and Wajcman (eds) (1985), *The Social Shaping of Technology*, London: Open University Press.
Mambrey, P., Paetau, M. and Tepper, A. (1995), *Technikentwicklung durch Leitbilder*, Frankfurt: Campus.
Mayntz, R. and Schneider, V. (1988), 'The Dynamics of System Development in Comparative Perspective: Interactive Videotex in Germany, France, and Britain', in R. Mayntz and T.P.Hughes (eds), *The Development of Large Technical Systems,* Boulder, Colorado: Westview Pr., 263-298.
Mayntz, R., and Hughes, T. P. (eds) (1988), *The Development of Large Technical Systems*, Frankfurt/M., Campus: Boulder: Westview Press.
Nelson, R. and Winter, S. (1982) *An Evolutionary Theory of Economic Change*, Cambridge, MA: Belknap Press.
Nelson, R. (ed.) (1993), *National Innovation Systems: A Comparative Analysis*, Oxford: Oxford University Press.
Nelson, R. (1994), 'The Coevolution of Technologies and Institutions. Evolutionary Concepts in Contemporary Economics', in R. England, (ed.), *Evolutionary Concepts in Economic Theory*, Ann Arbor, University of Michigan Press: 139-156.
Pflüger, J. and Schurz, R. (1987), *Der maschinelle Charakter,* Opladen: Westdeutscher Verlag.
Pinch, T. and Bijker, W. (1987), 'The Social Construction of Facts and Artefacts', in W. Bijker, T. Hughes and T. Pinch (eds), T*he Social Construction of Technological Systems*, Cambridge: MIT Press, 17-50.
Powell, W. (1990), 'Neither Market, Nor Hierarchy: Network Forms of Organization', *Research in Organization Behaviour* 12, 295-336.
Powell, W. and Smith-Doerr, L. (1994), 'Networks and Economic Life', in N. Smelser and R. Swedberg (eds), *The Handbook of Economic Sociology*, Princeton University Press, 368-402.
Rammert, W. (1993), *Technik aus soziologischer Perspektive,* Opladen: Westdeutscher Verlag.
Rammert, W. (1996), 'Computer Use at Home: A Cultural Challenge to Technology Development', in W. Brenner and L. Kolbe (eds), *The Information Highway and Private Households: Case Studies of Business Impacts*, Berlin, New York: Springer, 399-408.
Rammert, W. (1997), 'New Rules of Sociological Method: Rethinking Technology Studies', *British Journal of Sociology*, **48**, no.2: 171-191.
Rammert, W. (1998), *Ritardando and Accelerando in Reflexive Innovation or How Networks Synchronize the Tempi of Technological Innovation*, Working Paper TUTS-WP-7-2000 Technical University Berlin.
Rammert, W. (1999), *Inquiry into Innovation – A Pragmatist's Conception of Technological Change,* (unpublished paper). Madison: University of Wisconsin.
Rammert, W. *et al.* (1998), *Wissensmaschinen ('Knowledge Machines'). Soziale Konstruktion eines technischen Mediums. Das Beispiel Expertensysteme.* Frankfurt/M.: Campus.
Rip, A., Misa, T. and Shot, J. (eds) (1995), *Managing Technology in Society: The Approach of Constructive Technology Assessment,* London: Pinter.
Rip, A. and Schot, J. (chapter 5 in this volume): *Anticipation on Contextualization: Loci of Influencing the Dynamics of Technological Development.*

Rip, A. and van den Belt, H. (1988), *Constructive Technology Assessment: Toward a Theory*, Enschede: University of Twente.

Rosen, P. (1993), 'The Social Construction of Mountain Bikes: Technology and Postmodernity in the Cycle Industry', *Social Studies of Science*, **23** (3), 479-513.

Russell, S. (1986), 'The Social Construction of Artefacts: A Response to Pinch and Bijker', *Social Studies of Science*, **16**, no. 2, pp. 331-346.

Russell, S. and Williams, R. (chapter 3 in this volume): *Social Shaping of Technology: Frame works, Findings and Implications for Policy.*

Sørensen, K. H. (ed.) *The Car and its Environments. The Past, Present and Future of the Motorcar*, COST Social Sciences, COST A4 **2**, Luxembourg: Office for Official Publications of the European Communities.

Soskice, D. (1996), *German Technology Policy, Innovation, and National Institutional Frameworks*. FS I96-319, Berlin: WZB discussion paper.

Turing, A. (1937), On Computable Numbers, with an Application to the Entscheidungsproblem. In *Proceedings of the London Mathematical Society*, **42**, no.2.

Turing, A. (1950), 'Computing Machinery and Intelligence', *Mind*, **59**.

Turkle, S. (1982), 'The Subjective Computer. A Study in the Psychology of Personal Computation', *Social Studies of Science*, **12**, 173-205.

Turkle, S. (1984), *The Second Self. Computers and the Human Spirit*. New York: Simon & Schuster.

van den Belt, R., and Rip, A. (1987), 'The Nelson-Winter-Dosi Model and Synthetic Dye Chemistry', in W. Bijker, P. T. Hughes and T. Pinch, (eds), *The Social Construction of Technological Systems*, Cambridge: MIT Press, 135-158.

Van den Ven, A. and Garud, R. (1994), 'The Coevolution of Technical and Institutional Events in the Development of an Innovation', in J. Baum and J. Singh (eds), *Evolutionary Dynamics of Organisations*, New York: Oxford, 425-443.

Vergragt, P. (1988), 'The Social Shaping of Industrial Innovations', *Social Studies of Science*, **18**, 483-513.

Weizenbaum, J. (1976), *Computer Power and Human Reason. From Judgement to Calculation*. San Francisco: Freeman.

Woolgar, S. (1992), 'Configuring the User', *A Sociology of Monsters, Soc. Rev. Monogr.* **38**. J. Law (ed.). London: Routledge & Kegan Paul, 57-99.

Wynne, B. (1995), 'Technology Assessment and Reflexive Social Learning: Observations from the Risk Field', in A. Rip, T. Misa and J. Shot (eds), *Managing Technology in Society*, London: Pinter, 19-36.

PART TWO

New tools for analysing and intervening

7. Arenas of Development: A Space Populated by Actor-worlds, Artefacts, and Surprises[1]

Ulrik Jørgensen and Ole Sørensen

INTRODUCTION - SPACES OF INNOVATION

While commenting on the problems of managing technology and innovation in the Finnish consumer electronics industry, the CEO of NOKIA once said:

> When a manager of an electronics company in Silicon Valley opens his garage, he has a view to most of the world's innovators, suppliers and users of advanced electronics - when I open my garage, I view a pile of snow. Quoted in van Tulder (1988).

Besides the colourful illustration of the difference between being in the heart of the scene or in the periphery, this statement implies a huge number of assumptions and opens questions about the role of location, closeness, relations, excellence, and space for innovative activity and technology management.

But what is the nature of the locations and spaces of technology and innovation that have to be managed? In which type of space should the manager, the engineer, and other actors be looking or in broader terms searching,[2] when they are involved in, dependent on or studying the development of new technologies and the creation of innovations and structuring of innovative processes? And how should we map the field, how can change be enabled, and how can a re-structuring of relations be made possible?

In this paper, we develop a notion and understanding of this space and some of the locations that are involved. We assign the notion 'development arena' to the space that can contain these processes analytically as well as enable change management. We begin by setting the stage for our discussion by defining the notion and discussing it in comparison with existing theoretical concepts within innovation economy and sociology of technology. The development of the concept has been an interaction between existing theory describing similar phenomena and studies of developments within the communication and information technology industries.

We explore the notion by presenting contemporary developments in television and computing,[3] especially the process leading to the innovation of the next generation of television technology - often defined as HDTV (High Definition TeleVision).[4] One of the surprising elements of this case story is that the introduction and presentation of a working prototype of the technology by a number of relatively insignificant actors completely reshaped the space for the development of new television technology. The entrance of actors of different kinds and the fundamental reshaping of the arena is hard to capture using existing theory. This has been the starting point for this investigation.

DEVELOPMENT ARENA DEFINED

Innovation has been studied from within a number of different disciplines, and several aspects may have been caught in these approaches. Our experience shows, however, that developing new technologies involves a number of very dissimilar processes held together by various linkages and inter-dependencies. This has led us to seek a theoretical notion that includes all these various processes.

A development arena, in our definition, is characterised and delimited by a space that holds together the settings and relations that comprise the context for product or process development that includes:

- a number of elements such as actors, artefacts, and standards that populate the arena,
- a variety of locations for action, knowledge and visions that define the changes of this space, and
- a set of translations that has shaped and played out the stabilisation and destabilisation of relations and artefacts.

Our definition emphasises the different and dispersed elements of the space that comprise various localities of both a cognitive and physical nature. The 'development arena' does not generally have a specific locality or one single geographical space of existence or of central importance. However, a number of specific locations will form the stages for action in relation to each other. They do so without any pre-specified order of importance or set of relations.

Our epistemology implies that the cognitive definition – development arena – cannot exist in any meaningful sense without specific references to objects and situations having a locality and a material reference. As a pure abstract notion 'development arena' remains purely metaphorical.

Companies enter arenas and specific situations when they start developing technologies and products. They may already be in an arena as a result of earlier activities. Actors might unintentionally be present in an arena. They might be represented in reduced form as competitors or users. In this paper, we limit the discussion to the analytical perspectives, although our attempt may include, as a future perspective, a discussion of managerial methods and problems.

The role of the notion 'development arena' is to sensitise both researchers and managers towards processes of technology development that would fall through the theoretical 'grid' of economic and management theory. We hope thereby to initiate critical discussions about the role of management and the directions it chooses for the development of technologies and products. The notion 'development arena' should function as a frame and a mental space for a discussion that focuses on the relational, unstable and heterogeneous character of the development process.

We want to place notions such as markets, customers and costs into a new perspective. Customers, markets etc., cannot be stabilised before technological artefacts are stabilised as commodities. We advocate a more representational and process based focus than is often found in the literature on technology development. What we want to stress is the representational character of entities such as markets, customers, costs etc. Furthermore, we wish to provide a spatial notion for conceiving developments in competing and developing representations.

We attempt, in this chapter, to create a framework to characterise the space and thereby also focus on the boundaries for such processes.[5] We deliberately try to define the arena as an open-ended space, where certain actors and locations can be inscribed either by the actors themselves or by others engaged in the arena. The development arena should thus give us a frame for understanding and researching processes in which companies and other actors attempt to master technologies, products and markets. It includes both the static elements of locations, knowledge and artefacts, while it also frames a space for continuous action.

Having introduced the background and scope of this paper, we still have three remaining pieces of work: First, to argue why we think there is a need for a new notion and why existing theoretical frameworks do not cover it. Second, to establish a frame for understanding this notion. Third, to indicate how it works and how it may help to identify the spaces that should be looked into and the types of problems and relations in need of being managed.[6]

METAPHORS AT PLAY

There are several considerations to make when a metaphorical notion like 'development arena' is introduced. One fundamental concern is whether it is scientifically sound to use metaphors at all. We think this is an integral part of defining a new vocabulary to cover phenomena and processes not easily addressed by other scientific notions.[7] Therefore, it is an integral part of theory development and of making sense of empirical material (Weick, 1995). It is important to consider what kind of narratives or images the choice of metaphor invokes (Morgan, 1986). Metaphors do not entirely correspond to what they are supposed to describe. This is one of the frustrating and intriguing characteristics of metaphors (Morgan, 1980). However, the same applies, to a lesser degree, to symbolic languages (Leary, 1990). It is not enough to propose a new metaphor for something that may be described by existing categories or concepts. It is important to make sure that enough context and narration[8] is provided to convey the intended meanings to the reader. In this section, we discuss the proposed metaphors in their own terms. Below, we provide a case story to show the use of the terminology.

As described above, a 'development arena' is a metaphor for the space where political, social and technical performances related to a specific technological problem take place. It is a spatial imagery that brings together heterogeneous elements that seem distant in geographical and conventional cultural space. It resembles the idea of the 'patchwork' of technology stories. It uses the idea of partial connections and multiple stories (Law & Mol, 1995). In addition, it specifically addresses conflicting interests and contention about the space. Part of the imagery is that performances depend on the participation and enthusiasm of both actors and audience. The audience can be thought of as journalists, scientists, lay people, users, customers, victims, etc. The imagery is spatial, but it is not bound to any specific geographical location. The idea is that the metaphor makes it possible to visualise the many different heterogeneous elements that compete for attention and power in the arena.

By using the term 'development arena', two metaphors are used to illustrate a certain research perspective. One is the envisioning of change and shaping embodied in the word 'development' – but not all changes really achieve something new and not all processes of change are creative and shape new things. Some even have the purpose and result of destroying or forgetting technologies. The intent is not to focus solely on the early stages of innovation and development. The processes of market creation and user positioning, recruitment and interaction are equally important. Therefore, 'product arena' may cover some of the ideas in a better way. The whole process, however, is a process of construction, shaping and becoming, which cannot be limited by the specifics of established products. For this reason, the notion 'development' covers our perspective better than 'product'.

The word 'arena' comes from Arabic. It refers to sand both as the ground for activities and as the never settled character of this ground and its place – it is moving and the ground is thus eternally reshaped (Fink, 1996). The notion has been used in political science to characterise political arenas – spaces and locations where negotiations, conflicts and ideas are exchanged and developed. Activities that often have massive consequences for the society 'outside' the arena, both those watching the show and those not even invited as spectators but only presented the results of the processes. 'Viewed as a social institution the (political) arena is a special place for specially shaped events' (Fink, 1996).

Companies, standards, markets may belong to different locales with different characteristics associated with them. Products and processes may be differentiated between markets and localities. The development arena is a space where these different locales and objects are linked together.[9]

DEVELOPMENT ARENA AND EXISTING THEORY

This section discusses the relation between the notion of 'development arena' and existing theory that covers some of the same phenomena. Geographical and cultural influences on technology development are discussed in the literature on regional and national innovation systems.[10] Evolutionary economics touches upon technological and historical aspects of technology development.[11] Literature on sociology of technology has inspired the interpretational aspect of 'development arenas'.[12] All these theories include or refer to - partly similar, partly different – aspects of locations and spaces that the notion of 'development arena' is trying to capture. Some are more focused on mapping the involved components and institutions. Others are focused on the heterogeneous character of relations involved. Not all elements of the theories mentioned above will be discussed.

In the literature on clustering of industries and the uniqueness of the development in certain industrial districts, we have found a number of contributions that focus on specific aspects of the structuring of industrial development (Dahmén, 1988; Porter, 1990; Piore & Sabel, 1984). These notions focus on the inter-linked set of companies and institutions and their tendency to be closely located geographically in a regional and eventually national context. Although the nature of the proximity of these structures may change over time from being defined by closeness measured by distance to closeness measured by culture and knowledge, the basic idea is still the structuring of industries and competences in networks. The approach addresses a number of possible synergies that may lead to the shaping of such clusters or districts. They include lists of factors that may or may not feed this process, but the theoretical strength seems to be the retrospective analysis of industrial structures. They do not have a language or models that show the linkages and shifting importance of the cognitive domains and the different locales of action. The focus is more on industrial structure and strategies for sustaining established positions than on the study of specific emerging technologies and innovations.

The same is the case with the identification of the role of the national institutional system for the development of specific networks, product configurations, and priorities of knowledge (Lundvall, 1992). This type of literature shares a common interest in pointing out the limitations of standard neo-classical economic approaches for understanding industrial and technological change (Freeman, 1994). The fields of industrial economics, economics of innovation and evolutionary economics have all contributed to this critique of the lack of understanding of the incremental and institutional character of technological change (Lundvall, 1998). Although the networking, exchange of knowledge and cultural context of technology development are emphasised in the models of national innovation systems, the focus of the research is more on the policy consequences of these structuring processes than on the representational and organisational aspects of the innovation process itself.

One of the major weaknesses in these approaches is the lack of interest in how these networks, structures and positions were constructed and maintained. Material objects are left aside in the limited focus on companies, institutions and organisational relations. They are necessary specifics, but of less interest for the development of general models and notions. This literature provides us with a number of well-researched patterns of development and a number of questions of why things went this or that way. The argument may be satisfying as a critique of economic models, but it does not provide us with an analytical tool to understand things in the making (Callon, 1991). The understanding of interpretative flexibility and the role of technologies and products in the making is limited and mostly treated in generalised terms of 'learning' and 'knowledge' as omnipotent phenomena.

Models from the economics of innovation, such as user-producer relations, establish a focus on the problems of user configurations and abstract models of markets and user needs. Along with the interest for inter-firm networks (Grandori & Soda, 1995; Håkansson, 1987) in organisational theory, it fills a gap in the models from industrial economics for forms of market competition and the related types of firms. These models are inspiring and they can in total serve as the 'encyclopaedia' of economic activity, and thereby help to maintain a focus on the complexity of this field. However, the heterogeneous character of the knowledge, markets, product areas, and institutional set-ups is not part of these studies. Despite this they form a necessary corrective for companies wanting to understand their competitive surroundings and shape the basis for a product strategy.

Much can be learned from the economically oriented studies of technological development. Some of the influential theoretical models are technological guideposts (Sahal, 1981), paradigms and trajectories (Dosi, 1982), technological regimes (Nelson & Winter, 1977) and path dependencies (Karnøe & Garud, 1998). Although the main purpose of these models of technical change has been to show the path dependency and institutional embedding of technological change, guiding technological change in certain directions for longer periods of time, they can also provide some advice for management. But the focus is on the phases of development in which products and concepts have already found their initial shape, while the processes leading to such temporary stable development patterns are not in focus at all. Guideposts are the result of innovative processes. Trajectories are the result of institutionalised behaviour patterns of firms and other involved actors. These models serve as relevant correctives to the main theories of economic development that assume that change is an open-ended process.

In the literature on the management of technological innovation, competitiveness, and knowledge, we find a number of the above-mentioned models represented. They are usually followed by normative rules containing advice on how these models can be used as guidelines or checklists in future management situations.[13] We have chosen not to engage in a detailed discussion of this field of literature, as our starting point has been dissatisfaction with the normative approaches to management advice and the lack of analytical and empirical interest to be found in this literature. Especially, implicit assumptions of the unity of the company and the rationality of planning as the basis for strategic behaviour are a major problem from an analytical point of view (Araujo & Easton, 1996). There may, however, be a number of reflexive contributions to the discussion that we ignore by omitting it.

The discussion above has focused on economically oriented theories. Below, we take a closer look at theories working with constructions and interpretations of change. These are actor-worlds and laboratories from the actor-network approach, communities of practitioners from anthropology, and technological regimes and technological ensembles from evolutionary economics and technology studies. Like the discussion above of similarities and differences in scope, objects and context, the following will not so much advocate the 'development arena' as an alternative to the vast number of concepts proposed in the different fields of literature. On the contrary, we see the notion of 'development arena' as a supplement that may offer us a framework for analysis and eventually also a reference for understanding and ordering the other existing concepts.

LABORATORIES AND SOCIO-TECHNICAL ENSEMBLES

Originally, development arenas were conceived of, in our discussion, as a kind of 'laboratory' for developing technologies where especially the company would play an important role. The 'laboratory' as a metaphor has been elaborated in the literature on sociology of science (Latour & Woolgar, 1979). In this literature, the laboratory is seen as a space for reducing and ordering processes that are too complex to be understood in 'real life' (Law, 1992). When the processes are ordered in the laboratory, they can be translated and reproduced outside (Latour, 1988a).

One reason not to use the laboratory as a metaphor for technology development is that the studies of laboratory worlds seem to end up in a quite internally focused type of understanding of science (Brosveet, 1992). The studies tend to focus on the grandeur of already established scientific discoveries, or the impact of well-known and widespread new technological systems.

In the construction and delimitation of laboratories, certain actors engage in strategies to establish a hegemony or regime, both in terms of activities and in terms of discourse and knowledge. There is hereby a bias towards the heroic work of the entrepreneurs who had a vision and conquered the world through science and technology. The social studies of science and social construction of technology deliberately criticise these tendencies in technology studies. At the same time, the stories told implicitly support this view (Latour, 1993). This is another reason for being careful with the laboratory metaphor. We find it difficult to avoid this trap because of the limited help from notions and heuristics (Scott, 1991).

The notion 'communities of practitioners' also emerged from studies of laboratories. It was introduced to help understand how knowledge and competence must be understood as developed and translated in local settings (Brown & Duguid, 1997). This way of approaching certain problems in early phases of research and development is quite useful. It adds to the understanding of internal processes of negotiation and sharing of knowledge. But although several attempts have been made to explain the translation of knowledge from one local community to another,[14] these boundary-crossing activities seem in contrast with the basic role assigned to the locale as the setting and boundary for the meaning-creation of knowledge and community. This is emphasised through the role of practice and competence as 'tacit' knowledge, not bearing any meaning outside the community. The studies of tacit knowledge focus on the cognitive and internal aspects of technology and knowledge even more than the sociological studies of laboratories of science. The heterogeneous and relational aspects are potentially left behind.

Laboratories can be viewed as one of the most important types of 'strategic places' in the modern world (Latour, 1983). However, they are not the only ones. Corporate boardrooms and policy institutions strive for a similar position (Brosveet, 1992). Laboratories are like the fortresses of the medieval world, settings for the maintenance of power but not enough to keep control. Industrial development laboratories, technological standards, test sites for new products, and established product configurations (markets, specific forms of commodity constitution, etc.) are similarly important strategic places in the modern world of competition between technologies and between transnational corporations. However, they are not enough to characterise the development arenas in which the processes are unfolded. The re-configurations of uses and technologies, the interpretative flexibility, and the domestication of technology in every new situation of use are given too little room, if laboratory studies or communities of practitioners are taken as the overarching framework for research.

The hegemony of the laboratory and the deliberate construction of its reputation as the 'centre of calculation' also involves an agenda for marginalising actors not accepted as part of this system of expertise or experience. Users may enter the laboratory as objects of study or as representations in models and assumptions, but they are not allowed into the basic discourse of expertise in the laboratory. A representation of these actors in science and technology is a way of reducing them to a standardised and easier to handle set of relations. As a consequence, these actors later have the choice of becoming members and accepting the conditions to come inside and experience the strength of the interpretative flexibility that flows from being part of the centre of interpretation. Or they can choose to build another laboratory in order to enter the fight over standards and regimes as it staged by different centres of knowledge.

A development arena contains other elements than already constructed artefact and industrial laboratories. Also involved are real and virtual products, processes, and user configurations as the actors view them. This broader perspective could perhaps be covered by notions such as 'socio-technical ensembles', introduced in the literature on social construction of technology (Bijker & Law, 1992). This may be the notion that comes closest to the very idea of the 'development arena'. However, socio-technical ensembles also focus on the specific set of social practices and artefacts that assign meaning to a technology as something well known that can be understood in its complete social context as a socio-technology. This view keeps the attention on the heterogeneous character of the processes involved. However, it has a focus on the already established, on the 'ensemble', instead of the processes of becoming, shaping and structuring. In practical use, the notion of socio-technical ensembles has been used as a characterisation of certain inter-linked phenomena, to avoid a narrower focus on the functional elements of the artefact in question.

A similar generalised notion describing the institutionalised setting for technological developments and their structuring is found in the evolutionary economics literature under the heading 'technological regimes' (Geels & Schot, 1998). The term conveys the idea that technological change is guided into certain patterns regulated by a set of socio-technical factors and institutional settings. The focus is different from that found in the notions of 'technological guideposts' and 'technological paradigms' discussed in the previous section. Regimes are not based only on specific design or structures of knowledge. They have their basis in the inter-linking of different institutions, which support the dominance of that type of technology.

The concept of technological regimes has been used to describe processes of change – identified as regime shifts – by developing an understanding of what in the evolutionary vocabulary is seen as processes of niche-formation. Such processes can, e.g., be defined as a combination of 'coupling of expectations', 'articulation of problems, needs, and possibilities', and 'network formation' (Geels & Schot, 1998). They hereby add a micro-level of actions to the meso-level description of stabilising processes encompassed in a technological regime and framed in what at the macro-level is termed a 'socio-technical landscape' with connotations of a physical, multi-dimensional shape or structure in which change processes can move.[15]

For our purpose, the concept of technological regime has too much focus on the technology as a system that defines the backbone and core focus of the regime. The regime concept is a valuable contribution to understanding the institutions and mechanisms that sustain complex existing systems such as the road and car system of transport. It does not highlight the processes that may continuously break down and fragment such a system. It downplays the mechanisms that renew the structuring of the system. Furthermore, what may be looked upon as a more important problem, the focus of a technological regime is the already established technology and not the developing technologies. The core representation of the technology is seen as the centre of change and stability.

Laboratories, ensembles and regimes are essentially useful notions within their respective areas of application. The notion of 'development arena' aims at analysing and describing processes in the making. For this purpose, the other concepts focus on internal processes or on already established technologies. Furthermore, there is an element of heroic action that is inconsistent with our conception of development processes as complex, inter-linked and chaotic. The arena as a metaphor should capture the flux and changes in morphology of the conceptual space in technology development processes, which differs from the structured movements in physical space conveyed by a landscape image.

'DEVELOPMENT ARENAS' AND ACTOR-WORLDS

Sociology of science and technology provides some of the vocabulary we are looking for to describe the development arena. We have been particularly inspired by actor-network theory (Law, 1992). Actor-network theory provides a vocabulary for discussing the heterogeneous processes that constitute technology in particular and society in general. The concept of 'actor-world' (Callon, 1986) is especially useful in the discussion of the relation to 'development arenas'. Actor-world is a broad notion that covers a lot of different settings far beyond the scope of development arenas. An actor-world is developed around a certain set of situations and is thereby limited to what we here call a location in the space of a development arena. The laboratory as a metaphor is another way of analysing such complex localities. This is discussed below.

The notions 'actor-network' and 'actor-world' are often used interchangeably (Callon, 1986). The notion of actor-world indicates that central actors not only build an actor-network but also a complete set of narratives and translations for the purpose of holding together the different pieces of the actor-network. Actor-worlds can be seen as scenarios or as utopias created by central actors. An actor-network is mainly built and stabilised through juxtaposition, translations and punctualisations. In an actor-world, each part of the corresponding actor-network is assigned a role. 'It binds the functions of these roles together by building a world where everyone has his own place' (Callon, 1986:22).

'The actor-world not only determines the repertoire of entities that it enlists and the histories in which they take part. It also determines their relative size' (Callon, 1986:22). The entities of an actor-world are heterogeneous. They consist of humans along with organisations, electrons, machines etc. In the absence of one of these entities or in cases where one of the translations is resisted, the actor-world might break down. A fundamental point is that science and technologies have been successful in many cases in setting themselves up as obligatory passage points supporting certain actor-worlds (Latour, 1987).

Actor-network theory describes how actor-networks are built and maintained and how they break down. It describes what is included and what is excluded from certain translations. There are no good descriptions of how different actor-networks compete in building different actor-worlds. 'More than any other type of actor, technologists may be sometimes endowed with the capacity to construct a world, their world, to define its constituent elements, and to provide for it a time, a space, and a history' (Callon, 1986:21). This corresponds to the more fundamental proposition that space-time is created by the interaction between actors (Harvey, 1996). Callon considers an actor-world to be a construction of time-space. It is a partial construction, but the term 'world' makes the imagery quite closed. The relations and translations can be resisted from within, but the imagery does not easily include competition between actor-worlds.

In contrast, a development arena is a visualising spatial expression of processes of competition and co-operation. It should convey the idea that several actor-worlds are being construed within the same problem area. It depicts the idea that several actor-networks co-exist and interfere with each other within a certain problem space. A development arena is our attempt to bring together processes or entities that would otherwise seem to be dislocated. It can be seen as the place where actors relating to a certain set of problems meet and exchange ideas etc. It is a place relatively independent of geographical location but containing many locales through translation. Each hybrid actor in the arena has multiple geographical belongings.

When actors render a new area problematic and start to enrol other actors into an actor-network, they start building an actor-world. At this early point in time, one could say that the development arena and the actor-world are identical. But in competitive business, research or national environment, other actors soon start to try to enter the arena, possibly by introducing new actor-worlds and thereby reshaping the reference points for all other involved actors. Actors might also expand the arena by bringing together different development arenas or actor-worlds that had previously been seen as separate. This can be seen as a new actor-world incorporating all the others, but we would rather use a different concept to capture the conflicts between different actor-worlds.

However, it might happen that an actor is successful in enrolling and translating all other actors into this new arena. Thereby, the development arena and the actor-world are folded into each other. The case could be a successful company-based standardisation of a technology setting, a de facto standard for all other actors in the arena. It might happen that there is no audience to watch the play in the arena. This could happen when a technology becomes obsolete, or on the contrary is taken for granted.

Actors that are excluded from a certain actor-network might be overlooked in an analysis based on actor-network theory. With the concept of development arena, it is possible to incorporate actors that are excluded by the dominant translations. Thereby, actor-networks and development arenas can be used together. Actor-networks or actor-worlds are particular configurations populating the arena.

THE TYPICAL GENESIS OF A DEVELOPMENT ARENA

To manage innovation is to enter a development arena. Below, we sketch the stereotypical genesis of a development arena using the concept actor-worlds as an example. Typically, a company starts a development project and thereby creates or enters a development arena. In the beginning, most elements only exist in fragmented representations on paper, as mental structures and as emerging interactions. As the process moves along, pictures of customers/users and markets are explicated in reports, test presentations etc. Objects that eventually develop into products are transformed into new modes of representation, such as computer models, prototypes etc. More people and equipment are often enrolled to work on these configurations and translations. In this process, all elements are constantly informing and changing each other. An actor-world is being built in a virtual laboratory that is constituted by the development arena. Often, the meaning and materiality of artefacts and users change several times during the process.

One of the 'virtues' of the company as an entity is its ability to confine matters in order to keep them from broader public view. In other words, new actor-worlds can be hidden backstage until they have acquired a certain momentum. They enter the arena with a massive presence, when objects have materialised, tests have been performed, selected contacts to other actors such as key customers and consultants have been made.

When objects and configurations leave the confines of the virtual laboratory, they are being put to the test. Will users behave as they are supposed to? Will the objects? What is the reaction of other actors? In other words, the actor-world and all of the translations holding it together are exposed to the audience. Many spectators will find themselves represented. They might accept the translations and let themselves be enrolled. Part of the enrolment process is the building of infrastructures to disseminate information and objects, such as commercial information, products etc. During this process, the topology of the development arena might change radically. Translations are extended to accommodate certain regions and cultures. The material objects might have to change.

Studies of technology and innovation have shown that the nature of objects is often changed in the process of consumption. One reason for this is that the way customers are represented and positioned, both in theory and in practice, often limits or misguides development teams in relation to the perspective of user-centred product development. Another reason is that wants and desires change through use, and yet another is that it is impossible to envision all possible uses in the confines of a virtual laboratory. It is likely that other actors react after the exposure of a previously secret actor-world. They can do this by trying to redefine the arena by introducing other products. Another option is to restructure it by setting up regulations. Thereby, the actor-world is only permitted to grow in certain directions, or certain objects might have to be withdrawn.

The involved company and its local networks can extend the boundaries of the arena. The space can also be changed by other actors entering the arena in competition or by taking over the ideas in different ways. Companies, customers, information distributors, users are all part of this process and have to play their role in the arena. Companies may devise strategies to extend a local arena into the global sphere by making connections to other localities and translating them to be part of the arena.

Sometimes other companies are invited onto the arena and into the laboratory. In this way, there is a chance that they will not appear as competitors and that they can be mutually beneficial. In some cases, entrepreneurs cannot expand their actor-world from their place inside a specific company because 'central actors' refuse to be enrolled or oppose the project by putting constraints on time, people, money etc. Instead, they decide to start on their own, creating a 'spin-off'. From this new position, they might be able to change the arena much more efficiently than before. 'Entrepreneurs' are here understood broadly as persons or groups inside or outside established organisations (i.e. also intrapreneurs).[16]

For consumers in general, the 'successful' result of the stabilisation of actor-worlds in an arena is often commodities or tools 'ready to use'. They do not necessarily realise that they had been configured as part of the process. Commodities are only questioned if the networked configurations of heterogeneous relations break down (computer virus, floods, lack of operators, spare parts etc.), if actors resist translations, if new constellations enter the arena with new ways of doing things, or if spectators launch fundamental critiques or regulations (religious or environmental groups).

The broad view of the development arena makes it possible to simultaneously conceive of processes considered to be internal to companies and processes that cross such boundaries. Actor-worlds used in the description above are only one among many ways of describing configuration elements of a development arena. The company-centred focus of strategic management, management of technology, and knowledge management is questioned by the notion. Other centres of influence can be conceived of, developed and researched.

HDTV AS AN EXAMPLE OF A DEVELOPMENT ARENA

To show the variety of processes and locales involved in the development of a new set of technologies, we have chosen to exemplify the development arena covering contemporary innovations in telecommunication and computing, especially the unresolved problem of introducing a next generation of television technology.

One part of the unresolved problem is the potential merging of television and computing into a hybrid communication media based on digital technologies. For the last five years, this merging has been viewed as one of the hottest issues in most popular magazines, at conferences about new media, and at the big conventions where new electronic media products for the consumer market are presented. As part of this merging process, a number of attempts have been made to digitalise video systems. But also alternative strategies of differentiating the two media have been set in motion, to maintain television as a mass communication and film media and computing as a work place and play station. Another part of the problem is the stagnating markets for standard television products in Europe and the United States and therefore the quest for the construction of the next generation of television technology.

Five interesting stories – among others – can be told showing both the heterogeneous and unexpected elements that play a part in creating and transforming this space that holds the controversies about and development activities for a next generation television technology. Five events from television development illustrate in combination the very different configurations and locations that enable actions and are enacted in the process. The five stories also show the continuously changing borders of the high-definition TV arena, with new elements emerging and new types of settings being included.[17]

First Story - Initial Local Japanese Arena

The first event is located in Japan. The Japanese attempt to develop the next generation of TV in the MUSE system was launched as a full-scale experiment and included a great number of test users to identify the technical details to be dealt with and the consumer expectations to the television of the future. The new television was launched in a local Japanese test setting, a closed laboratory world that could enhance the capabilities of the Japanese electronics industry and prepare them for dealing with the world market of new television technologies. It entails all the complicated relations involved in developing new product generations. It shows the building of a heterogeneous network of technologies, producers, markets, users and content providers. Japan has had a long tradition for setting up large-scale experiments with new product generations. The arena for these activities has been national experiments. But the vision was from the beginning broader than just the Japanese market. It was to make HDTV, in the form of MUSE, an initiative that could bring Japanese producers of consumer electronics one step ahead in international competition.

Second Story - EU Enters and Extends the Arena

The second setting is different. It is initiated by the EU Commission together with the network of European electronics manufacturers. The Commission's policy ambition is to develop an industrial policy for Europe, based on a division of the economic and industrial power in the world amongst the rivals and partners of the 'triad'. The triad positions Europe, Japan and US as the centres of industrial development and economic power. Although the official ideology of the EU is neo-liberal and focused on free trade, the 'fortress Europe' idea involves a corporatist coalition of industry and the Commission in central areas of technology. Electronics especially has been in focus in the EU as a traditional stronghold for Europe for the telecommunications and consumer sectors, but also a source of constant pain due to the weak position of the European computer and semi-conductor sector. Since from this point of view a next generation of television technology could not be left to the Japanese, the EU launched a European alternative to MUSE in co-operation with some of the big players in consumer electronics, the HD-MAC standard.

In line with the Commission's scepticism toward national institutions, traditional national broadcasters were not invited, and instead the new commercial satellite broadcasters were identified as the partners for a strategic alliance for the implementation of the new television system. The solution should be a new generation of analogue TV systems[18] prepared for pay-TV and satellite broadcasting. The new arena first shaped in Japan (as described in the first story) was colonised by the EU, and a competition between technologies was initiated parallel to the potential merger of product visions and markets. An element of competition on technological standards and new tight networks and alliances were added to the now much more heterogeneous locations and elements of the arena.

Third Story - US Enters by Redefining the Scope

The setting for the third story is the US. Almost no consumer electronics industry remains in the US, or to state this more correctly, all the American producers are now owned by European and Japanese manufacturers. Instead, a widespread and very diverse industry exists that produces computer equipment with special focus on digital image processing hardware. Although European and Japanese firms also dominate a great part of the manufacturing of computer screens, they do not dominate the image processing products industry. The idea of integrating television and computers is proposed repeatedly as a result of the technology focus of the digital image processing activities. A real technology fix, one might suppose, but kept alive by a computer-focused industry and computer users, and continuously brought to practical test by the introduction of both inexpensive add-on cards for the computer and high end, costly product fusions of television and computers.[19]

Having no industry and no strong interest groups in the traditional television sector to serve, the US government decided to trigger a bigger step into the future. An open government invitation to tender for technological R&D was announced to support consortia that would develop the next generation of digital television standards and equipment. Without trying to construct any merger of the two sectors, a whole new set of future restructuring was enabled by bringing closer and eventually merging two different arenas of development: the television and the computer industries. No user groups or experiments with the new media are involved in this phase of development. The licensed consortia are closed. They limit their communication for strategic reasons, and their focus has shifted from specific developments of new products to technological concepts and standards for transmission and image processing.

Fourth Story - a Small Players Revolution by Prototyping

Meanwhile, the Danish national TV broadcaster, together with the Danish Telecom and other companies that include Swedish partners, had decided to by-pass all technological pre-occupations. By constructing a digital television set, they showed that digital television was possible at least from a technological point of view. The setting of the fourth story is local and almost national. Most of the partners had earlier experience in co-operation. They included a Danish producer of encoding equipment based on standards for live digital images in the consortium, because the main technical problem was the need for extreme data compression capabilities. This data compression, however, could reduce the convincing quality of live images. The product was called DIVINE TV. When presented to the public, DIVINE TV turned completely upside down the assumptions of companies and policy makers, and thereby heavily influenced the relations in the development arena. The EU and Japanese projects were doomed as outdated, and the US technological solution was much closer than expected. But in contrast to the Japanese experiments and the networks of the EU Commission, neither the DIVINE prototype nor the US consortia had any close linkages to users, emerging markets or alliances. The arena was left vibrant and confused after this simple demonstration.

Fifth Story - Companies Playing it Safe

This story is about the strategy of Philips – one of the big players in the European electronics industry. The setting is the boardroom and development departments of one company. But it is international, both because of the international reach of this company and because of the already internationalised arena. Philips was part of the EU project that developed the HD-MAC standard and equipment. It supported the idea that the EU should support the consumer electronics industry by setting a common European standard for the next generation of television sets, more or less forcing European broadcasters and consumers to invest in this technology. Shortly afterward, the Philips branch in the US teamed up with partners in two of the bidding consortia for developing the next generation of digital television standards and technologies and hereby gave its support to the switch from analogue to digital television. Strategic players seemed to be able to participate in different and competitive parts of the arena, mastering the uncertainty about the dominant solutions of the future.

One lesson to be learned from these stories is that traditional types of industry-government networks and cluster-like structures do not turn out to be as powerful as was expected when they were being set up and sustained. New technology visions and 'small' players end up in certain cases to be more important than the well organised and long-term co-operating companies and networks. Another lesson is that technological fixes in the form of technological expectations and visions still play an important role, and they turn out to be crucial for the priorities assigned to certain development strategies and configurations of technological solutions.

Thus, a number of lessons can be drawn from the five stories. As isolated lessons, they can also be found in the vast literature on innovation processes and technology management. All these lessons form a catalogue of 'war stories' that can help us define some of the mechanisms of change or paths of development, but they are often isolated and elevated as general lessons or rules for change. Instead of risking being locked into the problem of determinism and prediction, we will – as indicated in the introduction – follow a different path. We wish to use the notion 'development arena' as a frame to hold together the phenomena involved in the five stories, and use them to identify configuring elements in this space.

CONFIGURATIONS: ELEMENTS AND TRANSFORMATIONS

None of the stories can be understood easily by translating them into the language of single academic schools of thought and describing them by one type of scientific model. Even models that deliberately transgress some of the recognised limitations of single scientific disciplines dealing with technology, policy or strategy like the actor-network theory and the concept of socio-technical ensembles show some limitations.

From the five stories emerge new actors, new configurations of networks, new settings, and re-configurations of the technologies and visions involved in the process. Even translations emerge of the understanding of technology imposed by actors in locations that were not recognised from the beginning or that were excluded because they were not considered important.

We will now extract some of the main configuring relations and elements involved in the creation, extension and redefinition of a development arena. In other words, we want to introduce and characterise some of the structured relations, locations and translations that 'populate' a development arena.

The backdrop of the five stories and the arena for a next generation television is the already established actor set-up resulting from the development of television technology, production facilities, broadcasting institutions, and mass communication after World War II. There are basically three standards for television broadcasting in the world: the American NTSC system, which also is implemented in Asia, and the European PAL and SECAM systems. A number of international standardisation bodies and some co-operation among broadcasters have stabilised these systems and managed the transfer from black-and-white television to colour television. The saturation of markets for television products and the limitations of the analogue transmission technologies have raised concern in the television industry about a 'next step', but this has primarily led to intensified competition and concentration and established today's existing oligopoly of consumer television manufacturers.

This establishes a number of *institutionalised actors* comprising of companies, broadcasters, standardisation bodies, user configurations, and regulating authorities. It also leaves us with a defined set of *framing objects* such as: relatively fixed visions of the technology and its role in society; a discourse about the role of mass communication and its importance in the process of modernisation (shaping myths and expectations); a number of reference products of a shape and with operational capabilities that are commonly known (guideposts, exemplars); and a set of technological concepts distributed through the education of operators and engineers and sustained through the standards regulating the area.

These actors, relations, and objects make up the configuring elements of the development arena (the 'static' elements, one may add, as no action is invoked by just listing and relating these elements to each other). A number of *locations* are already involved at this stage of the description of the television arena: such as the geographical 'home' of national broadcasters, the standardisation bodies, test sites and company laboratories. A physical location that is easily forgotten is the placement of television sets in people's homes. This has important implications for consumption and thereby for how the technology is socially embedded and domesticated.

The idea of opening a new market for information dissemination and media competition has been one of the factors that made the television systems in many countries less stable. This has given rise to a number of new commercial television broadcasters that in competition with or just parallel to the traditional national broadcasters have made the media market grow but also stagnate from the point of view of content. The Japanese MUSE initiative introduced in the first story was a *transformation* of the whole setting of the television system. By establishing a test-site – a new location – that renewed the positioning of the television set and the content to be expected, the Japanese set the stage for a next step in television development. But at the same time, the Japanese context was still present. The technology was a continuation of the analogue system and the change incremental. The transformation involved both restructuring the locality for development and testing, and involved the users in an experiment that opened a new way of positioning. We could see this transformation process as recruiting actors by establishing a new setting for innovation, changing both the network of actors and the technological vision – a *strategy of resettling and inclusion.*

While the Japanese initiative was virgin, the actors were well aware of the potentials of being a first mover and focusing on identification of a new user concept. The European action – the second story – was not at all about testing new uses of television sets and not at all about consumer preferences and needs. The European action was re-active from the outset. By joining the important actors in industry with the satellite operators, they assumed they would be able to establish a strong policy network that could define the next generation technology. The transformation involved was the one of resettling and inclusion, but with a very different set-up of actors than in the Japanese case. The most important move was the construction of a competing network in the arena – a *strategy of extension and differentiation.* The result: a much more unsettled, competitive and heterogeneous arena.

The US was free from the binding relations to existing industrial interests. At the same time, they were assuming that a technological breakthrough would help American producers to regain positions by entering the arena. They did this by redefining the scope. The US government entered the arena by moving from the position of an observer to becoming a participating actor. They did this by establishing new technological locations through financing licensed collaborating laboratories – technology-focused test sites – thus building on the tradition of American procurement policies. For the first time, this re-framed the arena's focus on a prolonged period of analogue technology by differentiating the technology focus and introducing a digital solution as the future standard. This was clearly a technological fix without any clear user configuration besides the one developed in the computer industry. At the same time, it was a very different move than those of Japan and the EU and left them 'behind'. In addition to the strategies mentioned above, this exemplifies a *strategy of exclusion.*

The Nordic initiative of engineering a digital television prototype in fact changed the 'taken for granted' assumptions of most other actors. Digital television was expected to be a future technology, but meanwhile a next generation television would still have to be based on analogue techniques. The demonstrated and working prototype re-framed the technological expectations in the arena 'overnight', and translated the problem by illustrating a breakthrough by presenting a new guidepost. Instead of continuing a period of competing technologies, the newly established boundary object, the working digital television, ended the controversy. Digital television was to be the next step. The transformation involved a *strategy of re-framing.*

The fifth story about the multiple involvement of Philips illustrates the vulnerability of even transnational companies that dominate the world market and the technologies involved in the field. Together with a few other companies, such as Sony, Philips is involved in most of the de facto standardisation processes in the field. However, they are not, by themselves or in alliance with other consumer electronics companies, able to set the stage for a next generation television system. They are forced to engage in the locations for technological testing and change that are established, or at least have a very good understanding of these locations – a *strategy of multiple engagements.*

A type of transformation that is not outlined in the five stories is the process of competition and negotiation that leads to the construction of overarching networks of companies and other actors to try to settle the controversy between competing technologies and standards by creating new, integrated solutions. This involves the moving between locations and the assembly of new integrated locations – which could be called a *strategy of reduction and ordering.*

CONCLUDING REMARKS

The chapter presents the notion of 'development arena' to frame a number of complex and heterogeneous processes in technological change. The use of the notion and its implied modes and configurations offers the possibility to analyse new, emerging technological ensembles and their institutional embedding. By coupling elements and processes of importance into a development arena, it is possible to transgress some of the limitations in the theories and models of technology management, economics of innovation and even actor-network theory. It is suggested that the notions of 'development arena' and 'configuration' define a space containing certain orderings. It complements innovation economy and actor-network theory because elements from these theories have a place in the arena as configurations. It provides a space for conceiving and visualising the fight between competing attempts at translating and stabilising heterogeneous elements into concrete technologies and networks.

We have shown that a 'development arena' covers and includes a number of distinct strategies and processes that link together both competitive and co-operative types of relations between locations inside and between actor-worlds. The notion implies a mapping of the elements in the arena and the types of transformations that characterise the changes in the arena.

The role of the theoretical notion is to sensitise both managers and critical discussions about the role of management and the directions it chooses for developments in technology and products. It is not presented as an alternative to existing theory but a framework to assemble the locations and processes involved in innovation. As such, it complements other theoretical notions like 'actor-worlds' and 'socio-technical ensembles' in that it forms a specific 'mode of ordering' (Law, 1994).

The notion 'development arena' can be useful as a management concept in strategic thinking. It will especially be relevant in product innovation and technology development. It could complement conventional thinking about how the company is being positioned on markets. Using the notion of 'development arena' makes it apparent that markets and users are being constructed, as well as products and technologies. It might also help managers to broaden their view on the company in relation to other organisations, e.g. degrees of competition and co-operation. It conveys a flexible view of the space in which technologies are being envisioned and developed. It could be an element in activities such as scenario planning and technology foresight. Finally, it might help to indicate directions, from which potential threats are coming, because it concentrates on a problem space rather than specific competitors or markets.

NOTES

1 This chapter previously appeared as an article in *Technology Analysis & Strategic Management Texts* Vol. 11 No. 3, September 1999, published by Carfax Publishing, Taylor & Francis, (http://www.tandf.co.uk). An earlier version of this paper was presented at the 'Constructing Tomorrow: Technology Strategies for the New Millennium' conference, Bristol Business School, 14-15 September 1998. The paper was written as part of the research programme CISTEMA (Centre for Interdisciplinary Studies of TEchnology MAnagement) carried out at the Department of Technology and Social Sciences, Technical University of Denmark (DTU), funded by Danish Research Councils. The authors want to thank colleagues in CISTEMA and an anonymous referee who all gave useful advice.

2 The manager in the case of NOKIA is referring to the metaphor of the 'garage' as the development laboratory for high technology innovators, but if it were that simple, the title of the paper could also simply have been: 'The view from the garage'. However, he may also somehow be 'mobile', since he is expecting to enter a space of important relations after first having opened his garage.

3 The empirical material used in this case originates from a research project: 'HDTV og multimedier' on the social construction of electronic visual media, carried out at the Department of Technology and Social Sciences by Finn Hansen and Ulrik Jørgensen with grants from the Danish Social Research Council (See Hansen 1995 p.20).

4 Describing the complex set of processes involved this way (HDTV) implies that the resolution and quality of the TV picture should be the main focus of future development. This is only one of many elements.

5 This attempt is in some ways parallel to the ideas behind Law's introduction of the notion of 'modes of ordering' (1992), where he tries to create a new sociological vocabulary for discussing hierarchy, power and management and ultimately modernity. Or similar to Storper & Harrison (1991), who try to disentangle the notion of 'local agglomeration' from some of the specific and mutually conflicting theories it is used in. They try to do this by creating a language that makes it possible to discuss governance, network system and region as separate dimensions.

6 This last step also relates to the CISTEMA research programme from which the problem in this chapter has developed. The objective of this programme was to identify the role of networks and the role of locality and relations both internal and external to companies that stage the problems for technology management in a process of globalisation.

7 We will not venture into a detailed discussion about science and metaphors. In our view, science develops through expansion of language in combination with empirical inquiry. Applying metaphors is one aspect of this process. Science is already imbued with metaphors considered to have literal meaning, such as mass 'attraction' (Leary, 1990).

8 Mats Alvesson introduces the concept of 'secondary metaphors' to elaborate on the problem of context (Alvesson, 1993). He argues that it is not enough to state that an organisation works like a game. We need to know what type of game the author has in mind. There is quite a difference between conceptualising organisational processes as resembling chess or soccer. When a metaphor is introduced, an explanation is needed to show how the metaphor helps to understand some new behaviour or a given phenomenon in a new way. Although the concept of 'secondary metaphors' is enlightening, we prefer the notion of 'narrative'. We will provide some narration to explain our understanding of 'development arena' below. Furthermore, we will exemplify how a new notion can be used in specific empirical cases.

9 We imagine that the arena may be divided into inter-linked sub-arenas, provided that this analytical differentiation is fruitful. This is not explored in this paper. From the outset, we will develop the notion 'development arena' without going further into this aspect of differentiation.

10 Some of the key terms in this area are 'development blocks' (Dahmén, 1988), 'industrial clusters' (Porter, 1990), 'industrial districts' (Piore & Sabel, 1984), 'national systems of innovations' (Lundvall, 1992), 'inter-firm networks' (Grandori & Soda, 1995; Häkansson, 1987), and 'user-producer relations' (Lundvall, 1985).

11 Such as 'technological guideposts' (Sahal, 1981), 'paradigms' and 'trajectories'' (Dosi, 1982) and 'technological regimes' (Nelson & Winter, 1977).

12 Which is treated under the heading configuration below. Examples of key terms from this area are 'socio-technical ensembles' (Bijker & Law, 1992), 'laboratories' (Latour & Woolgar, 1979), 'actor-worlds' (Callon, 1986), and 'communities of practitioners' (Brown & Duguid, 1997).

13 An example of such a textbook is 'The Management of Technology' by Resen and Becker (1995). Its structure follows the common standard with chapters covering: 'management of visions', 'strategy, planning, and implementation', 'governance of research and development', 'management of production', 'use of information technology', 'quality management', 'environmental management', and 'global benchmarking'. A variety of themes based on a preconceived split of objects and actions. The textbook even includes the standard narrative of changes in focus in technology management from R&D, over technology and innovation, to knowledge management.
14 Similar problems apply in the case of transfer from one part of an organisation to another.
15 The complex of notions: 'niches', 'regimes', and 'landscapes' seem to address a number of the same problems as we do in this paper. There may also be a fruitful exchange of views with this approach in the future, but the strong evolutionary connotations in the vocabulary conveys rather fixed boundary conditions – the 'nature' of things – that we question.
16 Latour uses the Machiavellian figure of the medieval 'prince' as a trope for this type of agency (Latour, 1988b). We attempt to provide a metaphorical space to imagine where and how they struggle. There are many 'princes' in shifting configurations of co-operation and competition. The different 'princes' try to control and change the arenas.
17 The background material for the case comes from the 'HDTV og multimedier' research project by Finn Hansen and Ulrik Jørgensen (Hansen, 1995); see note 3.
18 Behind this choice was not a complete ignorance of the emerging digital video technologies and their use, especially in the professional area, but a deliberate forecast that there were too many technological barriers to a digital broadcast and television system.
19 Examples are the Philips CDI-system, computer boxes using television sets as monitors, television tuner plug-in cards for PCs, and computer capabilities built into high-end television sets. None of these have gained substantial market shares; they seem to be test probes from manufacturers to keep up with the continuously envisioned technology fix. As this aspect is not the focus of this chapter, it may be relevant just to mention that one of the problems with this 'fusion fix' is the total neglect of the social conditions for consuming television and working with computers. Interactivity is often mentioned as one of the differences. However, it is not only a question of new features. It is embedded in the social setting of the two technological ensembles.

BIBLIOGRAPHY

Alvesson, M. (1993), 'The Play of Metaphors', in Hassard, J. & Parker, M. (eds), *Postmodernism and Organizations*, London: Sage, pp. 114-131.

Araujo, L. & Easton, G. (1996), 'Strategy: Where is the Pattern?', in *Organization*, **3**, (3), pp. 361-383.

Bijker, W. & Law, J. (1992), 'Postscript: Technology, Stability, and Social Theory', in W. Bijker & J. Law (eds), *Shaping Technology / Building Society*, Cambridge Mass.: MIT Press, pp. 290-308.

Brosveet, J. (1992), *Laboratorietanken: En blindgate i aktørnettverksteorien?* (The laboratory idea: a blind alley in the actor-network theory?), STS arbeidsnotat No. 19/92, Senter for Teknologi & Samfunn, Trondheim: University of Trondheim.

Brown, J. S. & Duguid, P. (1997), 'Organizing Knowledge', Palo Alto, XEROX PARC. *Working paper* presented at the conference *Path Dependency*, Copenhagen Business School, 1997.

Callon, M. (1986), 'The Sociology of an Actor-Network: The Case of the Electric Vehicle', in M. Callon, J. Law & A. Rip (eds), *Mapping the dynamics of science and technology*, London: Houndmills, pp. 19-34.

Callon, M. (1991), 'Techno-economic networks and irreversibility', in J. Law (ed.), *A Sociology of Monsters*, London: Routledge, pp. 132-161.

Dahmén, E. (1988), 'Development Blocks in Industrial Economics', in *Scandinavian Economic History Review*, (1), pp. 3-14.

Dosi, G. (1982), 'Technological Paradigms and Technological Trajectories', in *Research Policy*, **11**, (3), pp. 147-162.

Fink, H. (1996), 'Arenabegreber - mellem ørkensand & ørkensand' (Notions of arenas - from desert sand to desert sand), in H. Fink (ed.), *Arenaer - om politik & inscenesættelse* (Arenas of politics and performance), Aarhus: Aarhus Universitetsforlag, pp. 7-22.

Freeman, C. (1994), 'Critical Survey – The Economics of Technical Change', *Cambridge Journal of Economics*, (18), pp. 463-514.

Geels, F. & Schot, J. (1998), 'Reflexive Technology Policies and Sociotechnical Scenarios', Twente, TWRC. Working paper for the conference: '*Constructing tomorrow: Technology strategies for the new millennium*', Bristol Business School, 1998.

Grandori, A. & Soda, G. (1995), 'Inter-firm Networks: Antecedents, Mechanisms and Forms', in *Organizational Studies*, **2**, (16), pp.183-215.

Hansen, F. (1995), 'HD-MAC - historien om det europæiske HDTV-system' (HD-MAC - the history of the European HDTV-system), *working paper*, Department of Technology and Social Sciences, Lyngby: Technical University of Denmark (DTU).

Harvey, D. (1996), *Justice, Nature, and the Geography of Difference*, New York: Blackwell.

Häkansson, H. (ed.) (1987), *Industrial Technological Development - a Network Approach*, London: Routledge.

Karnøe, P. & Garud, R. (1998), 'Path Creation and Path Dependence in the Danish Wind Turbine Field', *Papers in Organization*, Department of Organization and Industrial Sociology, Copenhagen Business School. Presented at the conference *Path Dependency*, 1997.

Latour, B. & Woolgar, S.W. (1979), *Laboratory Life: The Construction of Scientific Facts*, Los Angeles: Sage.
Latour, B. (1983), 'Give Me a Laboratory and I will Raise the World', in K. D. Knorr-Cetina and M. J. Mulkay (eds) *Science Observed*, Beverly Hills: Sage, pp. 142-170.
Latour, B. (1987), *Science in Action*. Mass.: Harvard University Press.
Latour, B. (1988a), *The Pasteurization of France*, Cambridge MA: Harvard University Press.
Latour, B. (1988b), 'The Prince for Machines as Well as for Machinations', in B. Elliott (ed.), *Technology and Social Process*, Edinburgh: Edinburgh University Press.
Latour, B. (1993), *We Have Never Been Modern*, New York: Harvester Wheatsheaf.
Law, J. (1992), 'Notes on the Theory of the Actor-Network: Ordering, Strategy, and Heterogeneity', *Systems Practice*, **5**, (4), pp. 379-393.
Law, J. (1994), *Organizing Modernity*, New York: Blackwell.
Law, J. & Mol, A. (1995), 'Notes on Materiality and Sociality', *The Sociological Review*, (43), pp. 274-294.
Leary, D. E. (1990), 'Psyche's Muse: the Role of Metaphor in the History of Psychology', in D. E. Leary, (ed.), *Metaphor in The History of Psychology*, Cambridge: Cambridge University Press, pp. 1-77.
Lundvall, B-Å (1985), *Product Innovation and User-Producer Interaction*, Ålborg: AUC.
Lundvall, B- Å (ed.) (1992), *National Systems of Innovation*, London: Pinter.
Lundvall, B-Å (1998), 'Why Study National Systems and National Styles of Innovation?', *Technology Analysis & Strategic Management*, **10**, (4), pp. 407-421.
Morgan, G. (1980), 'Paradigms, Metaphors, and Puzzle Solving in Organization Theory', *Administrative Science Quarterly*, pp. 605-622.
Morgan, G. (1986), *Images of Organization*, London: Sage.
Nelson, R.R. & Winter, S.G. (1977), 'In Search of a Useful Theory of Innovation', *Research Policy*, (6), p. 36-76.
Piore, M. & Sabel, C. (1984), *The Second Industrial Divide*, New York: Blackwell.
Porter, M. (1990), *Competitive Advantage of Nations*, New York: Macmillan Press.
Resen, L. & Becker, J. (1995), *Teknologiledelse* (Management of Technology). København: Børsen Bøger.
Sahal, D. (1981), *Patterns of Industrial Innovation*, London: Addison-Wesley.
Scott, P. (1991), 'Levers and Counterweights: A Laboratory that Failed to Raise the World', *Social Studies of Science*, **21**, (1), pp. 7-35.
Storper, M. & Harrison, B. (1991), 'Flexibility, Hierarchy and Regional Development: the Changing Structure of Industrial Production Systems and Their Forms of Governance in the 1990s', *Research Policy*, **20**, pp. 407-422.
van Tulder, R. (1988), 'Small European Countries in the International Telecommunications Struggle', in C. Freeman & B.-Å. Lundvall (eds) *Small Countries Facing the Technological Revolutions*. London: Pinter.
Weick, K.E. (1995), *Sensemaking in Organizations*, Thousand Oaks: Sage.

8. Spaces and Occasions in the Social Shaping of Information Technologies: The Transformation of IT Systems for Manufacturing in a Danish Context[1]

Christian Clausen and Christian Koch

INTRODUCTION

This article explores the social shaping of IT, emphasising the identification of the dynamics of spaces and occasions, where potential outcomes and risks can be analysed, addressed and politicised. In contrast to mainstream perceptions in policy and management, which tend to take technology for granted, the social shaping perspective views technological change as the outcome of social processes of negotiation through a complicated and heterogeneous network of diverse players. Social shaping studies demonstrate the choice and decisions concerning features of the technology as a global process, spanning a range of occasions and spaces (Clausen & Williams 1997). Our contention is that technology policy, including management of technology, need to be informed by an understanding of occasions and spaces open for negotiations on technology. A comprehension of technological choice as being social is not enough, we also need to understand how, where and when and under what circumstances the choice is taking place.

The paper uses the global shaping process of information technology for manufacturing as the overarching case and discusses three illustrative examples of occasions and spaces for the social choice of IT, namely:

- Segments of IT-suppliers and their customers;
- Company internal dynamics;
- Design for mass production of software and customisation.

These examples are drawn from case studies in the Danish CISTEMA research programme: 'Centre for Interdisciplinary Studies in Technology Management'. The first case analyses the socio-technical dynamics in the space constituted by four different negotiated (market) segments of supplier-user constituencies. More or less temporary levels of closure develop and include specific occasions and potentialities for customisation, workers' participation etc. The second case analyses the space of the internal company organisation. An account of the implementation of production management systems in three different companies highlights the political nature of the decision making processes in the companies and the stabilising role of the social systems of each company. The third case discusses the mass production of ERP-software in the later nineties, emphasising the two occasions of 'design' and 'customisation', which in this case is separated in time and space. The social choices made in the design stage have a strong influence in shaping the software packages. In contrast, however, the products seem open and negotiable at the enterprise level, because of the customisation facilities, features that can have impact on the company internal dynamics. Taken together with the transfer from the US, these examples present an overarching case of the transformation of IT systems for manufacturing in a Danish context over a 30 year period. Detailed case-studies cover the period from 1990-1998.

SPACES AND OCCASIONS IN THE SOCIAL SHAPING OF TECHNOLOGY

Our theoretical departure is the social shaping of technology (SST) perspective (Williams & Edge 1996). Literature and practices concerned with technology policy and governance, including management of technology often take their departure from ideas focused on the supply of technology and linear conceptions of the flow of technical artefacts and best practice knowledge from science and technology institutions, supplier companies and consultancies. SST implies a focus on the content of technology and incorporates a broader and more heterogeneous set of players taking part in a complicated play of negotiations across the development, selection and use of technology. The development of technology is seen as the outcome of social processes of negotiation where players have different commitments, perspectives or positions in the structure. The perspective emphasises the social choices related to the development and use of technology both related to the single artefact as well as to the wider socio-technical system or technological programme (Noble 1985). Of general interest is the identification of sites and situations where such a choice can be identified together with the factors and mechanisms' influencing the shaping processes (Williams & Edge 1996).

In this paper, we seek to chart the complexity of the social shaping process, characterised by different options for negotiations and different access by actors, through a focus on occasions and spaces, the 'where?' and 'when?' issues of the shaping processes. A space for shaping implies a social context, where socio-technical ensembles can be addressed and politicised. Some actors may be included in the space, leaving others excluded. Social as well as geographical distance thus characterises a space for shaping. A range of studies in the social shaping tradition have presented such spaces: Research laboratories, development departments and governmental technology promotion programmes as well as institutions for exchanging knowledge and best practice solutions, spaces established through supplier-user interaction and collaboration and spaces inside the company. It follows from this list that 'space' is mainly a result of social processes and not distinctly geographical. The shaping is however not a continual process comprising a constant level of negotiation intensity. Occasions or episodes of specific negotiations can be identified. An occasion is thus a time framed set of negotiations shaping technology. In the discussion we use spaces as ordering before occasions.

As SST reflects a rather broad church of technology studies, differences are found regarding the interpretation of the role of actors and existing structures, the conception of the technical and the social and the influence of discursive structures versus materiality etc. Accordingly, views on the malleability versus the obduracy of socio-technical ensembles vary between different approaches, whether they emphasise the reproduction of social systems and condensation of work practices, or point at changing interests and the construction of new actor positions (Clausen & Williams 1997).

Analyses taking their departure in supply side offerings and articulations may see potential choices varying over the history of a technology, unfolding more or less path dependent trajectories of technological templates, systems and their use (Clark 1997). Such approaches seemingly make good sense as the technology we are studying in several cases was strongly promoted by dominant suppliers. They may point at different occasions where technology can be interpreted or reinterpreted in different ways and a former 'closure' can be opened up. From this background we may expect such occasions to occur with changing contingencies and interests interacting with moments of increased 'interpretative flexibility'. But, linear models mapping technology out from supply side offerings seem unhelpful in analysing IT developments which often have a diversity of origins. Instead, social constructivists (Bijker 1995) devote special attention to the diversity of actor interpretations of the meaning and content of technology and strategies and the creation of actor positions as parallel processes to the shaping of artefacts and knowledge claims. Technology strategies are analysed as parts of transformations or translations of interests taking place as simultaneous processes with the co-development of technology and organisation in actor network building processes (Callon 1986).

Taking their departure in social studies of science, social constructivist approaches often emphasise the development of single artefacts or single socio-technical ensembles as a result of the closure of controversies. These analyses risk focusing too narrowly around technology and the presumption of closure as the main criteria of success for technological development (Sørensen & Levold 1992). Instead of focusing on the heroic engineer and attempts to mobilise actors in order to stabilise a single socio-technical network, Sørensen & Levold (1992) point at the complexity of socio-technological change including examples of permanent fluidity of technology. This, along with difficulties related to the transfer of technology across different institutional contexts, makes Sørensen and Levold (1992) point to the importance of studying the institutional embeddedness of technological practice. Similarly, Williams (1997) describes technology implementation as an interaction between two different socio-technical networks. On the one hand technology often develops around a template of work practices from the context of their earlier development and use, which form the basis for supplier offerings. The user organisation, on the other hand, often proves remarkably stable, where methods of operation becomes entrenched in working and management practices and in the cultures of the organisation. The potential choices vary as technology is moved across social settings and barriers of 'colliding' institutions of suppliers and user companies (Williams 1997).

The organisational side of our social shaping analysis is inspired by technology studies from industrial and organisational sociology. Studies have pointed at the importance of the company as an organisation and a social system where subgroups in management and alliances between these and groups of employees form social spaces and play important roles in the choice of management strategy and use of technology (Child 1972, McLoughlin & Clark 1993). Continuing this emphasis on the role of the different positions and players at company level, the concept of the company social constitution (Hildebrandt & Seltz 1989, Koch 1997, Clausen 1997) underlines the historically developed relations of conflict and consensus and their implications for different actors' interpretations of problems and potentialities related to the development of work and technology. Occasions for choice of technology and organisation are here related to mechanisms of 'freezing' respectively 'un-freezing' the company political agenda (Clausen & Olsén 2000).

Summing up, we expect to find varying negotiability and stability in specific spaces and at specific occasions through the transformation of IT. The transformation process as a whole cuts, as a historical process, across a number of these spaces with their different characteristics.

CASE-STUDIES: TRANSFORMATION OF I.T. FOR MANUFACTURING IN A DANISH CONTEXT

As our empirical case of the global shaping process we analyse the transformation of IT for manufacturing from early accounting systems and material requirement planning systems (MRP) to the Enterprise Resource Planning systems (ERP) of today. (As a shorthand we often use Computer Aided Production Management, CAPM as a general term for these IT-systems). Within the overarching case we present three sub-cases: the user producer segments, the company internal dynamics and the mass-production of software. The cases are presented here as illustrative examples of what a social shaping perspective can produce as insight into technological change processes. The presentations build on several research projects (Koch 1994 and 1997, Clausen & Lorentzen 1992, Clausen 1997). These projects have approached technological development covering several perspectives and main actors i.e. the management and employee perspective, the supplier perspective, the professional associations etc. Within these projects in-depth case studies have been carried out involving a range of interviews, participative observation in workshops and action research in interaction with a row of enterprises, software suppliers, professional associations and expert observers of the area. Together, they cover the period from 1986-1998. The empirical work enabled the identification of the three spaces and the related occasions. The broad fieldwork done tells us that the identified spaces and occasions of technological development processes we discuss below represent central and typical examples from the Danish context in the time span discussed. Moreover the social dynamics described and analysed build on in-depth studies of each type of space. Although we refrain from discussing alternative perspectives on the technological development process here, we contend that the methods adopted are instrumental in highlighting the negotiability of technological change as a central social dynamic for the management of technology. This does not exclude other perspectives, such as the learning perspective from being fruitful for another types of argument and results.

INTERNATIONAL TRANSFER

The transformation of accounting systems and material requirement planning systems (MRP) to Enterprise Resource Planning systems (ERP) represents a technological development process covering around 30 years. In a number of phases (situations) a central dynamic has been based on a vision developed in the US and incorporated in an IT-system, becoming the template for a technology transfer process into Danish companies. The possibilities for shaping and reshaping of these CAPM systems into the local social context, and hence the strategic options available can be related to mechanisms of flexibility versus stability of socio-technical ensembles. The CAPM-development process thus shows both contexts of stable developments as well as contexts offering increased negotiability.

The development of accounting systems and production and inventory control systems in the sixties has often been regarded as a specific product of the US-American manufacturing system (Clark 1997). Accordingly, the systems were patterned by contextual elements of American defence and industrial policy as well as specific features of the historically developed manufacturing system. Some of these elements include powerful visions of control, formalisation in planning, co-ordination and articulation of best practice manufacturing. This practice and articulation was sustained and driven by state supported interfaces between universities and larger centralised corporations with steep organisational hierarchies and a broader techno-structure including layers of academic trained professional groups sponsored by hardware suppliers (IBM) and major accountancy firms. This placed the US system in a leading position in the 1960s for providing Western firms with templates of best practice. Specific templates of software packages and best practice were strongly promoted to Japanese and European companies by US hardware suppliers with international consultancy firms and professional associations in a supportive role.

In Denmark, national actors like professional associations, governmental bodies, universities and employers federations promoted these visions and templates along with US IT suppliers (especially IBM). Until the end of the 1980s this created a rather unified supplier driven technological push towards user companies. This was the case, even if the Scandinavian countries in this period adopted the US visions in a less wholesale manner than in the UK (Williams 1997). In countries like Denmark and Sweden professional associations or similar bodies are less dominated by one particular professional group and suppliers and consultants play a less dominant role (Swan 1997). Here, according to Swan, a wider variety of practices may be more commonly discussed, reflecting the broader industrial culture. Visits to user companies may involve a wider array of actors, including e.g. labour representatives.

Similarly the formulation of the Manufacturing Resource Planning (MRPII) concept (Wight 1981) in the USA acted as a strong vision of total control over material flows within manufacturing (Webster 1990). This worked as a template for European companies and professional communities and impacted heavily on European manufacturing in the 1980s (Hildebrandt & Seltz 1989, Clark 1997, Williams 1997). The following Danish segment and company internal cases are all examples of Danish interpretations of MRPII. Finally the IT systems in focus in the software mass production case are Enterprise Resource Planning- systems (ERP supersite 1998, Davenport 1998). ERP-systems represent a merger of three visions of controlling a manufacturing enterprise:

1. Economic vision: The enterprise as financial entity with economic flows.
2. Logistics vision: The enterprise as material flow.
3. Information vision: The enterprise as information system and flow.

Visions related to other departmental and functional elements are also included in the systems, though these three dominate. Related to each of these visions are certain major discourses or technologies. Within the logistic vision the ERP software is offering a model for how to realise the full control system. It offers an interpretation of the main problems of manufacturing and of the tools and procedures needed to solve these problems. Thus supply chain management and more basically, the very ordering of product data in so-called 'indented bill of materials' and describing the production process for each product and sub product in routings are examples of central elements (which are to some extent still in line with the previous MRPII-vision).

In parallel to the material vision, the economic vision is related to discourses of accounting. Internal controlling of cost centres, activity based costing etc. are examples of this. Finally the information system vision is related to discourses on client/server systems, relational databases, object oriented programming, the NT-platform etc. Moreover, within each vision, templates and artefacts were developed over a long historical period. The basic accounting principles thus stem from the renaissance. Information technology is introduced into economy and logistic forming templates and artefacts of production planning systems and accounting systems.

CREATION OF A DIVERSITY OF SYSTEMS AND PLAYERS

The first illustration of local social shaping opportunities takes its departure where the monolithic, IBM-dominated MRPI push of the late 1980s changes into a more varied picture, showing a diversity of suppliers and systems. It is common for this period to describe how Danish companies with custom or small batch production faced serious problems in their attempts to utilise the US templates (Nielsen, Møller and Koch 1990). Stories flourished about the co-existence of a formal (edp) and informal planning reality and the use of 'surrogate data' to meet the needs of the system. These kinds of problems together with the new PC-technology and the development of new software tools formed a basis for many smaller national suppliers and system developers (Koch 1994). In this way, the earlier very stable US based technological situation was restructured into a number of new socio-technical ensembles. Some of the new suppliers and their systems were tailored to the situation of the Danish industrial companies. The functionalities of these systems were aimed at coping with a more varied program of products, handled through customisation of variants of the bill of materials program. Generally speaking, the new smaller software houses were established on the basis of entrepreneurship, even if a few of these managed to receive public funding to develop the first version of their system.

Incremental Development

The development of these software systems is to be understood as an incremental process. This technology is not developed overnight but is gradually shaped through a cyclic process in which one version, reflecting new visions of possible features, meets reality in the customer enterprises (see figure 8.1) This results in a new shaping and new innovation of elements that are integrated, or may be integrated, into the technology. This process gives rise to new learning, and a diversity of experiences that create new demands and in turn new visions in the supplier organisation. This cyclic interaction between supplier/developer and customers is clearly influenced by other social actors as well as biased in itself; the supplier organisation does not consider all learning and experience from the customers as relevant. Some experience never leaves the customer's organisation.

This observation is in line with several other studies mainly from a UK context (Webster 1990, Clark, Bennet et al. 1992, Clark and Newell 1993). These support the argument that technology is not developed through a clear-cut rationalistic process of innovation, diffusion and adaptation. The diversity of experiences connected with one or more versions of the technology opens up opportunities for other developers to emerge. If a developer/supplier is too lazy or too wealthy, others might be willing to take the risk. The development of shop floor scheduling systems is an example of such a process. In a period the dominant CAPM systems were mainly MRP systems (material requirement planning). At the same time the scheduling task in a large group of companies was getting more and more complicated since order based and customer oriented production was emerging. Thus, there was a need for IT tools to support the 'fine scheduling'. This task was taken up by a number of smaller developers and was even supported by some R&D establishments.

Figure 8.1 Incremental Development Inside a Segment

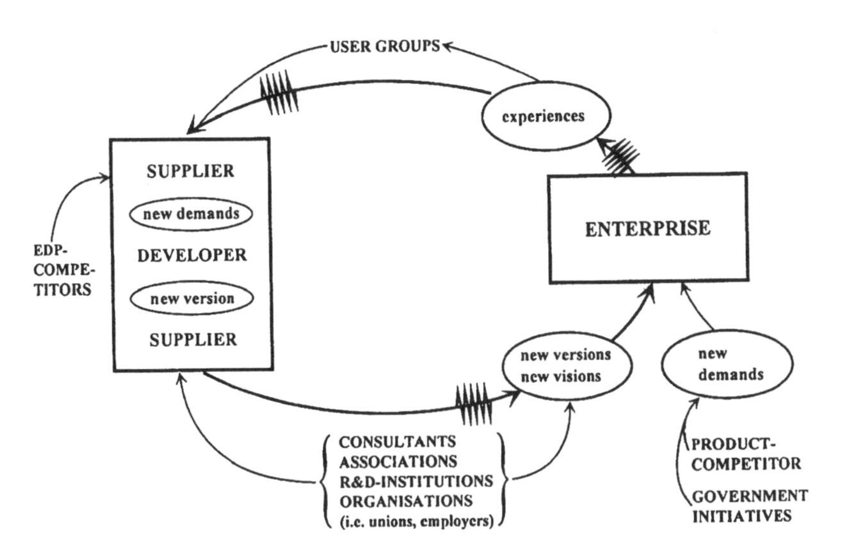

The CAPM market segments in Denmark

We have used the notion of segment to describe the frame for an incremental development. 'Segment' reflects the notion that developers/suppliers and their customers tend to build up an interdependency. A phenomenon that was prevalent in the early nineties. The customer has chosen the supplier's system and is therefore forced to follow the supplier in a period dependent on the 'weight' of the social and financial investment. On the other hand the developer/supplier often has about 50% of his turnover and organisational resources locked up with maintenance tasks for his customers (Koch 1994). Furthermore there is the tendency for enterprises to stick to a supplier even when the CAPM or other systems are completely replaced. It is necessary to discuss *the distinction between developers and suppliers*. Some of the enterprises offering CAPM systems are suppliers of systems with only limited capacity for customising. This goes for some of the internationally operating software and hardware IT enterprises. Others are operating as more or less independent enterprises connected with a developer; one developer in a network with suppliers etc. If an enterprise is linked to a supplier without development capacity it is less likely that customising goes as far as needed by the enterprise.

EXAMPLES OF SEGMENTS ON THE DANISH MARKET

As examples of segments on the Danish market the following will be briefly sketched and discussed:

- the IBM segment
- the SAP segment
- the PC segment
- the self development segment

The *IBM segment* was dominant on the Danish market. It was characterised by long term stability; generations of computers as well as CAPM software have been released. (A new version of MAPICS has been introduced almost each year since 1978). In Denmark IBM was mainly a hardware supplier in close alliance with software producing business partners. IBM/ Denmark in itself did not produce software for the CAPM market. The customers were mainly larger enterprises with the policy; 'We buy IBM'.

In a period in the late 1980s IBM was losing customers in the CAPM field due to its inactivity in software development. IBM was not responsive to enterprise needs and their products seemed 'frozen'. Since the beginning of the 1990s, however, IBM has followed the strategy of 'extended enterprise', meaning that software and hardware sales should be in balance. In Denmark this was realised by close alliances with 'business partners' – Danish software houses designing software for IBM hardware. As a result, the IBM segment was covered by a full range of computers, a range of CAPM software (fourteen different products, Koch 1994) and various suppliers. Hence *the segment became more open for social choice.*

The *SAP segment* is a result of IBM's passivity in the late 1980s. SAP is an internationally operating German software house. SAP/Denmark is a supplier without development capacity with customers amongst the 200 largest Danish enterprises. The SAP systems (R/2 and R/3) are standard systems with only small customising possibilities but with imposing features. The explicit strategy of SAP/Denmark is to find customers that have a 90% fit with the system characteristics. Furthermore SAP/Denmark tries to avoid designing specific software. Nevertheless SAP has been gaining market share and was able to double its R/3 customers in 1993 and sales have been rapidly growing since then. Hence, this segment is characterised by a rather closed situation, with limited options and choice.

The *PC segment* evolved very quickly in the 1990s. The main software house in this segment was Damgaard Data, an up till now independent developer operating with a network of suppliers. The product, Concorde XAL, is characterised by being inexpensive considering its features. Furthermore, customising was supported by development tools integrated into the software. The suppliers of this product were very different ranging from 'over the counter' sellers to large suppliers/developers who had large enterprises as customers. Some of the larger suppliers undertook their own development of supplementary modules. At least one supplier followed a *strategy of customisation*; The development tools were followed up by education, training and organisational measures. This opened up scope for a more detailed shaping of the CAPM. However, the PC segment in general was (and is) heterogeneous on this point.

The *self development* segment was the most heterogeneous in its character. This segment is often underestimated, but according to Christensen and Clausen (1992) it amounted to 33% of the Danish CAPM market. Some software suppliers offered development tools for this purpose (4th generation tools). Since the development was internal and started from scratch it can be *presumed to have been very open for social choice*. However internal development is governed by internal social forces; the company social constitution (see below) works as a filter for this development as well.

As outlined above, these segments differ in their stability and competition between segments has become greater - for example some enterprises have 'downsized' (or 'right sized') their hardware from mainframes to PCs or workstations. This opened up opportunities for more freedom of choice mostly due to the differences in suppliers' strategies since there was an extensive correlation in both the underlying visions of CAPM and in features of CAPM across segments (Koch 1994). This basically segmented *market* for CAPM was and is spanned by various other social actors and their spheres of influence. Examples of such constituencies are shown in figure 8.2.

Some of these constituencies are dominated by a single central vision, others are formed in order to promote a profession and others are promoting broad discussion amongst social actors. These constituencies involve numerous enterprises, represented by employees and/or managers. Each of them has a more or less evident impact on how production is organised or is developing answers to new problems arising in industry. For example there is a rather close link between the emergence of logistics managers and departments in enterprises and the 'production' of graduates in logistics from the business schools. The need for logistics originates in the need for systemic rationalisation (Hildebrandt 1990, Diess and Hirsch Kreinsen 1992).

Figure 8.2 Constituency Characteristics

Environment	Characteristics
The Japanese Production Principles-environment	Consulting institution Promotes J.P.P.
The logistics-profession	R & D institutions promotes a profession and a vision
The logistic association for engineers (bachelors)	Broad discussion centred in the organisation for engineers
User groups related to CAPM	Feedback to developers
The 'Technique and Data'-Environment	Monthly magazine Yearly exhibition with seminars
'Danish Industry'	Employers federation, heavy impact on R&D-programmes

DISCUSSION: SEGMENTS DETERMINING SOCIAL CHOICE?

As outlined above, it seems evident that the segments around suppliers constitute at least some level of closure for social choice. In some cases this closure is evident. In brief it is a question of whether the suppliers consider *customising as a surprise or customising as a strategy*. But it is more important to discuss cases where preconditions for open choice are good, but turn out to have only limited value: First, when focusing on suppliers which have a strategy for customising it might turn out that company internal forces actually 'constrain' the choice. Second, the internal development of adapted programs is only to a certain extent supportable by suppliers since their own development is costly and in competition with the internal development. Hence the social alliance between the supplier and internal actors will often be of importance. Third, the supplier might not support the necessary organisational changes.

These dimensions underline the contextual features of social choice; the segments are only self-determining to a certain extent. Other external social actors might intervene as well as the internal. In general, the opportunities for social shaping of IT systems in the space constituted through supplier-user segments are primarily related to *occasions* for social construction or reconstruction of the segment as well as changes in supplier or user strategies in the established segment. The most important occasions seem related to situations of: a) new customers entering the segment, b) suppliers changing strategies of customisation, c) changes in user needs and circumstances, d) changes in supplier-user interplay through adoption of new development tools or related to mechanisms of feed-back and learning and e) influence from contingencies as well as discourses of the surrounding constituency. Due to the contextual and heterogeneous nature of the segments, these occasions are often interrelated. Other occasions might lead to the destruction of the segment, such as corporate take over of the supplier, or the entering of competing suppliers.

COMPANY INTERNAL DYNAMICS

As indicated above, the possibilities for shaping and customisation of technology depend to a high degree on the context and interplay between supplier and user firms in the different segments. But, to which extent are these shaping possibilities used productively during adoption and implementation within a company? In this section we will discuss three examples drawing on the concept of 'the company social constitution' (Hildebrandt & Seltz 1989). This concept describes the historically developed concerted norms, rules and principles in the company which influence employees' behaviour, motivation and attitude. It emphasises the role of actors within the company while still recognising the structural patterns of corporate hierarchies. A company's social constitution develops through an historical process of conflict and conflict-solving activities responding to economic and market situations and technological possibilities. At the same time these norms shape the way problems and various solutions (the political options) relating to the production process and its technology are conceived. By conceiving the social constitution as 'frozen politics' of the company it becomes possible to explain periods of conservatism towards technological change combined with periods of debate and transformation.

All three case companies discussed here were facing new market situations and management saw broadly similar new demands for shorter lead times and customisation of production. At the same time the younger generation of workers were expected to have higher expectations towards the quality of work. The old traditional Tayloristic forms of rationalisation were seen as economically and morally 'worn out' by management as well as by shop stewards. On this basis there was a growing understanding among management of the need to mobilise 'human resources' in production. This new interpretation was developed into management policy through an earlier replacement of the production management.

The shop stewards experienced through this renewal of work policy an increased recognition of their importance for the development of the company and they were normally consulted before any change in production. A new union strategy was developing at the company level, which addressed new developments in management policy. This strategy was building upon and developing a mutual relation of trust between management and employees. The majority of shop stewards saw themselves as mediators of this relation of trust, guaranteeing the readiness to accept changes on behalf of the employees.

The change in work policies also became visible in work roles at the shop floor level. More diverse tasks were assigned to operators including scheduling and quality control. Higher responsibility for output within the workers or the work group and less direct surveillance and control from the foremen were part of a new compromise between management and workers. But in spite of many common trends in work policies and work norms and rules in the three companies we observed severe historical, structural and contextual differences between the companies. Differences in products and production processes, the work force and their culture and the company organisation had constituted differences in terms of the company's social constitution. Company sizes were small ranging from 250 to 500 employees (at plant level).

The machine builder

This was an independent company expecting high growth rates producing very sophisticated and high value machines mainly to export markets. Many skilled workers, a very low labour turnover and a long tradition of co-operation and mutual trust indicated the development of a stable company social constitution. On the other hand, management aimed at improving flexibility over working hours and job demarcations. Pressure was put on shop stewards to accept these new arrangements by establishing a new experimental CIM-factory on a green field site. Shop stewards eventually agreed this new arrangement including a shift in wage system.

Production was organised in three shops with three planning systems ranging from part production in large batches to one-off production in the final assembly. A traditional MRP-system operating centrally broke down orders into jobs for every machine/work station. The establishment of the CIM-factory was pursued by management during a lot of talk of visions of the fully automated factory, and plans were developed to implement IT technology at all levels of the organisation. After the implementation, ideas of the automated factory were revised in conjunction with problems of low machine utilisation and management eventually announced publicly a moratorium for the introduction of IT. The vision of CIM was substituted by a focus on organisation and utilisation of existing IT.

In practice many small changes and implementations of software were carried out as steps towards CIM. Shop floor terminals were linked to the MRP system and to the CAD/CAM system enabling shop floor work scheduling and real time data collection. Skilled workers developed some of the solutions. Management had a policy of 'letting some of the workers play with technology' and trying out different working solutions. Workers became responsible for detailed capacity planning and gained access to job orders 3-10 days in advance. In practice delivery schedules generated from the central planning department limited local autonomy. Instead workers experienced the planning system as insatiable and often worked overtime to adapt to planning requirements. The upper limit for 'a fair day's work' had become unclear.

The cable plant

The cable plant was part of an old traditional Danish company. Having produced electrical cables to the home market for many years in a protected market situation a hierarchical rule based and bureaucratic organisation had developed. Semi-skilled workers (men and women) dominated the workforce and a traditional Tayloristic management policy had accepted low educational levels. Due to the establishment of the European single market and tendencies towards deregulation, the company increased its efforts to develop a competitive position in export markets. This meant production of relatively small batches with short delivery times and increased quality management. In order to increase organisational performance, an ambitious educational project was launched aiming at improving job qualifications and attitudes towards work among the entire workforce. The majority of shop stewards were heavily engaged in this educational project which was seen as an opportunity for increasing job content. On the other hand some of the older workers and a few shop stewards were afraid they would not be able to meet the new demands.

The cable plant was in the middle of a process of implementing a so-called CIM-factory, where all processes should be controlled by computers. A first step was taken with the construction of a new factory based upon the new control concept. In the older factory the CIM system supplemented the old control systems. The CIM system provided features for down loading recipes for process control, operator access to work schedules and a real time data collection system concerning all relevant variables of the production process.

The work scheduling system was perceived as deterministic and centralist by local plant management as well as shop stewards. The design of the system was carried out by technicians in collaboration with the supplier. Operators and maintenance workers were involved in the design of displays and had improved the user-friendliness but had no influence on basic system features. The CIM system meant an increased level of automation based on a still growing control of the process at the level of the technical departments. Local autonomy at the shop-floor level was maintained in principle as operators maintained access to the control of process parameters. But work autonomy was combined with increased demands for responsible operator performances based on close management supervision.

The cooker factory

This case represented a smaller factory mainly supplying domestic markets. Being part of a large multinational company, important competition parameters as production costs, work in progress, lead times, quality parameters and so on were closely supervised and formed the basis for allocating resources between the different factories. A relatively simple production process was based on semiskilled workers and Tayloristic principles of work separation and repetition. Many workers were engaged in activities improving quality, work place design and the like through a comprehensive network of bipartite committees. Relations between shop stewards and production management had developed into a high degree of mutual trust.

Production scheduling was based on pure Just In Time principles and used a Kan-Ban card system. The Kan-Ban system was introduced and maintained by an 'enthusiastic and charismatic' manager. He saw the changing of workers' attitudes towards production as more important than introducing new technology and emphasised the simplicity and transparency of the production system. These qualities made it possible to delegate responsibility for the keeping of simple scheduling rules to the workers. This improved management supervision of production and the identification of potential bottle necks in the production system as a whole.

Shop stewards had a collaborative attitude towards the Kan-Ban system and saw possibilities for reducing direct supervision by foremen and thereby increasing work autonomy. A piece rate wage system was abandoned and semi autonomous work groups were established in combination with the withdrawal of the foremen level. But the new work autonomy was combined with increased centralised control. Through a new control system management collected data from the different work stations monitoring the progression of work orders. The production manager gathered supplementary information by walking around counting single pieces of work in progress. CIM visions were present in the mind of the production manager but his strategy implied a development in small steps where changes in the social system should precede changes in automation or IT-based solutions.

DISCUSSION: FREEZING AND UN-FREEZING POLITICAL AGENDAS

These three examples show how the implementation of CAPM is intertwined with decision making processes and politics in the company. CAPM systems are adapted to meet some of the needs and conditions of the companies. However the feedback processes informing suppliers' attempts to adapt CAPM concepts are complicated and imperfect. Thus, the introduction of new technologies and more fundamental organisational change proceeds slowly with numerous reversals. The (re-) shaping of technology is strongly patterned by the organisational strategies pursued by different actors in and outside the companies. Management strategies reflect the historical characteristics of the specific organisational 'filter' of the 'company social constitution' affecting interpretations of conditions in the outside world. Thus attention should be paid towards the specific interpretation of internal and external factors as an important element of the social shaping process. But, in the above described cases where management policy includes an open attitude towards discussions with shop stewards, CAPM and CIM belong to an area of 'hidden politics'. Not only management, but also shop stewards have a certain reservation towards bringing up subjects of design and introduction of CAPM/CIM.

The study of company level dynamics shows, that several strategies of centralisation and decentralisation are pursued simultaneously. On the one hand a potential control instrument for management is created at different levels that affects the balance of power between management and shop floor personnel. On the other hand decentralisation of responsibility for achieving quality, adaptation of NC-programs and local work scheduling point towards development of skills and control at the shop floor.

The Danish companies discussed here were not typical but were in the forefront in the modernisation of work practices and human resource management. However their size, the inclination to compromise, the informal approach of management and shop stewards can be observed in many Danish companies. Here, the 'company social constitution' points at the conservatism of traditional societal control structures and the struggle between old and new modes of control. The development of a new compromise is still being contested. Here we emphasise the social mechanisms for opening and closing areas for politics of work and hence for the social shaping of CAPM/CIM at the company level. The outcome of these political processes should not be taken for granted, as the very meaning of CAPM/CIM concepts vary between actors. Still, the non-deterministic character of CAPM/CIM concepts concerning basic features, skill-requirements, implications for work, and their implementation in small steps open up a considerable range of choices at company level.

Most important *occasions* for social shaping through company internal dynamics are related to situations where 'frozen politics are un-frozen': a) changes in management control strategies b) changes in worker roles and attitudes and shop steward strategies c) development of new management-worker relations and compromises, d) mechanisms of changing the interpretative filter concerned with company responses to the outside world (supplier offerings, product market trends, labour market trends) and e) influences through constituencies and broader societal discourses. Again these occasions are strongly related.

DESIGN FOR MASS PRODUCTION OF SOFTWARE AND CUSTOMISATION

Where the first two cases reflect upon empirical findings from the early 1990s, the next case focuses on the last part of the decade. From 1994 the ERP-market has changed significantly. A few ERP-suppliers gained market share by almost mass marketing of their systems; SAP, Baan and Intentia are all examples of this. All three suppliers grew around 50% a year, in contrast to the general growth of the IT-sector at around 8%. The general growth of the market did not necessarily enable smaller software-houses to survive (Steinmark 1996). Many suppliers still operate on the market, but roughly 80% of the sales are made by 20% of the suppliers.

In Denmark, five main suppliers covered most of the manufacturing companies: SAP, Baan, Intentia, Damgaard and Navision Software. The three first are international players penetrating the 'market' from 'above' (starting with large corporations), whereas Damgaard and Navision Software were local Danish players in the PC-segment up to 1994, developing from below. Both the PC-companies were characterised by a large network of resellers. From then on these networks internationalised and they are now global players.

The social shaping process now follows different patterns to the segment case discussed above. When Damgaard and SAP decided in the early nineties to renew their system, they both formed internal design teams with organised co-operation with a grouping of interested customer enterprises and other players such as management consultants. These companies were partners in the development of the R/3 and the Concorde manufacturing control modules. As partners and testers of the software a few companies were thus given influence on the software; giving them strategic advantage in exchange for their skills and competencies within the needed areas of software.

From 1994 on however, the generic software was offered to customers, who basically had to accept the main layout and content of the ERP-software. The basic design phase was over and the package was sold as a commodity. The construction of a market dynamic allows the supplier to gain capital to initiate their own relatively independent design and development processes, in contrast to the segment situation discussed above. And the manufacturing enterprise is construed into passive buyers of systems.

This could lead to a halt of the company level dynamics, if the systems had the same features as previously. The systems are however significantly more flexible in their basic design. Where the segments were characterised by processes of adding features as a joint process, the mass produced systems like SAP R/3 are characterised by '200%' facilities that the customers need to reduce to the requisite size and features.

In principle, this opens up scope for renewed company internal development: when choosing main and sub modules, setting the parameters of the system etc. The company actors can develop coalitions around a *customised* version of the ERP-package. By taking in the sales module the coalition might recruit the sales department etc. And the end-users can in principle be incorporated in the design of user-profiles, screens etc.

The extensive customisation possibilities are however not always used. The large implementation task and later upgrading of the system, lead to a much stiffer situation – one can label it the *power of default*. The basic settings of the system parameters offered by the supplier can easily become the main decision parameters for the enterprise actors. Furthermore, enterprise actors frequently refrain from programming additional features in order to cope with the upgrading. Finally, some modules and sub-modules are chained, meaning that if the company wants a full accounting system and or a full logistic control system it implies certain choices of module clusters. What should be extremely flexible is thus often not so very flexible. Moreover flexibility is dependent on organisational IT-resources within the manufacturing enterprises as well as externally. These resources cannot, while implementing the next versions of the ERP-system, be used for local appropriations.

The most important *occasions* for the social shaping of IT related to the space of the new mass produced standard software can be related to: a) early design of basic features through extensive producer-user co-operation, and b) later customisation of the offered standard software package to user requirements.

CONCLUDING DEBATE

The analysis of different moments of transformation of IT-systems for manufacturing systems from the general and powerful visions in the US context to the implementation in Danish companies has revealed important social shaping processes. Attention has been drawn to the choices which underpin the development processes of the socio-technical ensemble of CAPM or ERP systems. Occasions and spaces for shaping and reshaping technology were identified. These relate to the transfer from one social context to another, to the supplier-user interplay and to company internal processes in its implementation and use. These findings draw attention to the complex socio-technical reality of IT development and provide a strong argument against ideas of technological determinism and the linear dissemination of best practice solutions from supply to use. Instead, the spaces for social shaping of technology comprise a dynamic web of strategic opportunities, where technological choice alternates between occasions of stability and flexibility and situations of exclusion or inclusion of players and topics.

CAPM segments

Stability has been produced through reproduction of social systems, as has been the case in the articulation of US best practice in Denmark until the end of the 1980s and again after the restructuring of the supplier-user relationship and the development of a new set of what we call supplier-user 'segments'. Flexibility in the shaping of the socio-technical system is related to different actor-interpretations and their micro political strategies and varies with their specific social and institutional embeddedness or perceived contingencies.

The reshuffling of the user producer segments was enabled by major occasions such as the emergence and spread of PCs, new software tools as well as changing business conditions and related planning problems in small Danish companies. These events enable the establishment of new players in the shape of many smaller national suppliers and system developers offering new interpretations of relevant technology strategies. In this way occasions for the social shaping of IT has occurred where the former very stable US based technological situation was restructured or transformed into a number of new socio-technical ensembles. At least some of the new suppliers and their systems became tailored to the situation of the Danish industrial companies.

The opportunities for social shaping of IT systems *in* the space constituted through supplier-user segments are primarily related to two main *occasions.* First, when a customer company enters the segment, second through the continual realignment of supplier and user strategies in the established segment.

Company dynamics

The case study concerned with the company internal dynamics illustrates that suppliers' technological offerings often seem quite remote from the viewpoint of some of the actors involved in local political processes of organisational development. Our company examples show a common awareness of organisational development which parallels or precedes technological implementation. Though our examples are selected from a group of companies exercising high levels of mutual trust between workers and management, they indicate a broader experience: that what is taken up at company level is often more depending on management and workers' attitudes towards planning than is reflected in supplier strategies and offerings.

We have identified stable situations portrayed through the concept of the 'company social constitution' implying the existence of established compromises between diverse perspectives and interests. Technological stability or periods of conservatism towards technological change were described as situations where the company social constitution could be seen as 'frozen politics' of the company. These situations imply strong reservations among management and workers towards bringing up subjects concerning design and introduction of CAPM systems. Strategies concerned with these technologies are very seldom subject to an open discussion across management-labour or professional demarcations of the user company. Hence, the different perspectives of diverse groups are not taken into account and the interpretative filter of a dominant coalition will prevail.

Technological flexibility and potential renewal can according to the company level case studies be related to *occasions* where either new market challenges, strong supplier articulations of new technologies or changes in workers attitude promote a management change in politics. In such situations, an opening is possible for establishing new learning processes, new interpretations of problems and solutions and a subsequent formation of new alliances and coalitions in technology management processes. Still, CAPM design debates often belong to areas that are closed for a broader range of company internal actors.

New standard software based flexibility?

Technology 'in itself', understood as the artefacts consisting of hardware and software, does not seem to play a decisive role in the hardening of technology and work practices related to the modern ERP systems. After the initial design phase the systems are packaged commodities. The early openness in design of the basic facilities is thus converted into considerable stability in these facilities. The systems are quite flexible however in the sense that modules can be chosen, configuration choices can be taken and made to work according to local user needs and policies during customisation. These two *occasions*, design and customisation, are distinct and separated in time and space. The hardness of this kind of technology in the customisation occasion has to be identified in the single context. Here the importance of the supplier articulation of best practice versus the resources and politics applied in the single user company seems to play a decisive role. But, if one considers the short time horizons of operation of every generation of systems, users can expect difficulties in keeping up their skills in order to match supplier knowledge and experience. In this sense, the result can easily end up in a situation of suppliers pushing global solutions against the need of local user situations like under the early US/IBM dominance.

The main results regarding the three spaces and the occasions are summarised below (see figure 8.3). The *occasions* related to these spaces develop over time and relate to the actors' strategies and relations as well as market dynamics. In the context of the segment the rupture of established segment relations create a relatively open situation which transforms into a continual stream of smaller occasions for shaping. This is grounded in the power constellation of relatively evenly-balanced players within the segment. The continual stream of local supply options, however, are not available for all players within the participating organisations. As discussed in the case of company internal dynamics, the stability of actor relations can be considerable, leading to exclusion mechanisms and episodes of 'negotiation away', i.e. removing change options from the internal agenda. In the mass produced segment the early innovative phase is then transformed into the relatively diminished openness of pre-programmed customisation options due to supplier hegemony. Through all our cases, the character of the occasions are thus restricted and enabled by the strategies of the participating players.

Figure 8.3 Stability and negotiability of socio-technology in different spaces

Feature:	Stability of actor relations	Negotiability of technology
Segments	Considerable: - Customer - producer alliance	Good over long timespan: - IT-market dynamics opens - Moderate size of producer
Company internal	Heavy: - Long term compromises	Limited and company specific: - External pressure can be negotiated
Mass Production of software and Customisation	Considerable: - Supplier hegemony	Good in ruptures: - IT-market dynamics limits - New flexibility needs resources to be exploited

IMPLICATIONS FOR TECHNOLOGY MANAGEMENT AND POLICY

Our observations also point to more general implications for technology policy debate and practice, including the management of technology. First, the policy options and strategic possibilities seem to be shifting according to changing socio-technical constituencies and different contexts. CAPM or ERP are not defining one technology, neither when viewed in a historical perspective nor in a certain period of time. So, technology management should not be reduced to a question of selecting the best technology or to a question of implementation. Second, there is not one single player at the supplier level or at the user company level that had or has a natural position to design and implement CAPM or ERP systems. We have observed the involvement of a broad range of relevant players including different coalitions of top management, local management, professional or departmental groups, workers and unions involved directly or indirectly in the socio-technological change processes. Consequently, there can be no standard method for technology policy or management. The problem to be addressed, is that many of the political processes and coalition building processes hinder an open exchange of experiences and learning processes across diverse firms, groups and perspectives and, consequently, the development of a better technology. Important debates on CAPM experiences and solutions take place in different social spaces with only highly selected flows of information and hence a politically shaped and restricted learning across interests and perspectives.

This political dimension of complex socio-technical change evolving through a complex institutionalised socio-technical terrain is seldom reflected in current technology policy and ideas of the learning society.

The political dimensions can or should of course never be completely dissolved, as political processes also are of a productive nature. But, an alternative approach to technology policy should investigate the possibilities for improving learning that cuts across established barriers. Political dimensions of technology are often left to technology management strategies and limited to the perspectives of the single organisation or project. Cross cutting experiences and synergy between projects are limited to consultancy businesses, professional associations or more accidentally to knowledge transfer through mechanisms of the labour market for employees, managers, consultants or computer scientists.

The building of an overarching space or actor as a platform or institution for exchange of experiences and learning should be considered as a possibility. Such institutions could be developed on the basis of Danish experiences where professional associations are important although still imperfect spaces for exchange of experiences and learning processes and where government agencies as well as unions participate together with professionals and managers from supplier and user organisations. A long term emphasis on collective institutions like professional associations could develop these in order to promote debates across established organisational and professional boundaries creating real alternative learning spaces to potential dominant supplier companies or consultancy businesses. They could also form a learning forum for unions and employees that are involved with the new systems. Government technology policy could in this context promote a more experimental learning oriented approach by funding a more systematic collection of experiences and in this way support the development of understandings and concepts, facilitating the exchange of experiences and the social shaping of socio-technical ensembles of CAPM and ERP.

NOTES

1 This paper is a revised version of Christian Clausen and Christian Koch 1999, 'The Role of Spaces and Occasions in the Transformation of Information Technologies', *Technology Analysis and Strategic Management 11*, 3: 463-82 , published by Carfax Publishing, Taylor & Francis, http://www.tandf.co.uk. The paper is based on research carried out in the Centre for Interdisciplinary Studies in Technology Management - CISTEMA, funded by the Danish Research Councils.

BIBLIOGRAPHY

Bijker, W. E. (1995), Of Bicycles, Bakelite and Bulbs. Towards a theory of sociotechnical change, Cambridge Mass. and London: MIT Press.
Braverman, H. (1974), *Labor and Monopoly Capital*, New York: Monthly Review Press.
Callon, M. (1996), 'Some Elements of a Sociology of Translation', in: J. Law (ed.), *Power, Action and Belief: a New Sociology of Knowledge?*, London: Routledge and Kegan Paul.
Child, J. (1972) 'Organisation structure, environment and performance: the role of strategic choice', *Sociology*, **6**, pp. 1-22.
Christensen, J. & Clausen, J. (1992), *Product Innovation, Planning and Production in Danish Industry*, Lyngby: Technical University of Denmark (in Danish).
Clark (1997), 'Articulating and Transferring Best Practice: USA, Japan &UK', in C. Clausen and R. Williams (eds), *The Social Shaping of Computer-Aided Production Management and Computer-Integrated Manufacture*, COST A4 Social Sciences, Luxembourg: European Commission, pp. 63-84.
Clark, P. & Newell, S. (1993), 'Societal Embedding of Production and Inventory Control Systems: American and Japanese Influences on Adaptive Implementation in Britain', *The International Journal of Human Factors in Manufacturing*, **3**, Wiley & Sons.
Clark, P.; Bennett, D. et al (1992), 'The Decision Episode Framework and Computer-Aided Production Management (CAPM)', *International Studies of Management and & Organization*, **22**, (4),, p. 69-80. M. E. Sharpe Inc.
Clausen, C. & Koch, C. (1999), 'The Role of Spaces and Occasions in the Transformation of Information Technologies', *Technology Analysis and Strategic Management*, **11**, (3), 463-82.
Clausen, C. & Lorentzen, B. (1992), 'Har faglig ressourceopbygning en fremtid?', in: Clausen, Lorentzen and Rasmussen, (eds), *Deltagelse i teknologisk udvikling*, Copenhagen: Fremad.
Clausen, C. & Lorentzen, B. (1993), 'Workplace implications of FMS and CIM in Denmark and Sweden', *New Technology, Work And Employment,* **8**, (1), Oxford: Basil Blackwell.
Clausen, C. & Olsén, P. (2000), 'Strategic Management and the Politics of Production in the Development of Work: A Case Study in a Danish Electronic Manufacturing Plant', *Technology Analysis and Strategic Management,* **12**, (1), pp. 59-74.
Clausen, C. (1997), 'Social Shaping of CAPM/CIM and the Company Social Constitution', in C. Clausen and R. Williams (eds), *The Social Shaping of Computer-Aided Production Management and Computer-Integrated Manufacture*, COST A4 Social Sciences, Luxembourg: European Commission, pp. 171-86.
Clausen, C., & Williams, R., (eds) (1997), *The Social Shaping of Computer-Aided Manufacturing and Computer Integrated Manufacture*, COST A4 Social Sciences, Luxembourg: European Commission.
Davenport, T. (1998), 'Putting the Enterprise into the Enterprise System', *Harvard Business Review*, July-August.
Diess, M. & Hirsch Kreinsen, H. (1992), 'Structures of Technology Markets and the Development of Manufacturing and Control Systems', M. Dierkes, U.

Hoffmann (eds), *New technology at the Outset*, Frankfurt/New York: Campus Verlag.
Fleck, J. (1993), 'Configurations: Crystallizing Contingency', *The International Journal of Human Factors in Manufacturing*, **3**, Wiley & Sons.
Hildebrandt, E. & Seltz, R. (1989), *Wandel betrieblicher Sozialverfassung durch systemische Kontrolle*. Berlin: Edition Sigma.
Hildebrandt, E. (1990), 'Die betriebliche Sozialverfassung als Voraussetzung und Resultat systemischer Rationalisierung', in J. Bergstermann and R. Brandherm Böhmker (HG.) *Systemische rationaliserung als Sozialer Prozess*, Bonn: Verlag Dietz (Nachv.).
Koch, C. (1994), *The Technicians and Production Control*, Copenhagen: The Danish Union for Technicians, (in Danish).
Koch, C. (1997), 'Production Management Systems: bricks or clay in the hands of the social actors?', in Williams, R. and Clausen C. (eds), *The Social Shaping of Computer-Aided Manufacturing and Computer Integrated Manufacture*, COST A4 Social Sciences, Luxembourg: European Commission.
Manske and Wobbe (1987), 'Computerunterstützte Fertigungssteuerung in Maschinenbau', VDI Reihe 2, *Fertigungstechnik* nr. 136, Düsseldorf: VDI.
McLoughlin, I. & Clark, J. (1993), *Technological Change at Work*, Buckingham: Open University Press.
McLoughlin, I. (1998), *Creative Technological Change*, London: Routledge.
Nielsen, K. T.; Møller, K. J.; Koch, C. et al. (1990), *Japanese inspired production systems and working conditions*, Copenhagen: The Danish Technological Institute, (in Danish).
Noble, D. (1985), 'Social Choice in Machine Design: the case of automatically controlled machine tools', in MacKenzie, D. and Wajcman J. (eds), *The Social Shaping of Technology*, Philadelphia: Open University Press.
Ortmann (1990), *Computer und Macht in Organisationen*, Opladen.
Sørensen, K. and Levold, N. (1992), 'Tacit Networks, Heterogeneous Engineers, and Embodied Technology', *Science, Technology and Human Values*, **17**, 1: 13-35.
Steinmark (1996), 'Med Ryggen mod Muren', *Computerworld* DK, 33.
Swan, J. (1997), 'Professional Associations as Agencies in the Diffusion and Shaping of Computer-Aided Production Management', in C. Clausen and R. Williams (eds), *The Social Shaping of Computer-Aided Production Management and Computer-Integrated Manufacture*, COST A4 Social Sciences, Luxembourg: European Commission, pp. 85-108.
Webster, J. (1990), 'The Shaping of Software Systems in Manufacturing', Edinburgh: RCSS, University of Edinburgh.
Webster, J. (1993), 'Chicken or Egg?', *The International Journal of Human Factors in Manufacturing*, Vol. 3, Wiley & Sons.
Wight, O. W. (1981), *Manufacturing Resource Planning: MRPII - Unlocking America's Productivity Potential*, Essex, US: Oliver Wight Limited Publications.
Williams, R. (1997), 'The Social Shaping of a Failed Technology?', in C. Clausen and R. Williams (eds), *The Social Shaping of Computer-Aided Production Management and Computer-Integrated Manufacture*, COST A4 Social Sciences. Luxembourg: European Commission, pp. 109-132.
Williams, R. and Edge, D. (1996), 'The social shaping of technology', *Research Policy*, **25**, pp. 865 – 899.

9. Feminism For Profit? Public and Private Gender Politics in Multimedia[1]

Hendrik Spilker and Knut H. Sørensen

1. THE CO-PERFORMANCE OF GENDER, TECHNOLOGY, AND POLITICS

Presently, the traditional welfare state is under ideological pressure from liberalism. There is increased emphasis on individual freedom of choice, unfettered by government regulation. In this way, the market becomes the preferred regulatory institution where 'the invisible hand' is presumed to be able to cater for collective needs in a way that reduces the need for government action. For example, in the area of development of Information and Communication Technology (ICT) and the establishment of the Information Society, the prominent Bangemann report (High Level Group on the Information Society 1994) suggests that the European Union should pursue a policy that delegates as much as possible to market actors. On the other hand, most observers would agree that there is a substantial area left for government policy. From this point of view, the task of public policy is to supplement market forces and correct 'imperfections' (see, e.g., Weimer & Vining 1992).

An important challenge in this respect is women's rights and the presumed need to support equal opportunities in work and education. For a long time, there has been a widespread concern (not just amongst feminists, but quite generally) that women would not be part of the ongoing development of computer technologies. The task of getting women included on an equal footing in the Information Society has usually been placed with governments, according to the basic assumptions of so-called state feminism (Hernes 1987). State feminism represents a version of gender politics where, generally, the responsibility to act to improve the situation of women resides with the government. This view is well-established in all Scandinavian countries, and its basic tenets seem to be shared throughout the European Union as well.

There is a potential tension between state feminism and initiatives like the Bangemann report. Arguably, today we are facing a situation where the challenge of creating equal opportunities for men and women is becoming a shared concern of government and private industry. Thus, it becomes interesting and important to study how and to what extent public and private actors get involved in such goals. In this paper, we will analyse efforts to include women as users of ICT in order to characterise the underlying politics of gender. To what extent are feminist aims pursued in public and private initiatives, and what kind of social shaping processes of ICT may be observed?

For this purpose, the paper studies some recent Norwegian efforts. On the one hand, we analyse the strategy proposed by the Ministry of Education to get more girls engaged in the use of computers. On the other, we explore three private initiatives with a similar goal – of developing multimedia products aimed at girls and young women to promote their interest in and ability to explore new ICTs. The analysis will be focused on the conceptualisation of gender, technology and politics within the initiatives in question, with particular emphasis on their gender politics. A state feminist viewpoint would assume that the private sector cases are characterised by a subversion of feminist issues, displaying a basic ideology of gender neutrality and thus putting differences between men and women, power issues, etc. backstage. Public sector initiatives could be affected by the same ideology, but they may equally be objects of feminist politics. Thus, we assume that feminist politics will be more present in the government case, compared to the private sector cases.

On the other hand, as pointed out by Cassel and Jenkins (1998, p. 14 f), we need to be aware of the potential of entrepreneurial feminism to promote feminist ideas by targeting new markets consisting of women. Cassel and Jenkins base their argument on their analysis of efforts to make computer games for girls or women. If they are right, we need to be more careful in the standard assumption that state feminism will be more 'correct' or politically effective than entrepreneurial feminism. Thus, the task of this paper is rather to analyse similarities and differences between the gender politics of ICT among public and private actors and the way this shapes ICT initiatives to promote the inclusion of women in the Information Society.

In previous research on the relationship between women and computers we can distinguish three waves of concerns and conceptualisation (Spilker and Sørensen 2000). The first wave of studies of women and computers focused in particular on office automation, motivated by observations that women seemed to be excluded from the potential benefits of the new technology. A major cause was the widespread, misleading and oppressive image of female office work as low-skilled and easy to replace by computers (Lie & Rasmussen 1983, Cockburn 1985, see overviews in Probert & Wilson 1993 and Webster 1995).

The second wave, that emerged around 1980, was characterised above all by a concern with the lack of women in the design and development of computer technology and the gendered understanding of relevant computer competence. This was important given the presumption that a major cause of the oppressive effect of computers on women was due to male dominance among designers. Why then were there so few women in computer science, and what could be done to remedy the situation? One argument attributed the lack of women in computer science to the way computers had become embedded in mathematics and military interests (Hacker 1989, 1990). In Turkle's (1984) study of different computer cultures, the analysis of male computer hackers identified a style of life and work that put women off, in particular the tendency to prefer a close relationship to machines rather than people (see also Rasmussen & Håpnes 1991, Håpnes & Sørensen 1995).

We are now moving into a third wave of research on women and computers, fuelled also by changing images of the computer. ICT is no longer primarily about programming, systems, control and calculation. Increasingly, computers represent a gateway to communication and cultural activities, and they have become the backbone of multimedia technologies, of which the internet is the most prominent. Potentially, this may make the technology interesting to a much wider variety of users, women as well as men.

The main catchword in third-wave analysis of gender is 'cyborg', originally applied by Donna Haraway (1991) to conceptualise the strained, transformative, ambiguous, and ambivalent relationship between nature and technology, human and non-human, female and male. Typical of third wave analysis is also the critique of essentialist arguments about the meaning of gender, and the need to explore categories like male and female as dynamic phenomena.

To sum up, the first wave looked at women as victimised users of technology, critiquing gender bias and continuing efforts to suppress female workers. The second wave looked for ways to create space for women as designers, as well as users of computers, to identify strategies of reform. The third wave is an effort to focus on the performance of gender as well as computers. We will analyse the case studies to see if the thinking behind them may be related to these three waves of conceptualising gender and technology, and in the end, the related politics.

It should be noted that we have not tried to assess the relative effectiveness of the different initiatives. Nor have we studied how girls or women react to them. Thus, we are basically interested in the way the understanding of gender, technology and politics is shaped in the strategies of the cases.

2. CASE STUDY OUTLINE

Our empirical material consists of four cases, one from the public sector and three from the private sector. The public sector case is a study of the set of initiatives by the Norwegian Ministry of Education to integrate girls into the emerging Information Society over a period of 16 years. Due to the long time period, the relatively large amount of resources spent and the large degree of political attention given to these initiatives, we believe this case can broadly represent public actions in Norway towards such inclusion goals. The main sources of the case study are planning documents, supplemented by related material that focuses on the implementation and results of these plans.

The latter three cases have been chosen to provide insights into the importance of different market strategies. We present three commercial initiatives that were – to our knowledge – the first in Norway to design multimedia content targeted exclusively or mainly at women. They were all released in 1996 or 1997.[2] Moreover, they represent quite different sorts of initiatives. JenteROM (in English, GirlsROM) is a CD made by an entrepreneurial woman with clear inclusion motives. Libresse, a manufacturer of women's sanitary products has made a web-site and booklet that explicitly promote the inclusion of girls and young women. HjemmeNett ('HomeNet') is a web-site that in principle is aimed at women as well as men, and is thus an expression of an ideology of gender neutrality, but at the same time communicates explicitly with gender politics because women figure so prominently in the target group. JenteROM looked like a case of entrepreneurial feminism, HjemmeNet as a case of the subversion of feminism, while the Libresse case was selected to explore the possibility that feminist ideals could be made use of in the shaping of a market strategy.

Methodologically our three commercial cases are based on a combination of in-depth interviews with relevant actors and content analysis of multimedia products and relevant documents. We have interviewed the people most directly responsible for the configuration of the multimedia content. In the JenteROM-case basically we are relying on the interview with the entrepreneur behind it, Siw J. Henningstad. At HjemmeNett we have talked to the editor, Nina Furu, while at Libresse, we talked with product managers Åge Selseth and Merethe Slensvik.

Qualitative content analysis has been performed on the JenteROM CD, the HjemmeNett web-site, and the Libresse web-site. This analysis has been informed by the interviews, and we have compared interview results with our reading of the written and electronic material. Our reading strategy has not been to look for hidden meanings and intentions, but rather to identify the main structuring features and policy expressions of the material.

3. STATE FEMINISM: COMPUTER POWER TO GIRLS THROUGH EDUCATION POLICY MEASURES?

The concern for computers in education and the possibility of the exclusion of girls is by no means a recent one. The Norwegian Ministry of Education produced its first White Paper on the topic in 1984. This initiated the first programme to support experiments with computers in schools, which included equal opportunity goals and measures.[3] 2000 marked the start of the sixth consecutive ICT programme that included concerns about the potential exclusion of girls.[4] Thus, the Ministry has considerable experience in the area.

Interestingly, there has been considerable stability in the visions promoted by these six generations of plans. The computer has been envisaged partly as an object of mastery by pupils and students, partly as a tool to improve the quality of teaching and learning. There has been no real concern that computers might replace teachers. Probably this is due to the rather down-to-earth quality of the visions. Computers – with or without multimedia capabilities - are constructed as a tool that may open up new educational possibilities, new strategies for teaching, but the potential has been assessed conservatively. The ICT challenge is seen mainly as an issue of access and proper skills, with a limited potential for improving the quality of teaching. There are no traces of scenarios where computers are seen as a technology that might help to transform primary or secondary education in any radical ways, even if the latest plans have a stronger emphasis on the use of ICT as a tool for teachers.

Thus, logically, the six generations of ICT plans represent a steady, but modest effort. The main political concern has been access, measured as number of pupils per computer. However the central authorities have not provided additional resources for investments in computers. Such investments have been funded through the ordinary budgets.

Instead, the Ministry has been concerned with training teachers in computers/ICT and to provide them with proper software. In the late 1980s and the beginning of the 1990s, a major effort was put into the development of a tailored system for Norwegian schools called Winix. The effort was later terminated and described as a costly failure. Since, the Ministry has cautiously avoided any direct involvement in software development, even if some support is provided to outside initiatives.[5] As a result, little is done to shape educational software to become a tool to promote gender equality.

Notably, the latest curriculum plan for primary education, 1-10 grade, has only passing references to ICT and multimedia, even if the technology is considered important. It is discussed briefly, following short notes on traditional teaching tools and school libraries:

> The training shall contribute to pupils' developments of knowledge about, insight in and attitudes towards the development of the information society and information technology. Pupils should develop an ability to be able to use electronic devices and media critically and constructively, and as practical tools in their work with subjects, topics and projects. Information technology may make pupils able to use databases inside as well as outside their own country. Girls as well as boys should be stimulated to use information technology to counteract social and sexual bias in education.[6]

Quite traditional values and attitudes are front-stage in the curriculum plan, new technologies are not.

Thus, the gender politics related to ICT, performed by the Ministry of Education, appear as rather diffuse and limited by a focus on the agenda-setting role of the central authorities. In the Norwegian case, the agenda is dominated by the following factors:

- The importance of mobilising general interest in and enthusiasm for ICT in the educational system. In this kind of system, actors have to be motivated to do their part. To command them is difficult and expensive.
- The need for actors to be trained to be sufficiently skilled to participate in the implementation of the ICT programme. This is particularly important in the case of teachers, many of whom do not know the first thing about computers.
- A strong preference that the programme should achieve political correctness in terms of the need to argue that the use of ICT in education is compatible with Norwegian culture and values. Issues of equal opportunities have been placed high on the agenda. However the concern for cultural compatibility has probably been very important in shaping of the Norwegian ICT in education programme to be conservative in its visions as well as in its practice. No radical change through computers!

The Ministry is informed about the evolving situation regarding ICT in schools, including gender aspects, through large-scale national surveys (see Sundvoll & Teigum 1997) and annual reports from the regional offices of education.[7] They get statistical information about the stock of computers in different schools (including teachers' training colleges): about how many schools, municipalities, and counties that have made plans for the use of ICT in education; about the number of training courses held for teachers; and about how the extra money provided through the ICT programme has been spent. The survey also tells about the number of teachers who say they know how to use word processors, spreadsheets, and the internet. This provides information about gender differences related to a few variables like access.

Actual pedagogical achievements related to ICT are described in very general terms. We are told that it is important to teach children how to use word processors, spreadsheets, and the internet, and it is acknowledged that the multimedia computer represents an important resource for teachers. But what is achieved by employing this resource remains unclear. This vagueness also shapes the policy towards gender issues. The main feature of the strategy of the Ministry is a very general aim of equal opportunity.

The most recent initiative is a small booklet aimed at providing ideas for teachers to promote ICT to girls.[8] The motivation for focusing on girls is 'because it has been experienced that girls may be reserved towards the technology'. The main advice given is either to give girls priority to learn ICT, relative to boys, or to provide girls-only learning situations. The female approach to ICT is constructed in the following manner: 'Girls like to use ICT to communicate on equal conditions and to build networks. ICT meets the needs of girls to belong and provides self-confidence. ICT makes work more varied and real, something that particularly motivates girls'.[9] It is stated that girls and boys have different approaches to new technology, that girls tend to know less than boys, and that girls are more motivated by usefulness than boys. The implicit message is that ICT should be perceived as a serious matter where structured teaching is to be preferred to more playful, less systematic, approaches.

The gender politics of ICT displayed by the Ministry of Education seems well within the scope of waves one and two as described in the introduction. Basically, policy is shaped by an ambiguity between a goal that is untraditional – girls as users of ICT – and a rather traditional understanding of gender as something constant. Girls are assumed, essentially, to be serious and instrumental. Thus, the teaching of ICT should be serious and instrumental.

State feminism assumes that social movements, like feminist organisations, will be able to shape public policy. However, in our case, there are no clear traces of such influences besides the concerns voiced on occasion by Members of Parliament. The shaping of educational policy to include girls into ICT seems to follow a standard path of equal opportunities thinking, rather than a particular influence of political feminism.

From the perspective of entrepreneurial feminism (Cassel & Jenkins 1998), we would expect private actors to be more concerned with the entertaining and non-instrumental aspects of ICT. We now turn to the market-based cases to see if this holds true. Do private actors come closer to the understanding of gender and gender-technology practices of the third wave studies than public actors?

4. JENTEROM: A CASE OF ENTREPRENEURIAL FEMINISM?

In June 1997, the edition of the Norwegian young women's magazine *Det Nye* included a CD-ROM called JenteROM ('Girls' ROM'). The CD was marketed as a product suited to the needs of females between 15 and 30 years of age, with the expressed aim of attracting them to multimedia in general and the internet in particular. It was developed by a very small multimedia firm named *Blekkhuset A/S*, basically a single-woman company. However, *Det Nye* is owned by the biggest magazine publisher in Norway, Hjemmet-Mortensen, and is distributed with sales of 100,000 copies. The entrepreneur behind JenteROM, Siv J. Henningstad, expressed some of the rationale behind the product: 'It had for a long time irritated me that girls don't use data, and I wanted to find ways to attract them to the PC'.[10] This was to be accomplished by proving that the PC is a useful tool also for women. JenteROM was intended as a means to this end.

In what ways may JenteROM be said to be 'useful for girls'? Analysing the CD-ROM, we discovered basically three aspects. First, there is the diary-part that takes up most of the entry-page. Here we are invited to bring our secrets electronically. Privacy is to be secured by a password-lock. This part also contains a calendar and a time-manager. Second, from the entry page one can link to topical sections containing information about trends, sport, body, health, music, environment, food and IT. The topics, chosen on the basis of a survey of girls' interests, are presented in words and pictures. Occasionally, sound and video-strips accompany them. Invited 'special contributors' had written part of the content, i.e. people with a nation-wide reputation in their fields. Third, the CD tries to be useful by providing information about as well as giving access to the internet. Access can be accomplished directly from the entry page, and the Norwegian Telecom, Telenor, offers the user one month free internet access if they link up.

Our initial impression was that JenteROM is trying to appear as a source of knowledge and information as well as user-friendly. Thus, there is an emphasis on usefulness, indicating that the CD-ROM has been shaped by an educational concern. But JenteROM is not just an educational project. A sustained effort has been made to make the CD attractive to young women. There is an accent on intimacy, and the CD together with the web page act as gateways to topics and sources usually assumed to be of interest to the target group: pop music, fashion, etc. Clearly, different interests have shaped JenteROM. The initiative emerges from the cultural industry and a tradition of women's magazines to mix educational topics with entertainment. In this sense, it is typical of the so-called edutainment genre where the products are marketed as providing both learning and fun. But why make a CD-ROM to promote women's position as users of computers? And why would a magazine for teenage girls with a rather conservative profile in its gender politics agree to distribute it? And why would the sponsors – a rather mixed group of commercial enterprises – bother to finance it?

Henningstad told us that she got the idea for the CD-ROM in a computer store, suddenly realising: 'There's nothing for me in here! All these offerings belong to a boy's or a man's world!'. When Henningstad was to configure an alternative, she started by drawing on her competencies as a woman: 'And I thought, what on earth shall I make for women, what do women like? And so I thought, I had better start with myself'. The plans for the diary part came directly out of this reflexive monitoring. The diary represents a transfer of a well-known and familiar form into the new media, and in that way that is easily recognised. At the same time, the CD-ROM became linked to an activity that is widely regarded as being typically female.

The plans for the topics to be covered also came out of her introspection. Henningstad wanted the JenteROM to be a place where one could learn things. This part also embodies a clear reference to a recognisable cultural form, namely that of the thematic feature of the weekly paper. Still, here the intention was to use the electronic media to transcend the format: 'To put together themes ... So it's an advanced form of a weekly, but in a completely different fashion, where you also have sound and much greater stocks of knowledge on the different topics.' In Henningstad's opinion, a CD-ROM would also be a more 'quiet' and contemplative medium than the transient papers. When deciding on which topics to be covered, she did not rely solely on her own intuition. She used her contacts from the School of Marketing to carry out a market poll.

The decision to organise the CD-ROM according to themes allowed Henningstad to market the product as a 'knowledge CD'. 'Knowledge' should be viewed as an answer to the assumption that females want to use multimedia for useful purposes. This is also formulated in contrast to stereotypical male computer use. The 'male user' seems to be constructed around the hacker figure as a person who is technically fascinated and very much engaged in playing computer games. Reflecting on what a 'Boys' ROM' would have looked like, Henningstad suggests that: 'Men would have had 80 percent play and, maybe, 20 percent knowledge'. However, the concept does not only pertain to knowledge *about* multimedia, but also to knowledge *in* multimedia.

The third part of JenteROM, the provision of information about and direct access to internet, entered the project at a later stage. This part seems to have relied on the discovery of the technical feasibility of such a solution. However, it tied in quite neatly with the ambition for the CD. Young women should be made interested in multimedia by the use of JenteROM, but this should also serve as a gateway and a motivation to use multimedia in general. Moreover, the CD-ROM should be easy to use, and it should be easy to enter the Net. Again, we should note how this strategy contrasts to the stereotypical image of male computer use, in particular the assumptions about a closed brotherhood connected through secret devices and cryptic code languages.

Getting the idea of designing a CD-ROM to attract women is one thing, having it implemented is a different matter. Considering the fact the cost of producing JenteROM ran to around NOK 1 million, it should be clear that this is not just something to be done out of the blue. Henningstad needed to enrol other actors to support her and join her efforts of gender politics. Thus, the gendering of the JenteROM would be shaped by the need to translate (Latour 1987) other interests as well as her own scenario to come out with a proposal sufficiently attractive to elicit the support Henningstad needed.

In the introduction to the paper, we outlined three waves of research into women and computing. Referring to the concept of entrepreneurial feminism (Cassel and Jenkins 1998) and its relationship to games, we expected at the outset that the JenteROM initiative would draw upon the more open understanding of gender, typical of the third wave. However, as we have seen, the ideas behind the JenteROM have more in common with the second wave in its basic motivation: to get more girls and women interested in computers and multimedia. Henningstad's approach to the problem shares the assumption that women/girls are more concerned with usefulness than men, and above all that technology has to be designed consciously for women in order to be attractive to them. But the JenteROM is definitely a break with the standard assumption that technology somehow is neutral in terms of gender.

However, the JenteROM project is also an expression of some of the difficult challenges in the development of a progressive gender politics of technology. The main problem is the strong tendency to think in terms of dichotomies: women are like this, men are like that. The result represents a standardisation of the female gender, particularly in the process of inscribing femininity into the CD-ROM and its programs. In this respect, JenteROM is rather modernist, in contrast to post-modern efforts to view gender as more flexible and varied. This modernism is also reflected in the considerable optimism in Henningstad's belief in the potential of educational tools to bring about changes. She argues as if the main challenge is didactic and the challenge to make women interested in computers may be delegated to the computers themselves. Moreover, the design of JenteROM assumes that femininity may be inscribed in programs and information in a straightforward fashion, thus ignoring the translation problem presented by such inscriptions (Sørensen 1992).

In its basic assumptions about women and technology, JenteROM has much in common with the approach forwarded by the Ministry of education. However, Henningstad as a feminist entrepreneur had somewhat greater freedom in the shaping of the tool for inclusion. While she also insisted about seriousness in the approach to ICT, still she allowed herself more opportunities to exploit the entertainment potential of multimedia technologies, e.g. in the diary function. However, we have little reason to believe that the JenteROM case is typical to the way private actors approach issues of gender and ICT.

5. LIBRESSE: MARKETING FEMINISM?

Libresse is the brand name of a series of women's sanitary products, like tampons, pads etc., which are produced by the Scandinavia-based, multinational company Saba-Mölnlycke. The Libresse-products have a market-share of approximately 80% in Scandinavia, and are also – partly under other brand names – fairly big in the rest of Europe and in Australia. We have studied the Norwegian web-site for Libresse,[11] a separate investment initiated by the Norwegian marketing division of Saba-Mölnlycke, launched as early as spring 1996. It has later on been revised and extended several times, but the basic features and the philosophy behind have very much been the same since the beginning.

Libresse wants to be visible to women and reach out to women through all available channels. But the company is applying two distinct strategies to reach this goal. They have developed what we would call a 'broad' strategy, with general advertisements for their product. To this end, they use established media like television, cinema, and to some extent newspapers and weeklies. This is traditional marketing, in all respects, with soft, non-disturbing images of the benefits of the products.

However, they also have a second, more 'targeted' strategy. This strategy is aimed exclusively at girls between 11 and 16 years. Based on the marketing philosophy that these years are vital for the establishment of purchase habits, Libresse has over a period of 30 years made efforts to communicate to this group. In the beginning, the efforts were pursued by way of teachers and health staff in the schools, and consisted of 'objective' information material about menstruation and puberty. Over the last ten years, however, Libresse have also tried to address the girls directly, through a membership club called *Klikk*[12] *Libresse*. The company insists that subscriptions have been based on local recruitment, primarily girl-to-girl recruitment, and not the use of address databases. Still, they claim to have more than 80,000 members, or approximately 70% of their target group. Membership basically consists in receiving the free, quarterly Klikk Libresse magazine. It consists of various articles, short stories, competitions, letters from readers, an ask-the-nurse section, and some product information.

Interestingly, Libresse has come to associate their web presence solely with this second, targeted strategy, and especially the Klikk Libresse initiative. This means that the web-site addresses teenage and pre-teenage girls. The web-site is partly re-circulating Libresse's educational material, but they have gradually upgraded it to suit the new medium. Arguably, they have been able to retain the seriousness of the information, but the presentations have become more exciting and voyeurish, with virtual trips into bodily changes and other multimedia effects. The young girls are encouraged to learn about menstruation and puberty through invitations like: 'Look what's happening in your body' and 'What's going on down there?'

The web-site is still making use of many of the editorial ideas from Klikk Libresse,[13] but again they have been transformed by exploiting new technological possibilities. Thus, the horoscope has been replaced by a test-your-self quiz ('Find out what star you really are!'). More interestingly, the letters from readers were transformed into a discussion group (and later an uncensored chat line), while the ask-the-nurse section became interactive. This has made postings and personal feedback easier. Other features include a period manager and competitions.

Libresse has tried to market their web-site as well as to strengthen the link between the company and the internet in general through some other notable actions. One of them is the use of the now mandatory www-address in advertisements and commercials. But also in the welcome letter to the Klikk Libresse Club, attention is directed towards the internet: 'internet – only for girls! ... Haven't you used internet before? Get someone to help you! You'll soon find out that it's very easy – even if you have never done it before ...' The Klikk Libresse magazine has also been filled with internet related material in recent years – with articles like 'Best friends on the internet' and 'Girls in Cyberspace'.

The most striking effort, however, is the issue of a leaflet called 'Internet guide for girls'. The 8-page leaflet – first made in 1997 and in an updated version in 1998 – tries to encourage girls to go on-line: 'Internet is just a place for boys – you think. But that's wrong!' Of the 8 pages, half a page is reserved for information about Libresse's own web-site. The rest consists of arguments about why the internet is interesting for girls; information about how to use web browsers and search engines; a guide to Smiley's vocabulary; and several pages with addresses to other, supposedly interesting web-sites.

Apparently, Libresse is doing everything with and on the internet except advertising or selling their own products. Of course, this is part of a market strategy, where the ultimate goal is product loyalty. However, arguably, it is a strategy with altruistic elements. Instead of using the internet to market their products, the Libresse web-site (and the rest of the internet) is configured as a tool for bodily redemption and a helper in adolescence. Some of the features of the web-site may be said to follow 'logically' from the products of the company. As the product managers reflected: 'It is more logical that we have an interactive period manager than many others, to put it that way'. But the wider focus on bodily changes and puberty is a result of the company's long-term efforts to establish a long-term relationship with adolescent girls: 'They very much like to discuss menstruation and ... perhaps not necessarily types of sanitary pads and the like, but the whole set of problems of growing up and puberty and related stuff ...'.

The web-site is configured as a place for learning and inquiring about issues directly related to the female body. Thus, it offers a link between biological sex and a technological medium. Besides being an attractive way to present comprehensive and engaging information, the company's spokespeople stress that the internet has two additional advantages for adolescent girls. First, there is the interactivity. Since menstruation and puberty are personal experiences, there will be personal questions related to these issues that cannot be covered by standardised information. Through the discussion groups and the ask-the-nurse service – and the period manager as well – girls have a new opportunity of posing these questions. Secondly, there is the possibility of anonymity. Because issues relating to menstruation and puberty are sensitive, Libresse is encouraging girls to participate anonymously in the discussion groups and offering safety rules when making mail friends. 'Because it is extremely important that one can live without identity on a site like ours', states Slensvik, echoing issues raised by feminist internet analysts like Sherry Turkle (1995). Thus, the Libresse web-site is a particularly illustrative instance of what we can call the body paradox of the internet: internet as a technology that can bring you closer to your body by removing yourself from it.

Libresse does not represent entrepreneurial feminism, but rather a kind of strange mix of marketing strategy and enlightenment effort. The leaflet 'Internet guide for girls' has, in its intentions and rhetorical strategies, much in common with JenteROM and also with the kind of gender politics one normally would expect from public actors. It exemplifies a progressive marketing strategy, gender-wise, but it is nevertheless a marketing strategy. Maybe it represents a rare kind of marketing feminism, a case where a non-controversial feminism is used to reach a female audience.

This may perhaps be explained by reference to the particular character of the products of Libresse. When we look more closely into their construction of female use of ICT, we can see that Libresse does not construct women as the instrumental users like in the two previous cases. Rather, the female use of ICT is related to the women's bodies and the need to acquire knowledge about this. In this manner, the contrast between male and female use of ICT becomes an issue of biological sex rather than gender.

6. HOMENET: A CASE OF GENDER NEUTRALITY?

HjemmeNett is a joint investment by three large media actors Hjemmet-Mortensen, Orkla Media and Egmont. Hjemmet-Mortensen is the largest publishing house in Norway, and it entered multimedia quite early. Their overarching policy was that all their magazines should be on the Net. Already in the winter of 1996 all their most widely circulated magazines had their own homepages. In addition, niche-magazines like *Techno* and *PC-World* was regularly published with complementary CDs. The new web-site, HjemmeNett, was planned as an umbrella service that would receive material from all content providers inside the three media conglomerates, in addition to a limited group of external companies. Put simply, it was the biggest investment targeted at Norwegian households so far.

When it was launched on 25th September 1997, HjemmeNett carried out a giant campaign for its new service. No less than 800,000 promotional CDs were distributed to Norwegian households, bundled with several magazines and through mail. This should be compared to the total Norwegian population of 4.5 million. HjemmeNett was promoted as an internet-service aimed 'at the whole family'. The marketing emphasised that the service could be adapted to the needs of every member of the family. The editor of the new service Nina Furu's visions for the new service were immodest: 'HjemmeNett is to be the spearhead that drives the internet into "the thousand homes"'. By implication, however, this statement means that the ambition of HjemmeNett was to engage women in the use of the internet.

HjemmeNett represents an ambitious attempt to attract new groups of users to multimedia by directing efforts at private homes and non-work situations. According to Furu: 'The golden days of the 20 year old data nerd is gone ... Or, he might go on as he wants. But we want the mother and the grandmother and the grandfather and the little lad at four to come along.'

The ambition of HjemmeNett was to be able to give you 'all you'd wish to use internet for, at one spot' (their slogan). The HjemmeNett tries to accomplish this by linking the variety of content providers inside Hjemmet-Mortensen with selected external goods-, content- and access-providers. However, HjemmeNett seems to represent an alternative configuration of the (female) user than the one we identified in the case of the JenteROM and Libresse. The latter design embodies specific assumptions about its user, e.g. that she is knowledge-seeking, is interested in specific topics, and prefers quiet and private experiences. HjemmeNett tries to circumvent such assumptions by offering users the possibility to choose activities in safe surroundings. This is accomplished technologically with the introduction of a key-word subscription. Thus, HjemmeNett promotes a kind of liberalism – here you should be allowed to do whatever you want to do.

Another factor we will emphasise is the strategy of HjemmeNett to appear trustworthy. This is strongly connected to the problem of 'information overload', and the need to choose, digest and evaluate the available input. The key-word subscription found in HjemmeNett should be recognised as an effort to manage this problem. However, the idea of trustworthiness above all applies to the external world of the internet. The main assumption is that users want and need someone to trust. This is the rationale behind Furu's concern to gain control of what is performed in the name of Hjemmet-Mortensen:

> We want to tell that this (i.e. what's found on the HjemmeNett) is material delivered from serious, respected content suppliers, not from three teenagers from a small village town that is pulling out some rubbish ... . We have respected, established brand names that have a certain trustworthiness, right? That trustworthiness we wish to bring with us into the Net.

HjemmeNett is seemingly an effort to contradict the need for special initiatives to get women interested in multimedia. The main strategy seems to represent a return to the idea that technology is neutral. When one adds web software that facilitates the choice of applications or information area, it is argued, this home page may cater anyone's interests.

In this manner, HjemmeNett represents a rather traditional gender-neutral strategy that breaks not only with state feminism, but also with entrepreneurial and market feminism. However, we should be careful about dismissing this kind of initiative as just traditional and an example of the unwillingness or inability of market actors to address gender politics. When listening carefully to Nina Furu's arguments, she is making quite strong statements about the gendered nature of the internet. On the one hand, there are the male enthusiasts, the nerds. They care neither about usability nor trust. On the other, we find the reluctant users, some men and most women, looking for an accessible and trustworthy internet. They are the target group of HjemmeNett. Thus, HjemmeNett claims to be based on an inclusion strategy aimed at all reluctant users, but – in a world of a dichotomous gender – constructing its users from a stereotypically female model.

7. THE GENDER POLITICS OF COMPUTER INCLUSION: PUBLIC CONSERVATISM VERSUS PRIVATE PLAYFULNESS?

We began by juxtaposing the idea of state feminism, privileging public actors with political correctness, with entrepreneurial feminism, which argues a role for private actors as well. However, our findings are by no means this clear-cut. In particular, our case-studies show that private actors may pursue quite different gender policies.

The strategy of the Norwegian Ministry of Education could be characterised as a pursuit of structured teaching, shaped by an instrumental understanding of ICT and a willingness to give priority to the training of girls. Arguably, the understanding of gender as well as of ICT was quite traditional.

JenteROM was a multimedia experiment aimed at young women. In order to construct this new audience, the innovator (Henningstad) had to invoke new conceptions of multimedia technology and its relationship to users. What happened with JenteROM is a clear-cut example of a mutual reconfiguration of technology and gender. On the one hand, we see an attempt to translate the technology into a feminine artifact, and, on the other, an effort to change some aspects of the definition of femininity – for example to overcome the assumption of computer-phobia (Lie & Sørensen 1996, Berg 1996). The end-result also represents an effort to regulate and 'freeze' a particular conception of the relationship between women and multimedia.

JenteROM was clearly a product of entrepreneurial feminism. In that respect, the initiative shows that such opportunities for pursuing feminist goals may be offered by the market, even if it is a rare phenomenon. Libresse and their Klikk magazine is an example of a different approach. Here, we find a company which tries to persuade girls to become users of the internet in the hope that this will make them become users of their sanitary products. We characterised this strategy as marketing feminism because they clearly promote some feminist ideals in order to configure girls as future consumers of Libresse products.

The Libresse web-site is definitely one for women, in the sense that the topics covered and the style of communication reflect the explicit intention that women are the main, if not the only, users. The mutual shaping of gender and ICT occurring here shares many characteristics with JenteROM, like the emphasis on quietness, utility and knowledge-seeking.

In the case of HjemmeNett, the construction of gender is different. While JenteROM and Libresse construct women as quiet, knowledge-seeking but technically not-so-competent users, a constituency that values usefulness over play, the latter assumes that women are seeking trust, broad access, flexibility, and individualism. The design of JenteROM was based explicitly on an assumption that gender is dichotomous. HjemmeNett constructs the feminine and the masculine as the end points of a scale.

The above analysis has called into question traditional presumptions about the exclusion of women from ICTs by reference to ongoing private sector efforts to make the use of multimedia, in particular the internet, more attractive to women. The potentially entertaining and pleasurable qualities of multimedia applications may encourage women to get involved with ICT (see Nordli 2001), and this involvement seems to be important to emerging commercial players. Moreover, the commercial initiatives seem to supplement public sector strategies to bring women into the Information Society. This calls for a widened understanding of the policy mechanisms available and suggests that public policy should seek to be aware of and build upon this kind of embryonic development.

It is important to note that commercial initiatives display different gender politics. When looking at our cases, the feminist reader will probably be more sympathetic towards Henningstad than towards Furu. S/he will recognise a definition of gender in Henningstad's thinking that is more in line with feminist analysis than what is found in Furu's statements. In this case study, we have not elicited the opinion of the intended core group of users: women. However, we would guess that Henningstad as well as Furu might find a lot of supporters for their respective strategies and gender definitions. This would be in line with a basic 'third wave' assumption: the heterogeneous quality of gender.

Nevertheless, taken together, JenteROM and HjemmeNett indicate a profound change in the gender dynamics of computers and multimedia. Primarily, their existence suggests that the previous symbolic identity between masculinity and computer use has been broken. Recent studies of Norwegian girls' relationships to computers support this view (Nordli 1998, Gansmo 1998). In particular, Nordli shows how girls, when becoming interested in computers, are pursuing the avenues of exploration made available by JenteROM as well as by Libresse and HjemmeNett. In fact, they might not even see the differences between these initiatives in the way we have presented it in this paper.

A further analysis across the four cases suggest that we may observe two different strategies of getting women to use multimedia and the internet. The educational equal opportunity case, the JenteROM and the Libresse web-site were all constructed on the basis of a *women-in-particular* strategy. The underlying strategy represents an effort to inscribe a standardised female pattern of use. Girls as users of ICT are perceived to be different from boys. In contrast, HjemmeNett was constructed from the idea that the problems of using the internet are the same for women as for men. Thus, here they used a *women-and-everybody-else* strategy, based on a standard, unisex pattern of use. Thus, we have identified two ways of supporting the inclusion of women in multimedia, even if the latter strategy is not explicitly concerned with the female gender.

The strategy of equal opportunities for girls in learning ICT pursued by the Norwegian government was mainly concerned with equal access to PC's among boys and girls. In addition, it suggests that girls may need more attention in this field than boys, and that they will appreciate a more structured and less play-oriented teaching. In many ways, this strategy still reflects the concerns from the first wave of research on women and computers: how to avoid the exclusion of women. Its measures are fairly conservative, with a few local initiatives as exceptions, and the thinking behind them is mainly defensive. Girls should get equal opportunities to learn about computers, but the reason why this is so important is more or less taken for granted. The official rhetoric is focused on rather abstract possibilities in the future labour market, with rationalist and instrumentalist arguments.

This contrasts with the ideology of the commercial experiments we have described here. They are much more on the offensive, struggling with the concern of inclusion rather than the avoidance of exclusion. This point is underscored when *Libresse*, the producer of sanitary products for women, and not any educational authority, is issuing the 'Internet guide for girls'. In a way, the commercial experiments, supported by a wide range of different forms of advertising, appear to be more progressive that the public education sector.

However, when we compare JenteROM with HjemmeNett, we see that the story is more complicated. At the end of the day, the commercial efforts have difficulties in pursuing a feminist politics of technology. The JenteROM case shows that it is possible to make a commercial product that is quite progressive in terms of its gender politics. However, we also observe that a more large-scale and commercially important effort like HjemmeNett is shaped by a traditional and conservative strategy, that of gender neutrality.

In some sense, it is a paradox that Siv Henningstad failed to get public grants to support the making of JenteROM. In many ways, this is a product that is in line with a way of thinking familiar to many teachers as well as policy makers in the education sector. However, it was not perceived as sufficiently 'serious' by government actors, partly because of its commercial orientation, partly because of its entertainment-like profile.

This situation is indicative of a kind of division of measures between the gender politics of ICT by government agencies and by commercial actors, respectively. Commercial actors will look for features sufficiently attractive to a sufficiently large audience to be able to profit. Presently, the main inroad to the design of software seems to be to make it entertaining, even though the concept of entertainment may have some flexibility – as demonstrated by the case of JenteROM. Governmental actors, including those found in the educational sector, are expected to be more 'serious'. This may mean that they lose in the competition with commercial products. At least, education faces considerable challenges from 'edutainment'.

NOTES

1 This paper draws upon research done as part of the EU/TSER project 'Social Learning in Multimedia' and the project 'From cult to culture: The making of the future internet-user', supported by the strategic university programme 'Information technology, communication and culture' at NTNU.
2 By the time of writing (late 2000 and early 2001), we have witnessed a 'second boom' of commercial web-sites for women in Norway. The last year as many as 10 – 12 new web-sites – or 'portals' as they now are most commonly named – have been launched in rapid succession. See www.jentenett.no, www.baby.no, www.kvinneguiden.no, www.trendmagasinet.com, www.kvinneveven.no, www.jenteporten.com, www.barnimagen.com, www.femme.no, www.klikk.no, www.jenteklubben.no.
3 See Stortingsmelding nr. 24 (1993-1994) Om informasjonsteknologi i utdanningen, p. 16f.
4 *IKT i norsk utdanning. Plan for 2000 - 2003.* Oslo: Ministry of Education. See also Aune and Sørensen (2001) for a more detailed analysis of these efforts.
5 See *Stortingsmelding nr. 24 (1993-1994) Om informasjonsteknologi i utdanningen*, p. 26-27.
6 *Læreplanverket for den 10-årige grunnskolen*, Ministry of Education, Research and Church Affairs, Oslo 1996, p. 78-79.
7 See http://odin.dep.no/odinarkiv/norsk/dep/kuf/1998/publ/index-b-n-a.html
8 'Et klikk og du er hekta. Et inspirasjonshefte om jenter og IKT', http://odin.dep.no/kuf/proj/it.
9 'Et klikk ...' (note 8), p. 5.
10 Britt Wang Løvik: 'Nedlåste betroelser', http://www.idg.no/datamagasinet/0297/jenterom.htm.
11 www.libresse.no
12 'Klikk' means gang, but the word also means 'click' – which became a funny coincidence when the phenomenon of internet surfing came into being.
13 Exceptions are the articles and short stories, which are saved for the paper magazine.

BIBLIOGRAPHY

Aune, M. and Sørensen, K.H. (2001), 'Teaching Transformed? The Appropriation of Multimedia in Education: The Case of Norway', in Marc van Lieshout *et al.* (eds), *Social learning technologies. The Introduction of Multimedia in Education*, Aldershot: Ashgate, pp. 159-189.

Berg, A-J. (1996), *Digital feminism*, Report no. 28. Trondheim: Centre for Technology and Society.

Berg, A-J. & Lie, M. (1995), 'Feminism and Constructivism – do Artifacts Have Gender', *Science, Technology, and Human Values*, **20** (3), 332-351.

Cassell, J. & Jenkins, H. (1998), *From Barbie to Mortal Kombat. Gender and Computer Games*, Cambridge, MA: The MIT Press.

Cockburn, C. (1985), *Machinery of Dominance: Women, Men, and Technical Know-How*, London: Pluto Press.

Gansmo, H. (1998), *Det Forvrengte Dataspeilet*, Report 36/98. Trondheim: Centre for technology and society.

Hacker, S. (1989), *Pleasure, Power, and Technology*, Boston: Unwin Hyman.

Hacker, S. (1990), *'Doing It the Hard Way': Investigations of Gender And Technology*, (edited by D E Smith & S M Turner), Boston: Unwin Hyman.

Haraway, D. (1991), *Simians, Cyborgs, and Women: the Reinvention of Nature*, London: Free Associations Books.

Hernes, H. (1987), *Welfare State and Women Power: Essays in State Feminism*, Oslo: Scandinavian University Press.

Håpnes, T. & Sørensen, K.H. (1995), 'Competition and Collaboration in Male Shaping of Computing: A Study of a Norwegian Hacker Culture', in Grint and Gill (eds), *The gender-technology relation. Contemporary Theory and Research*, London: Taylor & Francis, pp. 174-191.

High Level Group on the Information Society (1994), *Europe and the Global Information Society: Recommendations to the European Council*, (Report of Bangemann Group), Brussels, 26 May 1994.

Latour, B. (1987), *Science in Action*, Cambridge: Harvard Univ. Press.

Lie, M. & Rasmussen, B. (1983), *Kan Kontordamene Automatiseres?*, Trondheim: Institute for social research in industry.

Lie, M. and Sørensen, K.H. (eds) (1996*), Making Technology our Own? Domesticating Technology into Everyday Life*, Oslo: Scandinavian University Press.

Nordli, H. (1998), *From Spice Girls to Cyber Girls? En kvalitativ studie av datafacinerte jenter i ungdomsskolen,* Report no. 35. Trondheim: Centre for Technology and Society.

Nordli, H. (2001), 'From Spice Girls to Cyber Girls? The Role of Mulitmedia in the Construction of Young Girls' Fascination for and Interest in Computers', in Marc van Lieshout *et al.* (eds), *Social Learning Technologies. The Introduction of Multimedia in Education*, Aldershot: Ashgate, pp. 110-133.

Probert, B. & Wilson, B.W. (1993), *Pink Collar Blues. Work, Gender and Technology*, Carlton: Melbourne University Press.

Rasmussen, B. and Håpnes, T. (1991), 'Excluding Women from the Technology of the Future? A Case-study of the Culture of Computer Science', *Futures*, **23** (10), 1107-1119.

Sørensen, K.H. (1992), 'Towards a Feminized technology? Gendered Values in the Construction of Technology', *Social studies of science*, **22** (1), 5-31.

Spilker, H. & Sørensen, K. H. (2000), 'A ROM of One's Own or a Room for Sharing', New Media and Society, **2**, pp. 268-285.
Sundvoll, A. & Teigum, H.M. (1997), IT i skolen 1997. Del 1, Tilstandsundersøkelse i skolene: Hovedresultater og dokumentasjon, Notater 97/42, Oslo: Central Bureau of Statistics
Turkle, S. (1984), *The Second Self: Computers and the Human Spirit*, London: Granada.
Turkle, S. (1995), *Life on the Screen: Identity in the Age of the Internet*, New York: Simon & Schuster.
Wajcman, J. (1991), *Feminism Confronts Technology*, Cambridge: Polity Press.
Webster, J. (1995), *Shaping Women's Work. Gender, Employment and Information Technology*, London: Longman.
Weimer, D.L. & Vining, A.R. (1992): *Policy Analysis: Concepts and Practice*, 2nd edn., Englewood Cliffs, N. J.: Prentice Hall.

10. Governing Measures: User-stories and Heat Pump Subsidies[1]

Pål Næsje

INTRODUCTION

Most policy analysis relies heavily on an eclectic mix of macro/micro economics for problem definition and a strong case-orientation on the problem in question (cf. Weimer & Vining 1992). The goal of policy analysis - indeed of much public policy – is to secure the most efficient market in terms of resource allocation in society. If key actors are profit-maximising firms and utility-maximising consumers, it should, in theory, be possible to move towards a Pareto-optimal distribution of available resources. When this does not happen, we have a market failure, and this is the central concern of policy analysis.

Perceived market failure is central to the case that is to be presented here as well. Bear in mind that the different types of market failures that are of interest to the policy analysis tradition all relate to a basic economic competitive model. The causes for such failure must be found empirically, but in the case of energy savings it seems clear that the central culprits would be natural monopolies and/or information asymmetries. In the former, the inherently high investments involved in energy production combined with high price elasticity hinder a higher social surplus – i.e. energy prices are set higher than they need to be or important external costs are not reflected in energy prices. In the second case, information asymmetry hinders the implementation of alternative energy technology solutions – i.e. consumers do not have information or knowledge about the solutions available.

Policy analysis differs from traditional academic social science research in that it has an applied focus, where real-life recommendations are an integral part of the work. In many cases, policy analysis will engage a public problem by estimating the consequences of alternative decisions. The *modus operandi* is to take the factors and facts that are involved in a problem as a base for analysis. This means that *how* information is constructed relating to a problem is not important for the analysis. For policy analysis, the *quality* of data is important, but as soon as satisfying quality has been secured, the analysis can commence.

Thus, the production of data pertinent to the case and the ensuing definition of problems resulting from this production is, by and large, neglected in policy analysis. I will argue that such an approach leaves out important aspects of the analysis of public policy (cf. Sørensen, chapter 2 in this volume). In a number of studies a constructivist approach to policy problems has been used (notably Hajer 1995). These studies show how problem definitions and data are, rightfully or not, skewed in specific directions. In the case I will present here, the importance of how a problem is understood and defined with respect to the public policy used, challenges traditional policy analysis.

THE CASE

In November 1990, Finn Kristensen was appointed Minister of Energy and Industry in Norway. Being a man of action, he wanted to rejuvenate Norwegian energy policy, and especially energy savings policy. At this time, 15 years of research and development efforts in energy savings had done little in getting actual savings. Kristensen was, unlike his predecessors, willing to use direct measures in the energy market, directed towards the owners of the buildings. In short, he saw the need to subsidise energy saving technologies. This was due to several factors.

First, an energy market deregulation act had just been passed in parliament. The political importance of sound energy savings measures was heightened by this new market situation: if electricity prices stayed low after market reform, alternative energy technologies would be left in an untenable competitive situation, and other important societal goals, e.g. protecting the environment, would be hurt. The presumption of this argument is that price and return on investments are the major factors that influence the choice of energy technology, and are more important than knowledge diffusion etc. So to avoid negative consequences of market reform, alternative energy technologies needed subsidies.

Second, the alternative energy technology of choice at this point was heat pumps. According to surveys of energy use, heat pumps had a large, but unrealised, potential. Exploiting this potential would be advantageous for several reasons: resource allocation in the national energy sector was seen to be less than ideal. Furthermore, unemployment was relatively high, in the Norwegian context, and according to Keynesian logic, government spending would create jobs, and generate more spending and even more jobs. Therefore state funds could and should be used. Anyway, the potential for heat pump installations combined with the political importance of a sensible allocation of funds made heat pumps attractive.

A subsidy of 40% of investment-cost and other energy economising measures was introduced to the public on March 16th 1992. In a publicity stunt, the Minister was interviewed while putting up posters in central Oslo for the State's Energy Economising Campaign (*Aftenposten Aften*, 03/16/92, p.4).

Not surprisingly, the installation rate of heat pumps rose quickly. Many small firms started to sell heat pumps to private building owners. There were many installations in apartment complexes and office buildings. Interest in heat pumps was overwhelming. At last implementations were happening, at an unprecedented scale. Those who had worked with heat pumps over the previous decades were thrilled. This was, as they saw it, a breakthrough for heat pumps, and an opportunity to realise an energy supply based on thermodynamics, i.e. where a low-quality energy need such as heating would be met with extraction of low quality energy (ambient air, seawater etc.). At last, heat pumps would take on their rightful role in the national energy supply.

But the energy economising campaign turned sour very quickly. The sheer monetary amount granted made the campaign visible in the political landscape. The idea of directly subsidising private actors' technology choice had a problematic ring to it in political circles, even if the cause was good. Consequently, the campaign became a target for political attack. Moreover, this attack was soon bolstered with critical comments from the Ministry of Finance.

The central question was, then, were the subsidies well spent? Internal discussions at the Ministry of Energy and Industry, concluded they were 'well spent' if installations met a number of criteria:

- installations had, of course, to save energy. This was not as unproblematic as it might seem. On an average, it was held that 40% of gross energy savings in residential buildings was '*lost in raised comfort and heating of more rooms, after the measure is taken*' (St.meld. nr. 41, 1992-93 p14). In industry these losses were held to be less than 5%.
- installations had to be cost effective for the state *in* a market. Now, how do you define that? Within the market logic the actor is or should be profit maximising. Based on this, actors were not to receive subsidies for installations that were clearly cost-effective for the actor in the (unsubsidised) market, with short payback time (5 years or fewer). If actors did not choose such solutions by themselves, they were deemed ignorant and undeserving of subsidies. Of course, the state never intended to fund actors' ignorance. So subsidised installations that were economically viable without subsidies, were *free riders*. On the other hand, the state wanted to support installations that were cost-effective in societal or macro-economic terms, avoiding those not cost-effective by

> any standard. What was deemed cost-effective for the state were all installations that broke even within 12 to 15 years. These installations did not yield a high enough return on investments for a private actor to undertake them, but were high enough for the state to support such installation as they were favorable on a macroeconomic scale.[2]

Hence all installations that were in the band between being barely – economically sound for private investors and cost-effective over the longer run, would be legitimate objects for investment subsidies. If the 40 percent subsidy was to be deemed 'well spent', the ratio between free riders and legitimate installations would have to be favorable.

Now, subsidies were used for heat pumps and other energy savings measures in three areas; in state and municipal buildings, in industry, and in residential buildings.

Figure 10.1 Calculated gross energy-savings for the different areas of subsidy up to March 1, 1993

	No. of accepted applications	*Energy Saved (calculated in GWh)*	Investment per saved kWh
Municipal/State buildings	862	246	ca. 2 NKr
Office/Corporate Buildings	400	175	ca. 2 NKr
Industry	356	1019	0.90 NKr
Residential Buildings			
-Houses (<5 living units)	2471	27	ca. 3.75 NKr
-Apartment bldgs. etc	99	49	ca. 2.50 NKr

(from St. meld. nr. 41, 1992-93).

Figure 10.1 presents figures for energy savings measures (this includes other measures e.g. isolation, installation of thermostats, and more energy efficient industrial processes, though the bulk of subsidies went into heat pumps installations). The calculations indicate that the subsidy was most effectively used in industry. Here, the investment costs were lowest, and the savings largest both in total and on average (the average installation saved 2.9 GWh). On the other end of the scale was the 'houses' category. The number of applications accepted was large, but costs were high (almost 4 NKr invested per saved kWh) and size was small (mean savings per application was 0.011 GWh=11 000 kWh).

In economic terms, such small installations seem to be unattractive. The average yearly energy consumption per household in Norway was 16,000 kWh in 1996 (Department of Oil and Energy 1997), but this is an average across apartments as well as villas. A large villa could use over 50,000 kWh per year. Furthermore, it is not uncommon to have 2 and 4 living units in a house. Now, a quick estimate would indicate that the mean investment to save the 11,000 kWh in the 'small houses' category was 41,000 NKr, which at an energy price of 0,50 NKr/kWh saved 5,500 NKr per year. If the investment is depreciated over 10 years (a heat pump's approximate life span), and the internal interest rate is set to 7%, capital outlays alone are 7,000 NKr the first year, declining towards 4,300 NKr the last. So, over 10 years, in this example the total capital costs would be about 57,000 NKr, while energy saved would add up to 55,000 NKr. On top of this, a heat pump would need servicing, which would add to the costs. This makes the average, small installation marginally cost-effective, if at all. Furthermore, the installation is vulnerable to external factors. Higher interest rates, shorter depreciation periods, and lower energy prices all make the installation less cost-effective.

In addition, central heat pump actors held back or were overtly negative about subsidising heat pump installation in small dwellings on the grounds of the poor economy of small heat pumps, low quality, and experiences of unfortunate configurations in a number of residential installations in the 1980s.

However, little actual data existed on the scope and distribution of free riders and other problems. Accordingly the 1993 budget proposal included an assessment of means as part of a general overhaul of different state funded subsidies and other related measures. The goal was 'to secure a better foundation for the development of goal-effective means and subsidies' (St.prp.nr. 1, 1992-93, p.72).

The main findings of this assessment was reported in and was a foundation for the government White Paper on 'Energy-economizing and New Renewable Energy Sources' (St.meld.nr. 41, 1992-93), submitted in April 1993.

It is important to notice, however, that the detailed assessment only focused on: Municipal/State buildings, Office/Corporate Buildings, and Industry. Residential buildings were not included in the assessment.[3] Thus conclusions on free riders are valid for the former areas, but *not* for residential buildings.

The White Paper described the overall results of the state's policy as positive. The subsidies released energy savings in the order of 1,200 GWh (net savings). The environmental effects were very positive as well, CO_2-emissions were cut by 700,000 metric tons and 'the pressure for' hydropower developments were eased. Moreover, subsidies were macro-economically sound, saving 80 million NKr. Energy awareness was raised, even among actors who did not implement any energy savings measures, due to the analyses energy users would undertake in the wake of the subsidies.

Nonetheless, negative aspects loomed in the background. The assessment found that 15% of the companies implemented energy savings that they would not have undertaken without subsidies, 50% of all companies sped up energy saving implementation – i.e. undertook measures earlier than planned. But 35% of the companies 'did not have a real need for the subsidy given' (ibid. p.16), meaning that the pay-back time was so short that the installation would have taken place anyhow.

According to this logic, the number of free riders was very high indeed. When subsidies speed up installations by less than 2 years, the gain was 'not macro-economically sound when related to the cost of the measure' (ibid.). All in all, a full 70 percent of the users were categorised as free riders in the report, based on the notion that what is economically sound for private actors does not merit State support. Hence, the terse conclusion was that 'the number of free riders exceeded expectations' (ibid.).

The political translation of this was, not surprisingly, that the subsidies had to be terminated. But the argument went through a surprising turnaround.

The issue of free riders was central to the evaluation. It was reported that a full 70 percent of installations were free riders according to the definition. What is clear, however, is that the number of free riders was largest in the most cost-effective installations, where there was an incentive to energy savings measures without subsidies.

Equally clear is that the saving measures implemented in small residential buildings were marginally cost-effective, if at all. The small residential buildings were never free riders, as defined in budget propositions and committee statements. In fact, residential heat pump installations were not evaluated at all. But the subsidies, it would appear, *improved* the cost-effectiveness in small installations and therefore did exactly what these subsidies were meant to do, namely to get macro-economically sound installations underway.

Thus, the policy problem was shaped by the particular context, emerging as a way of 'sweetening the pill' in negotiations to secure the deregulation of the energy field. This resulted in two non-reconcilable positions that had to be solved. Either follow the principles laid down on subsidies and prolong subsidies for small heat pumps with relatively good macro-economic viability *or* break these principles and prolong subsidies for large heat pumps with good private economics.

The first alternative, to prolong subsidies for small heat pumps, was antithetical to the views held by central actors at research institutes, at the research council, and in the Norwegian Energy Agency (NVE). Among these, small heat pump installations were seen as fundamentally unsound. In this paper, I will try to show how these views gained primacy, formed part of a revision of the conclusions from the evaluation and, in turn, held up small installations as the most problematic economically.

CONSTRUCTION OF POLICY PROBLEMS

It has become something of a platitude to say that policy problems are socially constructed (cf. Hajer 1995). In this case, nevertheless, such a perspective throws light on the problems and solutions encountered. The process of choosing and changing energy policy measures is a political process, informed by socio-material factors, but also informing these socio-material configurations. Simply put, I will try to show the correlation between measures chosen and understandings of users in the energy field.

The field in this case has some similarities with what has been defined as a *development arena* (Jørgensen and Sørensen, chapter 7, in this volume). The case moves beyond the narrow confines of either policymaking or the laboratory, to include also information on users. The boundaries are not clear-cut, however. Arenas are populated by a heterogeneous set of actors, with strategic agendas. Likewise, the information used in arguments is broadly collected, but strategically used. It seems important to stress that a central part of the policy formation discussed here concerns the inclusion and exclusion of who is allowed to speak on the topic. This actually reduces the number of relevant actors significantly.

The way problems are defined, the context of the problem and the overall perspective on actors' motivations and rationality are all part of the process of making (energy-savings) policy. On the more general level this is described, e.g. by Jasanoff (1995), as a co-production of policy and action. We can talk of 'co-production' where technological knowledge and political order 'are simultaneously created and recreated so as to sustain each other through complex rituals of interdependence' (ibid.: 527).

From the point of view of the 'state', the broad reorganisation of the energy market constituted a social experiment. The experiment was, basically, how to induce market logic in an area which previously was monopoly-based and where prices were set by-and-large by the state's dictum. The basis for the reorganisation was that resource allocation is improved in a well-functioning (but not necessarily perfect) market. To make this work, users would have act more along an economic rationale and reduce costs wherever possible. Hence, macro-economic goals are intrinsically linked with micro-action compliance.

Of course, some market imperfections were expected, particularly as far as alternative energy technologies are concerned. Energy saving technology was seen to need some support, which would make such material configurations more attractive. In short, alternative energy technologies were seen to be *more* exposed to market imperfections than the already existing energy structures.[4] Moreover, the lack of energy savings knowledge had been seen as a central obstacle to implementations for most of the period. Subsidies would be used, then, to keep these two barriers in check and make deregulation more attractive among policy makers. Of course, subsidies were negotiated results of diverse interests and concerns, including industrial development, energy system soundness and environmental concerns.

At the base of these issues was the compliance or noncompliance of users to what can be called market logic. The knowledge needed for making and revising a subsidy decision, where insights about users is central, could be constructed in several ways. One way is to conduct a survey. But, in the evaluation of subsidies, larger installations were surveyed and smaller installations were excluded. Therefore, knowledge of small heat pump usage came from other sources.

The prime source of such knowledge was, therefore, the experiences of heat pump practitioners, including those in research institutes and related organisations. As experts, the latter had considerable leverage vis-à-vis the Ministry. Actors in that group constituted a link between the micro-actors, the small heat pump installations, and macro-goals, the policy makers. This makes the user stories told on this level especially interesting.

STORIES AND POLICY

As it turned out, during interviews with such actors, quite a number of stories on user-behaviour were told (Næsje 2000). I will briefly outline the organisational and policy role of stories, before turning to their roles in this case.

The use of stories and story-telling has been studied quite extensively in organizational settings, as a part of narrative processes in organisations (Gabriel 1998, Orr 1996). It seems uncontroversial to hold that stories are a part of organisation's symbolism and culture, and can act as vehicles for communication and learning (cf. Orr 1996). It is held that stories are one of the artefacts that generate and sustain meaning and, more important here, underpin world-views and rationality in organisations. As such, these stories can also play an important, if neglected, role in policy-making processes. We have seen in a number of other works (cf. Dierkes *et al.* 1996) the role of viable visions as guiding images for general technological development. I will argue that the use of stories is an important aspect of policy-making and technology as well.

The difference between fact and stories is not a clear-cut one. In my case, stories are told with the assumption of having factual substance. The stories told are *true*, but something is added. This something is narrative power. For example, a lengthy technical report on heat pump successes and failures (e.g. SINTEF 1992) is factual and true, but never moves beyond description. Such reports are terse and to the point, but lack narrative power outside its intended group of readers.

Terse accounts are not suited as vehicles for world-views. The narrative power of a good story, on the other hand, makes it fitting to this end. A good story *sticks*, hence it can bridge from one specific discourse to another. Simply put, a story can be retold due to its well-known structure and good points.

Then again, stories told in a specific setting, such as the energy discourse, are oftentimes not so interesting outside of the discourse. Gabriel (1998) points at the combination of terse narration and narrative deskilling as the prime reasons for the languidness of organisational stories,

> ... hardly a single story bears repetition outside its home territory as a 'good story'. ... organizational stories have the tendency to shrink into codified signifiers devoid of narrative. (Ibid: 95).

Furthermore, as a field becomes saturated with different kinds of information, the good story loses ground as it is seen as imprecise and lacking deductive force compared to other types of information: e.g. the survey central to the energy savings White Paper. Hence the story-teller's authority is stripped and the space for stories reduced. Likewise, some strategies for story-telling are excluded, especially in terms of credible generality and scope.

On a discursive level the stories told are one of the parts which constitute the discourse. In the analytic frame suggested in the late works of Foucault,[5] a discourse functions as an interpretive mechanism, allowing some statements and excluding others (cf. Foucault 1972a). The mechanism is defined through *exclusion* (prohibition, rejection and falseness), through *internal rules* (commentary, authorship and discipline), and through the *rarefaction* of the speaking subjects (ritually, in fellowships and doctrine) (Foucault 1972b).

The rarefaction of speaking subjects is relevant in this case, where, as we will see the stories are told by people who have some status within the 'fellowship' of energy savings discourse, being policy-makers or practitioners. Hence, stories are kept in circulation in policy relevant settings.

To briefly sum up: stories are a part of the rationalisation of policy. They are an integral part to the construction of meaning. They have more latitude in areas where little 'factual' information exists. Stories are suited to encapsulate knowledge and meaning, and bridge from one domain to another. The relevance here is that knowledge about energy savings subsidies in residential buildings was lacking and, more important, knowledge about the use of such technology and adoption of the market deregulation was low. Therefore, the operational space for such stories was large.

THE STORIES TOLD

> The plot must be of ... a length to be taken in by the memory. ...
> The perfect plot must have a single ... issue; the change in the hero's fortunes must be ... from happiness to misery; and the cause of it must lie not in any depravity, but in some great error on his part. (Aristotle, Poetics)

The perfect plot for a tragedy is easy to remember and has one point, according to Aristotle. The pivotal cause in the persuasive tragedy is the deployment of hamartia – the tragic flaw – as opposed to the deployment of moral failure. Tragedy caused by moral failure is not tragic; in that case, the hero got what he deserved, hence he is a villain.

The plot in the stories told about the use of energy does not qualify as great literature. The plot is, nevertheless, compact and singular, trading literary quality for mobility and effect.

Let us, then, turn to the material. By defining stories as having a singular punch-line and energy use as a theme, and keeping within the confines of what was conveyed as cases experienced, I found 30 stories in my material. These spanned from a couple of sentences, to 630 words over 11 paragraphs (in transcribed form). The choice of pieces constituting a story was done somewhat eclectically. Especially those interlaced with other arguments or those mentioned in passing, are difficult to identify according to rigid standards. All stories, however, relate, directly or indirectly, to some argument.

The shortest stories often presuppose a familiarity with the field. Failed heat pump installations is a case in point. Let me give one such example:

> We succeeded with many things, but there were many things, which did not go that well. We were only able to introduce [heat pumps] in

> residential buildings, which is the big area. But it did not go far enough. The Heimdal-project said: Heat pumps are not suited for residential buildings. (*informant AB, KKT, Interview transcript, emphasis added)*

The Heimdal project is well known amongst energy savings practitioners in Norway. It is not well known in the larger field of energy politics as such, but the argument is. The point in this story as told here is very compact, making it less mobile. It presupposes that the person involved in the conversation knows the case (something I, as an interviewer, did). On the other hand, the story can easily be extended in terms of mobility, by adding the different experiences from the project. Here experiences were those of a flawed installation, with construction failures, mechanical malfunctions and lack of user attention. By adding these experiences, reasons for not having heat pumps in residential buildings can be made clearer (or more opaque), and it is possible to identify a culprit (or avoid doing so). In essence, the story told here is open to interpretation, but the lesson learned – that heat pumps in residential buildings were not successful from the outset – stands firm.

Now, the overall impression is that it is possible to place the stories told in my material along an axis, based on where the focus lies. The stories have a hero (or villain) who is between two extremes; either based on a very tight coupling between social and material *or* a very marked decoupling between social and material.

Let us first look at stories which focus on social behaviour where the material configuration is decoupled from behaviour.

A major uncertainty in the attainment of energy savings concerns how users actually use energy and energy technologies. As we have seen, small heat pump installations and other measures for residential buildings were not surveyed in the evaluation of subsidies, so there was a considerable knowledge gap on the policy making level. However, in several policy documents and energy debates, users are held to waste energy: e.g. 'Users waste a lot of energy, because we have so much of it' (Reidar Due, chairman of the standing committee on Energy and Industry, debate in parliament; St.forh. April 2, 1979, p. 2884). As energy is cheap and abundant, a lot of users waste it. The view held by PF, IFE reflects this:

> Well, if you had a cheap meter to record energy use in an apartment, you would not encourage increased energy use, even if the energy system [such as a central heat pump] were collective. But we know, it has been proved, that all things which are shared indiscriminately, are also used a lot less conscientiously. ... I lived in an apartment complex with shared car-engine heating. And the bill for this was shared equally per head. We

> were annoyed by this one neighbor who left this thing on all night, you know ... why should we mess with timers and stuff, when he didn't care, so after a while, all of us did like him ... when a small meter would have reduced the energy use by, what ... 70% *(PF, IFE, Interview transcript)*

The point here is that free or cheap energy use (as in this case) provides no incentive to save energy. On the contrary, (small) economic gains are easily outweighed by the extra 'trouble' of changed behaviour. Furthermore, the lesson is that users have a low general awareness on energy use. Even as environmental issues become more important for people at a more general level, when it comes to actual behaviour it is difficult to see an impact, or so the story goes. Another, similar example is this:

> I'll give you one example, here from the city. It was this place [...] which had a ground source heat pump... people here used water from this to boil their potatoes, that is, they put the potatoes in a kettle under hot running water in the morning, and that water held like 60C. And when they came back in the afternoon, the potatoes were cooked. An extraordinary splurge! ... *(JS, KKT, Interview transcript)*

The problem readily identified is that energy is cheap and the energy cost is shared. In such a situation, people waste energy, as simple as that. The culprit as it is told here, is low price. Obviously, a material explanation could be offered. For example, some statistical data suggest that material factors, like the size and type of dwelling is the biggest (only) factor which impacts upon energy use (c.f. Aune 1997:95). Such an explanation is rebutted in the continuation of the story:

> Our apartment is connected to the district heating net, and it is superb. Thermostats in every room, you don't use more energy [than other energy systems] *(JS, KKT, Interview transcript).*

His experience is that modern energy systems do not invite extravagant energy consumption. The material configuration of the system delegates the control of energy use to thermostats, making it more difficult to use more than necessary. *Overuse* is thus avoided. What is not avoided is *waste*.

The introduction of a market is, then, one way to increase energy-use awareness. If economic gains are possible, the user will adjust his use accordingly, it is thought. This is even more so if energy prices rise and bigger gains are possible. The low energy prices, which decreased even further after the deregulation of energy markets, were held to be temporary: 'in the longer run, it seems clear to me that prices *have to* rise' (ED, NVE,

Interview transcript). A number of informants made such points. If this was not to happen, a number of the interviewees argued for energy taxation as a possible mean to induce energy awareness.[6] In any case, price and cost are seen as the consummate moral enforcer.

These stories underpin arguments that users are not as interested as they should be in energy conservation. These arguments tends to focus on the connection between what they see as too low energy prices and extravagant use, either by luxurious habits (too high temperatures in living quarters) or outright squandering, as in the two stories I reported. To moderate such use, the invisible (moral) hand of the market is considered to impose higher standards. Now, Weber has argued that protestant asceticism was pivotal in developing capitalism, with its contempt of luxurious consumption and appraisal of the 'rational and utilitarian use of wealth [...] for necessary and practical things' (Weber 1958:170f). In the modern society this is by-and-large naturalised, so we can have asceticism without Protestantism. But control and adherence to such values are, according to Weber's pessimistic views, not only internalised, but also frozen in 'the technical and economical conditions of machine production which determine the life of all [...] with irresistible force' (Ibid.:181). This argument quickly runs into problems, though.

In modern economies, the weight placed on free markets and *laissez-faire* does seem to be increasing. Countering the assertions of Weber, the subjects in free markets are moved to subscribe to the principles of the market, namely individualism, competition, and self-interest, in order to make it function (cf. Scott 1996). It is not the material fixations of every-day life which further capitalism, but the recreation of selves as rational consumers. It is in this process that the user stories make sense: how to assure rational energy-use (normatively) when the tools for correction of social behaviour are increasingly subtle? The behaviour of the rational egoist is easy to predict, as attention to costs and profit-maximising choices will prevail. From this point of view, rational egoists are the best subjects of the political economy in late capitalism.

Now, the reality is never this clean cut. A free energy market without any intervention from the state was thought to lack, at least to some extent, the ability to secure wider societal goals, like protecting jobs or the environment. So while a rational consumer would be nice in theory (or to such a theory), in the real world such rationality had to be fostered and managed.

The stories mentioned above underpin a rather negative or pessimistic view of social behaviour. There were some positive accounts of user behaviour as well. A typical positive story would often have the following structure. First, the story-teller would characterise the user's motivation.

Most often the motivation would be a concern for the larger environment (green values). On the other hand, a number of stories told that the user was concerned with comfort and/or indoor-climate. The first type is characterised by a surprising turn: heat pumps are installed with good (but not profit-maximising) intentions, but turns out to be economic anyhow! Installations for reasons not relevant for a free market or economic viability perform well on the latter issues. Good intentions (social) give good outcomes (material). Here the technology – the material fixations of behaviour – secures the outcome. The heat pump technology is thus vindicated and order prevails.

Stories where the other motivation – based on comfort – is at play, does not end that well. They relate to the role of door-to-door salesmen, from small firms, salesmen which, according to the same stories, had a field day when subsidies were available. Here the lack of technical competence combined with oversell is portrayed as important reasons for unsuccessful material fixation of energy consumption patterns. Moreover, these sellers 'give heat pumps a bad name!' (JS, KKT, Interview transcript). Thus wrong intentions combined with lack of knowledge pervert both the technology *and* the rational energy market, by playing on the irrational consumer – or as Veblen would put it, his conspicuous consumption.

Let us sum up these types of stories and their rationales before moving on to stories that couple the material and the social. We have seen that these stories keep material configurations out when they explain consumption patterns. Three faults are at play: the lack of knowledge; the lack of interest; and the moral inclination of user's behaviour. These constitute the base for wasting energy. But these three faults are easily altered. Therefore, the most important lesson in these stories is that *it is not difficult to save energy.* The issue is how to stop using it unnecessarily. Many consumers display less than optimal energy consumption patterns. And the construction and internalisation of the free market will reduce such use and secure the best possible usage of societal resources. The prime mover for attaining this goal is the price mechanism.

In most cases, installations transcend the specific artefact (e.g. a heat pump) into a configuration of several parts, with interlocking inputs, outputs, and use. In a number of stories, the configuration of social and material parts was the central theme. These stories establish close ties between material set-up and social behaviour, on several levels. Thus the stories carry a coupling between material arrangements and users.

There is one basic type of such stories. It configures technical interest and heat pump artefacts. These I will call 'The Citroën Configuration'. Let me give the seminal story:

PF: Well, [Heat pumps] are probably for the specially interested, technically interested, and among heat pump users in the residential market, if you go out and ask what they really are, they may very well be normal people, you know, but have a technological ... technical interest. Or are engineers. If you put up statistics on what type of cars heat pump owners drive, you would probably find a large share of Citroëns, don't you think? ... That's the type of constellations you can make to characterise the people who actually bought heat pumps.

I: So, do you find that to be positive or negative?

PF: No-no, it's positive, because the technically interested drive Citroëns. ...

I: But [Citroëns] are known to be unreliable, aren't they?

PF: Just as heat pumps! ... Technically advanced, but so-so with respects to secondhand value and reliability. Those who drive Volvos, they won't have heat pumps! *(PF, IFE, Interview transcript)*

The configuration here relates the technology to other social components. The technically interested, lay or professional, like heat pumps. Other stories, in a similar vein, focus on other social characteristics, like the story of a plumber installing a 'homemade' heat pump, often with other energy saving ad-ins. The latter configuration, according to those stories, performs poorly.

Either way, the stories hold individuals and their motivation as the prime mover for installing heat pumps. The point in the story is then that heat pump technology is special, not (based on the story-teller's experience) ready made, for ordinary users. They are not black-boxed (Latour 1987). Heat pumps are for those with the time and knowledge to pursue special solutions. And more indirectly, these stories question the very base for heat pump installation. The technically interested do not choose heat pumps out of economic self-interest only. They are seduced by the artefact itself, and therefore place too little weight on rational, economic calculations.

When it comes to pedagogical points, these stories individualise energy saving measures. The lesson learned, then, is that interest is indispensable when either behavioural or material changes are wanted, as personal motivation is the only factor which will move users to undertake fairly complicated installations. Moreover, users that have undertaken such installations are marked with the insignia of 'special interest'. On the other hand, users who lack political interest also lack motivation, and no change will occur.

The story told is then a story about the *tragic* energy economising effort, that was destined to fail, because such interests are for eccentrics – those '*who drive a Citroën*'. The coupling between material setup and social makeup clearly reduces the potential for energy savings. As solutions are individualised and made a part of the social makeup of the user, any energy economising effort aimed at a substantial change in energy consumption patterns will fail. Now, one alternative could have been to allow for subsidies even with a large proportion of free riders. Such subsidies could develop people's interest in technical solutions. But this was at odds with the energy discourse, with its credo of rational and attentive users, to the extent that when the subsidy was securely connected to free-riders, it was killed in parliament with only minor regrets.

To sum up, we have seen that the stories told can be classified along two axes in terms of storyline and rationale. First, the stories operate with different views on how the social aspects of energy use relate to material installations: are use and installations closely linked or not? Second, the stories differ in how easy or difficult it will be to change energy use. These findings are summarised in figure 10.2.

Figure 10.2 Storylines and the relation between social and material aspects of energy use. Summary of findings.

	Coupling between social and material:	
	Tight	Loose
Easy to change energy use	Storyline: *Citroên-owners* Positive interest spurs installation	Storyline: *Wasteful behaviour* Much use is unnecessary
Difficult to change energy use	Storyline: *Volvo-owners* No interest equals no installations	Storyline: *Rational use* Use meets real needs
Strategic Question:	Distribution of interest	Distribution of moral makeup

This also has implications for the strategic question of whether policy has to change interests or the moral make-up of its citizens (the latter a difficult task indeed).

If any singular story prevailed, these stories would bring different outcomes to the life of the heat pump subsidy. The next part will discuss the subsequent life of the stories, here it would suffice to mention that the *rational use* storyline found few allies and had never any central role in the energy discourse.[7]

Configuration of stories and measures

As indicated in the stories reported above, the actors I interviewed - energy policy-makers and heat pump technology experts - had quite negative views on users and small-scale installations. To make such views known, accepted, and integrated in decision-making processes, several different mechanisms could be at play. It is important to keep in mind that no reified knowledge existed on how subsidies were used in small installations, as only larger installations had been surveyed in the 'evaluation of means'.

Let us then look at the decision process in parliament and the arguments used in policy documents and debates. It is important to notice that there was low interest in these questions from the mass media, so they did not provide material for the debates. Points to substantiate the arguments were, as we have seen, based on the stories that have been discussed above. But in political rhetoric, the use of stories, and the type of knowledge such stories represent, will often be challenged by a call for 'hard facts'. If stories are made more authoritative by being connected to 'hard facts', there was a gap in argumentation in this case, as no 'hard facts' existed. This gap was filled by the stories told, but carefully reshaped for use in political debates. The role and authority of stories will differ corresponding to the subject's position. Thus, an argument put in parliament has to rely on different rhetorical devices than, say, in energy researchers' brainstorming sessions. For the sake of argument I will hold on to this distinction (albeit problematic) between hard facts (social/economic data) and stories.

In parliament, the leitmotif was the role and distribution of free riders (St. Forh., June 7, 1993). The arguments regarding this were substantially based on circumstantial or indirect claims. Let us take one example: when discussing the distribution of free riders the representative Eva Finstad (Conservatives) put forward the following argument:

> I am very satisfied with the change of the non-specific subsidy of energy usage. ... We Conservatives are in full agreement with the State Secretary Gunnnar Myrvang *[Labour Party]* who in an editorial in the Arbeiderbladet *[Labour Party-friendly newspaper]* wrote: ' ... an analysis of the energy saving subsidies shows that there are too many free riders, more than the State can accommodate. ...' *(Finstad, St.f. June 7, 1993, p4200)*

Now, what Finstad calls 'a change' is the deletion of energy savings subsidies. It is interesting to examine the basis for this argument, which involves three distinct modal layers.

First, we have the evaluation of energy economising subsidies in 1992. As pointed out before, this evaluation did not survey small heat pump installations, and its main conclusion was that large installations were free riders. The argument here is, nevertheless, stretched to cover all energy usage.

Second, the editorial piece is invoked as holding real information on the subject. The truth-value is enhanced by pointing out and using a piece which is written by a Labour Party politician, the principal opponent of conservative politics in Norway, and where the 'fact' presented can be turned into a point substantiating the conservative's standpoint. So when the article argues that the subsidy was flawed, without mentioning results/lack of results for different user types, this is because the information is tuned into a support for the argument which they want.

Third, the use of the editorial piece neatly adds distance between responsibility for 'fact' and the argument. The 'fact' portrayed is used as such, but the responsibility for its truth-value is not Finstad's. A story of the kind presented in the previous section would here be easy prey. Whereas this argument's value is secured by pointing to the broad distribution of the (free riding) phenomenon, a *story* plays on the special and the singular for effect. Thus a story may be a vehicle for translation and transport of world-views and knowledge of the field, but has less to offer in a political debate. In this specific case, the truth-value of stories is supported by the use of rhetorical devices such as authority: it is presented by the State Secretary, in an editorial in the well-known Arbeiderbladet. All these modalities could be changed and it would change the effect and impact of the argument.

In important ways, the argument presented effaces the fine-drawn socio-economic arguments put forward in the evaluation. But in-so-doing, it aligns the 'facts' presented in the report with the knowledge presented in stories. There are very few references to socio-economic data and numbers in the debate (and none to the residential energy saving measures, of course). I would suggest that this might be due to the integration of two stories: 'that small installations are not worthwhile', with the results from the evaluation: 'that subsidising large installations is not worthwhile'. The terse referral to 'fact' opens discursive space for knowledge based on stories.

The result of the debate was that the government White Paper on energy savings and its recommended means – cutting back of subsidies – was approved. Promised subsidies would be given, however, which secured a relatively large grant in the budget allocations for the next year. After this, only minor funds were granted with negligible funds for heat pumps. Some other technologies were redefined as being 'new renewable energy technologies', like biomass energy and solar energy and were the only technologies to be granted direct funding.

POLICY RESHAPED

> Oh no, not that story again! We'll have to slave with it forever! (TS, NOVAP, Interview transcript)

The cutting back of energy technology subsidies can be seen as weaving together several factors. In this paper, I have tried to depict the outlook of practitioners and decision-makers on small installations and their impressions of user rationales. One additional factor in the equation was the institutional framing of subsidies.

As mentioned, the Minister of Energy and Industry at that time, Finn Kristensen, leveraged large funds in direct subsidy for alternative energy technologies and energy savings. The backdrop for this was energy market deregulation, where state involvement was seen as necessary to foster good energy usage and promote energy solutions which were beneficial at a societal level, if not on the individual level. The individual user was seen to lack a long-term approach to energy related investments. This was thought to be worsened by a deregulated energy market, as calculability is made more difficult with fluctuating energy prices. When budgets are tight (which would be the starting point of any budgetary discussion for the Department of Finance at any time), the effectiveness and success of all state funds, especially subsidies, are reviewed. As we saw, these subsidies were targeted for a pilot evaluation of state means, which concluded that a large proportion were given to free riders.

Thus the subsidies were under pressure from the Department of Finance. Some energy technology proponents lamented this, arguing that the evaluation was given too much weight. One proponent, the organisation for heat pumps in Norway (NOVAP), argued that the evaluation was too limited in scope and that its conclusions were tentative: 'this tiny report from Energidata' (TS, NOVAP, Interview transcript). Nevertheless, it was used to its full capacity, as the death sentence for a flawed state measure. This was only possible with the direct interaction between stories and evaluation; these were mutually reinforcing with the result that subsidies were cut.

More generally, the case illustrates how the definition of problems and construction of data relevant to the problem are informed and shape, by a set of factors mostly neglected by traditional policy analysis (e.g. Weimer & Vining 1992). The use of stories is an integral part of policy making as a reservoir of easily mobilised bits of information. This information makes political sense, or sense out of policy.

Let me then return to the notion of co-production. As argued by Jasanoff (1995: 527ff), today we are experiencing a 'shifting yet self-perpetuating "system" in which scientific and political order is created and recreated'. In important ways, we can broaden this assertion to include socio-technical systems in the place of science. The relation between energy policy and energy technology is also about negotiating order. But this negotiation was never one-way, from policy makers stating means to diligent practitioners, on the contrary, the stories presented among practitioners were active vehicles of information transfusion. They allowed the evaluation of means to encompass areas not surveyed. However, it would seem reasonable to expect that the stories were altered by the evaluation as well: the centrality of economic free riders indicates this.

Therefore, the process was a reordering of the field, where diffuse political means were aligned with diffuse technological set-ups and brought in accord with Norwegian energy discourse. That this meant that both measures and technology were terminated was – alas! – never intended by heat pump proponents.

NOTES

1 An earlier version of this paper was presented at the 1999 meeting for the Society for Social Studies of Science, San Diego, CA, October 28 to 30; session 'On Kilowatts, Carbons, and Cars: The Politics of Energy and Environment'. I would like to thank Jane Summerton, Hans Glimell, Knut H. Sørensen, and Robin Williams for valuable comments on the paper.
2 Cf. Governmental white paper on energy economizing, St.meld. nr. 41, 1992-93 pp.14
3 Letter from the Ministry to the Parliamentary Standing Committee on Energy and Industry, dated June 1, 1993
4 Private communication with Senior Officers at the Ministry of Petroleum and Energy.
5 Oftentimes, the work of Foucault is seen to have changed focus from the earlier (e.g. *History of Madness*) to the later works (e.g. *History of Sexuality*), from archaeology to genealogy. Though problematic, in the later phase, the relation between knowledge and power in modern society can be said to be the main focus (cf. Dreyfus & Rabinow 1983: 104ff).
6 In current energy debates, the use of energy taxes is hotly disputed (e.g. *Aftenposten* 10/11/00, p.2 'Budsjettforhandlingene').
7 This position could be named 'the social-scientist-position' as it has been most strongly advocated in social science research. (Cf. Aune 1998).

BIBLIOGRAPHY

Aristotle (1920), *The Art of Poetry*, I. Bywater (tr.), Oxford: Oxford University Press.

Aune, M. (1997), *'Nøktern eller nytende'* 'Energiforbruk og hverdagsliv i norske husholdninger,' Dr.polit. dissertation, *STS report* 34, Trondheim: Center for Technology and Society, NTNU.

Dierkes, M., Hoffmann, U. and Marz, L. (1996), *Visions of Technology*, Frankfurt: Campus Verlag.

Dreyfus, H.L. and Rabinow, P. (1983), *Michel Foucault. Beyond Structuralism and Hermeneutics*, Chicago: University of Chicago Press.

Foucault, M. (1972a), *The Archaeology of Knowledge*, New York: Pantheon Books.

Foucault, M. (1972b), 'The Discourse on Language' p215-237 in *The Archaeology of Knowledge*, New York: Pantheon Books.

Gabriel, Y. (1998), 'Same Old Story or Changing Stories? Folkloric, Modern and Postmodern Mutations', p84-103 in D. Grant, T. Keenoy and C. Oswick (eds) *Discourse + Organization*, London: Sage Publications.

Hajer, M.A. (1995), *The Politics of Environmental Discourse*, Oxford: Oxford University Press.

Jasanoff, S. (1995), 'Science, Technology and the State', p527-531 in S. Jasanoff, G. Markle, J. Petersen and T. Pinch (eds) *Handbook of Science and Technology Studies*, Thousand Oaks, CA: Sage Publications.

Latour, B. (1987), *Science in Action*, Cambridge, MA: Harvard University Press.

Næsje, P.C. (2000), 'Pumps and Circumstances: The Political Configuration of Heat Pump Technology in Norway,' *STS-report* nr. 46, Trondheim: NTNU

Orr, J.E. (1996), *Talking about Machines*, Ithaca: Cornell University Press.

Scott, A. (1996), 'Bureaucratic Revolutions and Free Market Utopias', *Economy and Society*, p.89-110, vol.25(1).

SINTEF (1992), *Erfaringer fra prototyp- og demonstrasjonsanlegg for varmepumper*, SINTEF report STF11-A92017, Trondheim: SINTEF Energy.

Weber, M. (1958), *The Protestant Ethic and the Spirit of Capitalism*, New York: Charles Scribner's Sons.

Weimer, D.L. and Vining, A.R. (1992), *Policy Analysis: Concepts and Practice*, Englewood Cliffs, NJ: Prentice-Hall.

11. Restructuring the Power Arena in Denmark: Shaping Markets, Technology and Environmental Priorities

Ulrik Jørgensen and Lars Strunge

INTRODUCTION

Liberalising the power sector in Denmark has turned out to be a much more complicated process for the Danish government than initially anticipated by the politicians and ministries involved. The process that started in Denmark in late 1996 with the decision in parliament on law no. 486 is involving still more elements and engaging a growing number of actors with different backgrounds.

The fascinating aspect of this restructuring process is the complexity of the policy formations involved and how radical the changes in configurations of actors and priorities seem to be. Heterogeneous elements are melted into this policy process, including: economic expectations about the results of liberalisation, traditions of public involvement in power supply companies, opening access for investment capital, environmental priorities, and support for innovative activities in power supply technologies.[1]

From the perspective of technology and environment studies, the interesting aspects of this policy process are the roles and performances played by both technology options and environmental protection measures. Or to rephrase the problem: how are technological and environmental options constructed and translated throughout a radical re-configuration of a supply sector? These constructions will often be controversial and imply radical shifts of inclusion and/or exclusion of both the objects of policy and the actors involved, as new, temporarily stabilised configurations are achieved in a transforming policy arena. This implies the stabilising of new actors, routines, practices, and objects of regulation.

Two questions can be regarded as central and can be used to pinpoint the implied impacts of the restructuring process:

1. How does the distribution of property rights and institutional structure influence technological change and innovation in power technologies, especially concerning the environmental priorities?

2. What is the potential for regulating the power sector regarding the objects and means of regulation conditioned by the structure of companies and markets?

These questions are important from a technology studies perspective, although they do not cover all possible policy perspectives involved in this process. By pursuing these questions, the rather spectacular changes in the policy process are highlighted, showing the very different policy perspectives that dominate the different phases of change. The technology studies perspective pays serious attention to the role of technology in societal development. It also criticises policies based on simplified and determinist ideas of technological change. Although both technology implementation and policy formation can be viewed as ways of imposing change, they cannot be viewed as fundamentally different or separate processes. Technology formation and priorities include specific policy perspectives, and policies concerned with the structure of the power sector include views and implications for the role of technology. In other words: technologies imply politics and politics imply technologies, even if this is not part of their expression.

SETTING THE STAGE: LIBERALISATION AND RE-REGULATION

Traditionally, large publicly owned power supply companies or licensed private companies have dominated the power sector in almost every European country. They have operated as 'semi-public monopolies' and have been closely regulated by government agencies with regard to supply and demand, price setting, environmental issues and ownership. But in line with the general political aspiration to liberalise the economy in this and similar traditional and closely regulated infrastructure sectors such as telecommunications, air transport, railways etc., the 'semi-public monopolies' of the power supply sector in the European Union (EU) were to be transformed into private companies competing at a marketplace.

Following a EU Council Directive in December 1996, all member states were supposed to open the power market for competition by February 1999. According to the EU directive, the idea of competition among the companies was to give power consumers – especially public and private companies and perhaps even individual households – a free choice of supply in the marketplace. On the basis of transparent prices, consumers should become empowered to choose between different power producers and distributing companies. The overall benefit was supposed to be lower prices and a more efficient production of power. This schematic reasoning about the transformation process delegated the principal role to market competition, which was expected to be the most effective mechanism for distribution and optimised process renewal (Collier, 1998). Although this 'neo-classical' rationale can be and was questioned by a number of actors, it constituted a very strong point of reference for those actors in the policy arena who were able to stage the transition. This delegation to market forces of the ability to produce benefits in terms of efficiency and continuous technical innovation creates a simplification and excludes an explicit discussion of implied technological and environmental priorities.

Besides being phrased as a liberalisation of power supply, the process has also been labelled as a de-regulation effort, although it eventually and more precisely can be described as a re-regulation of the power supply system. The liberalisation effort does not *per se* create a non-regulated marketplace for transactions for power, but it imposes a new institutional framework and constitutes new avenues for action as well as new conditions for public regulation. Also, the possibilities for setting environmental performance criteria for the power sector (e.g. concerning the international CO_2-measures outlined in the Kyoto-protocol) will be influenced by these new regulatory conditions. This quite complex policy process creating the new development arena for the power sector will be investigated.

The answers and solutions to liberalising the power sector have in fact been of very different kinds – as will be shown in the chapter. They have ranged from very limited changes, opening competition only on restricted markets and for restricted groups of customers, to a 'laissez faire' liberalisation opening competition on all levels, including production, distribution, network infrastructure, company structures and sales of power to consumers. By setting new boundary conditions, the traditional development arenas in the Danish power sector are challenged and eventually made obsolete, and the power sector is opened for new actors that completely reshape the policy arena and change the objects that are significant for policy. Thus, the ability of actors to enter their perspectives in new constellations and to set in motion and translate their interests in the arena is important for consolidating or gaining influence. The translation of policy objects also leads to the exclusion of positions and actors, regardless of the significance of their earlier status.

THEORETICAL FRAMEWORK: ARENAS OF POLICY AND OF DEVELOPMENT

The analytical framework used in this chapter is inspired by developments in political science, where new theories dealing with the framework for political decision making and interest articulation have been developed, known as policy-networks and sector approaches (Benson, 1981; Rhodes & Marsh, 1992). In line with these developments, the concept of policy arenas, as introduced by Fink (1996), is used. The point here is that the arena is not only the stage for political performances, but it defines the space where policy is constituted, reflected, and enacted.

The notion of 'development arena' is used to identify a framework for analysing the emergence and stabilisation of new products and technologies (see Jørgensen and Sørensen, chapter 7 in this volume). It hereby extends the scope and reach of approaches developed in technology studies, especially in actor-network based studies. The metaphorical notion 'arena' was borrowed from political science, where it has been used to define the spaces where certain policy actions and discourses are located. It was also used to provide a framework to understand both the structured and limited characteristics of those discourses, as they seemed to be in flux and potentially open for restructuring by the entrance of new actors and new policy perspectives.

What arenas of development and policy arenas have in common is that they enable the study of patterns of action performed between heterogeneous networks of actors engaging in the creation of change in policy formation and technological innovation. Arenas frame spatial and temporal ordering processes. The framework also provides an opportunity to deal with the boundaries of staged spaces of the arena and thereby helps to identify the very different sets of actor-networks and actor-worlds that they inscribe themselves into.

The arena metaphor creates an associative imagination and language that enables the analyst to deal with the relatively open character of an arena. And it enables the study of restructuring of the arena based on new entrances and strategies of recruitment on the one hand, and the exclusion or pushing to the margins of actors, artefacts, and visions on the other. What for a period of time seems to be a stable actor-network, forming an arena of strongly related and well-positioned actors, and defining a set of taken-for-granted routines, development initiatives, and sense-making interactions, can be radically changed by new entrances or changed alliances and mediations. The emerging new actors constructing new networks and defining new objects of policy and development are judged by themselves and other actors in the arena for their ability to stabilise their presence and practice. This implies a restructuring of the actor-networks and their actor-worlds that are already represented in the arena.

The notion of machination has been used to cover this ability to transform an arena (Latour, 1988), whereby the focus is on the actors' performance in setting other 'actants'[2] in motion and translating the mediations in the network. Another way of analysing the performing aspect of actions is to study the role of delegation, in which actors delegate objects or other actors to carry out the transformation and to secure inscribed transformations for the future. This is the case when actors delegate and trust economic institutions, like markets and competition, to carry out the inscribed job of securing lower prices and maintain the future technological development in a sector; or when a certain established regulatory regime, based on close and established relations, is translated to work in a different configuration of ownership relations and company structure. Expectations delegated to objects or processes are not sustained throughout their successive translation in the network as immutable capabilities to put other actors in motion, but are dependent on continuous re-mediation by other actors. One central example in this context is that technological developments resulting from a regulated energy sector cannot be expected to be the same as those resulting from the competitive arrangements in a liberalised market setting, even though the policy goals of regulation stay unchanged in the process.

Especially the porous nature of the boundaries of an arena framing the loosely coupled actor-networks for specific temporal engagements motivates the use of this framework to identify the rather often weak actor-relations and provide the process with continuity.[3] By making the messy boundaries conceptually visible, they can be made into explicit objects of study. Unlike sector boundaries, which are often constructed through functional differentiation and established institutional relationships, policy arenas form a space where actors operate and create new relations by managing uncertainties, re-structuring relations, and recruiting other actors. The boundaries of arenas are fluid and mutually shaped by the actors engaging themselves and being recognised in the arena in question. The empirical question is: who cares and what do they want to do about it? This is similar to the 'follow-the-actor' maxim established as a basic guideline in technology studies.

It is therefore quite obvious to try to cover the number of different controversies in policy by introducing the arena notion and to use this framework to link together perspectives and results from technology studies with studies of policy formation and the involved number of different rationales and institutions. Technology studies have developed an understanding of technological change under the umbrella of 'social shaping of technology', but the impact of these findings on technology policy and policy in a broader perspective has still been limited. This is partly due to the relatively new emergence of this approach in technology studies, but it is also because technology studies have often included a too limited understanding of the specific conditions of policy making. Although individual case studies may reflect a rich understanding of specific policy processes, an overall framework to bring these fields together still has to be established.

In certain periods of time, very different actor-networks may be identified in an arena that is opening for different configurations of the arena. These configurations can build on both the character of the involved actor-networks and their linkages and on boundary objects that might be created (such as the role of wind energy) or obligatory points of passage (such as the importance of meeting the minimum measures for liberalisation). The configurations may also differ depending on the existence of centres of authority either defined inside separate actor-networks or established as common centres of translation (Callon, 1986). Actors in an arena may have multiple engagements in several of the actor-networks that assign them to the role of co-ordinator, but in certain situations also as translator of an outcome of conflict.

Compared to the set of strategic moves that have been discussed in chapter 6 on development arenas, it can be questioned whether there is a need to extend the list of strategies that actors can try to engage in. This list includes strategies of: 'resettling and inclusion', 'extension and differentiation', 'exclusion', 're-framing', 'multiple engagements', and 'reduction and ordering'. Especially the strategy of 're-framing' has to be understood as the action of certain actors to enrol what is so far a marginal or external actor-network by presenting a set of schematised inscriptions and delegations that are given authority by the included actor-world. This is the case when e.g. market and competition are introduced as mediating entities in the power sector.

A successful translation can be the outcome of an enrolment, whereby the actor-world and, what may be termed specific 'rationales', provide a repertoire of well-established rules of thumb and taken-for-granted translations and exclusions for the actors. Rationales are hereby understood as an internally consistent and ordered conceptual framework of arguments that enables the actor to filter information and arguments and at the same time align certain actions and arguments (Strunge, 1995). Rationales bear similarities to both the enabling features and the simplifications involved in the actor-world concept, but they are especially of interest when analysing the background for taken-for-granted operations by actors. Rationales are also of importance as schematised arguments in boundary work, leading to stabilised and simplified networks. In this way, actors inscribe certain enabling features into the objects of action and policy, creating a local context of meaning. The creation of policy objects and the enrolment of certain rationales in policy discourses are thus important steps in the policy formation process in an arena.

Especially, the role of policy rationales included from economic policy and theory are contrasted with strategies of technology development and environmental protection measures. The delegations and eventually implied technology priorities in economic models and metaphors are hereby analysed within the framework of technology studies emphasising the differences in societal support and priorities for technological change.

Using this framework, it is possible to highlight how different actors can enter the arena and stage very heterogeneous entities and policy perspectives and thereby change the possible outcome. Their involvement is also important in defining policy agendas and establishing rationales based on the inclusion of established actor-worlds. This may sometimes lead to translations of elements in the involved actor-networks and the construction of new stable boundaries and regimes, sometimes leading to contradictory strategies and long-lasting conflicts in the arena.

THE NATIONAL BACKDROP: EVOLVING DANISH ARENAS OF POWER

Until recently – and before liberalisation took off – the power sector in Denmark was regulated and operated as a public monopoly, in spite of private or semi-public ownership of most of the contracted operators.[4] The principal operators in the Danish power sector have been the distributing companies at the local level and the power utilities, which mostly work on a regional basis. The distribution companies were owned and run either by local communities, by consumers in co-operative societies, or in one case, by a limited private company, the majority shareholder of which is one local community north of Copenhagen. The distribution companies had a government-contracted monopoly on selling power to consumers in their local community and on distributing power through local networks. Prices were regulated by government-negotiated agreements and through the representation of consumers or community owners in the distribution companies. The Power Price Board, one of the Danish authorities responsible for competition control, has supervised the negotiation process. Since the power sector was considered a non-profit business, prices had to satisfy the principle of 'only paying the costs', which included the need to accumulate profits for re-investments, performance improvements, and capacity enhancements in long-term planning of supply.

The supply of power has traditionally been maintained by a dozen large power utilities distributed around the country. The typical situation was that the local distribution companies owned the power utilities in their region, and consequently were obliged to buy their power from that same power supplier. It is in fact a distinct feature of the Danish power sector that all the large power utilities have been closely vertically integrated with the distribution companies – and hereby also indirectly controlled by the consumers. In comparison, in other countries like Germany and France, the power utilities controlled the distributors (Collier, 1998).

As a result of the growing centralisation of production facilities after WWII, the power utilities have based their power production on large-scale, steam-turbine technology. In the beginning, they were fuelled by oil but have switched gradually to coal. This has led to even bigger power producing facilities, as a result of the growth in the size of power generators internationally. Following the growing centralisation of the power utilities and their co-operation in the two regional structures, the number of actors has been limited in this part of the arena, dominated until the late 1970s by the government and the utilities and distributors. The main objects of policy were the continued negotiations on prices, efficient production units, and regulation of investments in the necessary surplus capacity to maintain growth and meet peaks in power consumption. The large power utilities have also been running the national power network that connects the local distributors. Hereby, they both regulated access to the network and had responsibility for securing the balance between supply and demand in the national power network. To efficiently maintain this balance, the Danish power network has been linked to both Norway and Sweden and has exchanged power with companies in the other Nordic countries.

The Danish power arena was quite stable for a number of years, starting in the 1960s, with regard to the objects of both policy and regulation. The technology development arena has also been rather stable, dominated by questions of techno-economic optimisation and growing attention to the environmental impact of power production. This did include continuous change concerning fuel optimisation, power production and distribution technologies, and a growing international integration of power networks, but the changes happened according to a quite stable set of engineering rationales. These rationales were developed and maintained in an actor-world defined by power planning engineers and energy officials who made the forecasts and planned capacity, growth, and prices. The changes were negotiated within a policy framework with a stable configuration of actors that included the ministry, local municipalities, engineering departments of the utilities, and the power company boards and their public representatives. While there were no incentives for improving production facilities and the organisation of the power companies, the professional standards and concerns of management (also mostly engineers) and regulation enforced a relatively cost-effective production of power.

The oil crisis in the early 1970s prompted several new policy options to enter into this seemingly well running and closed power arena. The need for national self-sufficiency in the supply of energy, the focus on power-saving measures, and the interest in renewable power sources were the new objects of policy to be dealt with.[5] A number of different measures were taken that opened the power arena for new entrances and reshaped its boundaries. Some of these rather sustained the existing actor-networks in the power arena, but others opened opportunities for new actors and more radical changes. One of the sustaining elements was to speed the transition from oil to coal. Another was to re-enact the already existing, special feature of the Danish power utilities in bigger cities, where co-generation of electrical power and steam-transmitted heating for buildings had been a way to optimise the energy efficiency of the utilities. The Ministry of Energy made a major effort to establish a heating plan for Denmark, focusing on reduced energy consumption for heating. Besides the renewed focus on co-generation, the plan established support schemes for insulation and better control systems. The renewed techno-economic rationale was to maximise efficiency in fuel utilisation and minimise dependency on imported oil described as 'integrated resource planning'. As a supplement, the consumer-owned distributing companies were made responsible for campaigns and advice to consumers to reduce power consumption by changing habits and buying energy-efficient equipment. This obligation was phrased 'demand side management'.

The slightly changed power production system was, although still based on large-scale technology and optimised power management, now also meeting the new demand to localise utilities to enable the distribution of steam-based heating for buildings. At the same time, more efficient electrical equipment and the results of government plans for more efficient building insulation, supported by strict building codes regulating the construction sector, started to slow down the growth in power consumption. Instead of just expecting continued growth in power consumption, a need for better power consumption prognosis was evident. All in all, the established pattern of continued scaling up production facilities was discontinued, as in other countries (Hirsh & Serchuk, 1996). Although these new measures bracketed the growth in energy consumption and thereby also slowed down the continuous renewal of production facilities, it was still possible for the dominant engineering actor-world to sustain its visions and planning rationales. The new energy sources were integrated into the overall planning and management schemes, and the growth scenarios exchanged with scenarios for plant renewals and environmental protection schemes for filtering smoke.

Fitting nicely into the scale-intensive power production system, the power utilities originally intended to build at least one nuclear power station in Denmark according to a plan presented to the public in the late 1970s. But due to an overwhelmingly strong popular resistance, the Danish parliament passed an act in 1986, which *de facto* put an end to plans for nuclear power plants in Denmark (Petersen, 1994). Instead, there has been strong and growing public support for developing renewable energy sources to substitute the use of limited oil and coal resources and to reduce emissions from power production. These changes in visions and alliances in the power arena were dramatic and first led to the building up of expertise in the atomic power plant construction which was later given up by the utilities, but this phase of the developing power arena is not in focus in this study.[6] As a result, instead of a continuation of the existing technology regime with inclusion of atomic energy, which would also imply the exclusion of the energy technologies based on renewable sources, the arena underwent a more radical restructuring.

DISTURBANCES: ENERGY CRISIS AND RENEWABLE ENERGY SOURCES

Denmark has focused particularly on wind power. However bio-gas production, wave power, and solar energy have been promoted by public procurement and R&D-funding by the government. Following the turmoil after the oil crisis in the mid-1970s, this focus on renewable energy sources and development of energy technologies has entered and restructured the power policy arena. New actors entered, including grassroots entrepreneurs, small investors in new energy technologies, and energy and environmental movements. Technologies that had been deemed non-viable in earlier phases of Danish energy policy were now taken up again by small producers, forming, in the first phase, local development arenas around the new technologies, which were only loosely linked to the national power supply system. But they resulted in extending the boundaries of the power arena to also comprise these new actors and a number of new policy objects, including small power-generating installations, network access, support measures for new energy technologies, and environmental priorities. In this process, not only were the new wind turbines becoming part of the power production system, but they were also translated from being local power producing units integrated in self-supply networks of the green movement to becoming 'just' small power utilities connected to the grid.

As a result, the power arena now consisted of at least two partly connected, partly separated sub-arenas, each of which referred to still separate actor-worlds of visions and priorities. And in contrast to earlier situations, where most of the equipment for power generators had to be imported except for the Danish stronghold in the production of insulated pipes for co-generation systems, the new arena also included a major industry, its consultants and spokespersons. The group of actors in the power arena has been extended to include turbine owners, industries, environmental policies, and a growing and diversified number of new energy-producing artefacts and connection management tools.

While energy efficiency and co-generation were merged into the traditional techno-economic rationales of the existing regime in the old power arena, the growing importance of wind-turbine based power production was seen as a major threat in the engineering actor-world of the power sector viewing the new energy technologies as small and unreliable. This was the case even though it was supported by an emerging new group of environmentally concerned engineers and politicians. The traditional rationale of growth, planning, and centralised power production was attacked and, if not rendered obsolete, then at least de-stabilised. The power arena had undergone a reconfiguration implying both an extension and a differentiation in visions and core artefacts.

Under heavy opposition from the traditional power companies, the technologies based on renewable energy sources even received strong political and financial support from the government. Offering positive tax and price incentives for individuals to build and run small-scale wind power facilities was one of the support measures initiated. Another was that the power utilities were forced to purchase excess power from the private wind turbines at a negotiated price, which was typically higher than the production price from the big power generators. This created a supply driven development arena for renewable energy technologies, and helped the environmental policy measures introduced by the energy movements to become a legitimate object of policy and an obligatory passage point in the energy debate in Denmark.

As one result of these struggles, and the creation of a sub-arena for renewable energy technologies, Danish wind power technology is today at the forefront of innovations, and the Danish wind power companies are market leaders on the global scene. Although this was not the result of a deliberate industrial policy of the Danish government, the demand-driven approach to the development of wind turbines to produce power has had this indirect outcome (Jørgensen & Karnøe, 1996). The conflict in the power arena between the traditional energy actors and the new innovative companies has established a continued process of innovation, forcing the wind turbine producers to optimise and up-scale their products and become involved in strict testing procedures. Renewable energy actually plays a growing role in the power supply. In 1996, renewable energy accounted for three percent of the total production of power in Denmark, and the actual installation plans will increase this to account for 15 percent in some years.

A major shift in the power arena has occurred recently, following the growing distribution of wind turbines in the Danish landscape. This shift allows only few possible localities for wind turbine placements. By establishing the visual qualities of the landscape as a criterion for allowing the localisation of wind turbines, the local communities have placed tight restrictions on the area utilisation plans, forcing government and other actors to seek alternative solutions to secure the planned capacity of installed turbines. This has led the government to use its regulatory power over the utilities to force them to engage in installing wind turbine parks, first on land but in the coming years with a large capacity to be installed offshore, where the wind regime is optimal.

During the last couple of years, a growing number of small-scale combined heat and power production facilities have also been built locally. And as mentioned, there are also relations between the power arena and the gas arena, now having local co-generation systems as boundary objects. They are not controlled by the large power companies, but directed by local communities. Legislation required the large power companies to assure access of the power produced by these decentralised power-heat facilities to the power distribution network. This has also introduced a new group of actors involved in energy planning and added one more sub-arena to the contemporary configuration of the Danish power arena.

THE EUROPEAN COMMISSION: A (NEW) PLAYER CREATES (NEW) BOUNDARY CONDITIONS

The EU directive on liberalisation of the European power industry was passed in December 1996. It prompted the reshuffling of the national arenas for energy policy and energy technology development. The directive outlined some new guidelines concerning the definition of the open market, criteria for inclusion of policy goals concerning renewable energy sources, a separation of the network system and the suppliers, and some other regulatory measures regarding approval of new production capacity, third-party access, fuel sources etc. By doing so, it delegated a number of features and policy concerns to be taken care of by market forces, whereas they earlier were taken care of in the energy plans and development activities of governments and energy suppliers. At the same time, the focus was shifted from integrated planning to secured competition and market access.

The delegations are based on assumptions and rationales developed in economic theory and sustained in the growing policy arenas that focus on privatisation as the standard institutional reform for governments to pursue. The European Commission hereby re-framed the existing networks of power arenas in many countries. Not that they were alike in their constitution and actor-worlds, but in most countries, specific structures of ownership and regulation had been developed to reduce competition and secure planning visions and technology priorities specific to the energy policies and established networks in these arenas.

The directive stated that by February 1999, at least 22 percent of the national demand in the member states was to be traded in an open market, and this share was to be extended to 33 percent in the year 2005. The directive also allows member states to decide how to open the markets. The only limit is that every power buyer (consumer) using more than 100 GWh should be allowed to act as a 'privileged consumer', which implies the right to buy power for its own consumption both nationally and abroad. Accordingly, the distribution companies would also typically be allowed to act as privileged consumers in the new marketplace for wholesale energy.

The directive allows a certain degree of calculating costs of Public Service Obligation (PSO) and adding these to the price of power. As types of PSO, the directive points to extra costs from utilising renewable energy sources, waste and combined power-heat, as well as costs derived from securing provision of power, environmental protection measures, and the protection of consumer rights.[7] The PSO-based costs have to be allocated across all consumers through obligations to purchase e.g. a part of PSO-production at higher cost than the most competitive energy production prices, or to pay an extra fee to be re-allocated to secure the PSO. In practical terms, the directive hereby allows for other policy issues to restrict competition, either by dividing the market into separate sub-markets or by differentiating fees to be paid for power of different origins. Environmental protection measures and prioritisation of energy sources and technologies are to be translated into price measures, market segmentations, and fee structures to be accounted for by legislation establishing a framework of institutional structures.

The directive demands a separate operator of the network system to be constructed and monitored by public authorities. The system operator has to be either a company or a public institution that has the task to operate, maintain, and develop the national power transmission network and to ensure balance between demand and supply, eventually by exchanging power with other national system operators. The purpose is to secure all suppliers an equal access to the network and transparent prices for transmission. The system operator must be independent of all other interests but transmission, and a separation of accounting between production, transmission, and distribution, despite the company structure or institutional set up, shall avoid cross subsidising between the different functional parts in eventually integrated companies. But there is no requirement for separate ownership. Here, the re-framing goes to the heart of former engineering planning and monitoring. What had been an integral part of the negotiation of management procedures in the old power regime, maintained by the producers in co-operation, now demands an institutional separation to maintain the 'freedom' of market forces without changing the need for balancing the system.

Finally, the directive describes two different procedures for establishing new production capacity. The first model is to base the building of new capacity on licenses given by the authorities in a country making possible capacity regulation (limitations) and the direction of new capacity to use special fuels, as the Danish authorities have done hitherto. The second model implies that new capacity would be the object of a public tender procedure according to the general procedures in EU. No matter which model is chosen, discrimination against foreign providers may not be employed.

The preparations for this directive have been going on for several years. It defines a potential new avenue for EU regulation that demands new institutional constructions and positions the EU in the power arenas of member states. In doing so, EU regulation has also entered traditional national infrastructure policies that are seen as vital for national economic development and national supply policies. While international linkages, new technology options, and sector convergence seemed quite obvious as arguments for re-regulating the telecommunications sectors, the re-regulation of the power supply sector has had a much more far-reaching political impact. It also allows for completely new entrances in the policy arena, and can therefore be seen as opening up possibilities for radical shifts in the actor constitution, planning rationales, and focal points of this policy arena, thereby de-stabilising existing coalitions and established separate development arenas and regulatory tools. As an obligatory passage point it forces national governments to translate strategic concerns by moving them from a typically centrally regulated institutional regime into an indirectly regulated set of large corporations constructed in ways assumed to support future national strategic concerns.

PREPARING THE DANISH POWER REFORM

The Danish process towards reforming the power sector, converting it from a publicly licensed monopoly to a competing marketplace, was initiated before the passing of the EU directive on liberalisation of the European power industry. During the fall of 1996, the Danish Energy Agency initiated an open working process directed toward reforming the whole energy supply and its infrastructure in Denmark. Before the preparations for the reform were launched, and because of the complexity of the reform, the parliament passed temporary legislation[8] to fulfil the requirements of the directive to be enforced by February 1999. The legislation (law 486 of 12th July 1996) assured that the legislative foundation was in line with the requirements of the opening of the market, the independence of the system operator, and the request for separate accounts for production, transmission, and distribution.

In fact, nearly all the requirements of the directive were formally met in the temporary legislation, which gave consumers of more than 100 GWh the possibility to purchase power, including the distribution companies. Hereby, the market opening was in principle more than the 22 percent required by the directive. The temporary legislation also mentioned PSO-production, defined as power from small-scale co-generation plants and power producers using renewable sources, wastes, or recycled materials as fuel. Costs of establishing such facilities and R&D costs were also defined as PSO-costs. Furthermore, power from co-generation plants, if sold at or above production costs, could also be accounted for as a PSO-cost. In late 1997, the Danish interpretation of the directive got approval[9] from the European Commission.

Six industrial power consumers using more than 100 GWh per year were identified and together with the 50 distribution companies that account for more than 90 percent of consumption, the opening up of the market was almost complete. But the limits imposed by the PSO actually restricted the opening of the market, as 77 percent of the Danish market would be considered PSO-protected, leaving only 23 percent to open competition (IDA, 1998). This would fit the requirement of 22 percent by the beginning of 1999, but not automatically the 33 percent required in 2005. On the other hand, if all the power production derived from co-generation plants was also sold at market conditions and without subsidies 'outside' the PSO, there would be no problem in meeting the requirements of the directive.

KYOTO AND THE REDUCTION OF CO_2 EMISSIONS: ANOTHER BOUNDARY CONDITION

To further the process towards fighting the growing greenhouse effect, the Danish government and parliament have made strong commitments obliging Denmark to cut Danish emissions of CO_2. According to the Kyoto climate summit, and according to the distribution of the burden in EU, Denmark accepted to cut its emission of greenhouse gases by 2008-2012 to a level corresponding to 21 percent less than in 1990.

Given the fact that it has not yet been possible to achieve common means to cut emissions at the EU-level, such as quotas or tariffs on emission of the gases, the Danish parliament decided to impose a top level for the emission from power plants operating in Denmark. Danish power plants play a key role in the campaign against emission, since the power sector, due to its wide use of coal, emits 40 percent of CO_2 and 33 percent of the other greenhouse gases. In spite of the reluctance to do the same in other European countries, it is agreed to impose CO_2 emission quotas on each production facility in Denmark. The quota will slowly decrease in the years to come from 23 million tons to 20 million tons from 2000 to 2003. The practical implication for the different companies is that they will have to observe their quotas during the four years, although there is also a possibility for them to trade quotas between companies in Denmark.

Trading quotas might be extended, first to the other Nordic countries, and later to other EU-countries. The system may later pave the way for a market for tradable emissions permissions. But if the quota regime is to be imposed only in Denmark, power producers will eventually have to limit their exports of power, since the production of power based on coal is quite CO_2-intensive.

This has been the industry's major argument against the Danish policy and international position in the CO_2 debate. As long as no international agreements on the cross-border trade of quotas exist, the Danish power utilities will be restricted in their competitive actions. As Danish power plants operate more efficiently and with less pollution than e.g. several plants in Eastern Europe, it seems unfair not to be able to extend production for exports. That this position goes hand in hand with a more general scepticism regarding CO_2 regulations on the part of power-intensive industries and transport companies does not come as a surprise.

STAGED PERFORMANCES: ACTOR WORLDS, RE-CREATION OF NETWORKS AND BOUNDARIES

Fuelled by the interests of the actors involved and stimulated by the evolving public debate on the nature of the Danish adjustment to the EU directive, existing actor-networks were re-enacted and new ones engaged in active recruitment strategies to stage their presence in the evolving and re-structuring policy arena. The openness of the process was seen when the different parties in the energy sector – industry, researchers from universities, experts from sector research centres, energy companies and others – were invited to attend conferences and workshops on the subject. Also Danish unions, grassroots organisations with special interest in energy policy, consumer groups, professional societies, and companies dealing with consumer goods took part in the debate that unfolded during 1998 and 1999.

Competition and market opening for lower prices

Economists from universities and government agencies dealing with energy policy, mostly working within a neo-classical theoretical framework, and a number of other actors welcomed the liberalisation as a huge step forward in gaining lower prices by creating this new market. Not surprisingly, they got support from the Danish Federation of Industrialists (DI - *Dansk Industri*), the Danish Competition Agency, and companies within related markets like oil and gas, spotting a potential new market. For these groups, the EU directive identified a standard path of transformation of traditional government-regulated public monopolies and, according to the assumptions the new markets, would provide benefits for both private investors and consumers. This group of actors was only weakly represented in the old power arena and was more connected to actor-networks focusing on a possible outcome of a market opening for lower prices for large industries or for new entrances into the power arena as brokers and investors. Liberalisation provided an avenue into the arena that otherwise was not open.

To support the arguments for potential price reductions, reports from both DI and the Competition Agency argued that the installed power utilities had created over-capacity in power supply and also that network maintenance and service was much too costly (Konkurrencestyrelsen, 1998). Both resulted from lack of pressure due to the monopoly status of the utilities and distributors, which provided no incentives to keep capacity at a minimum and enhance the efficiency of the employed staff. The over-capacity was evident but was the combined result of reduced growth in power consumption and the investments in wind energy and small, local co-generation plants. Phasing-out of older power plants had not occurred due to potential export possibilities, resulting in the over-capacity.

Consumer influence and ownership

Another rather new group of actors emerged that widened the policy process to also include the question of local and democratic influence. Researchers dealing with energy policy and the promotion of alternative energy systems advocated that liberalisation would undermine an environmentally friendly, integrated energy system, based on optimising utilisation and saving resources. They formed an alliance with grassroots organisations that advocated renewable energies and the delegation of this obligation especially to consumers in control of the distributors and the power companies. At first, they opposed the idea of liberalisation as a whole, but later they argued for a minimal Danish adoption of the directive maintaining the local community influence on the companies (Hvelplund & Lund, 1995). This entrance into the arena re-enacted the more or less vanishing role of public participation and ownership.

A second focus of the new actors was the distribution of the accumulated values in the companies. If they were to be limited companies, the distribution of wealth between consumers and shareholders would become an issue in the debate. To what extent would the owners – local communities or consumers – gain status as ordinary shareholders and be able to eventually strip the companies by selling their shares? Or even worse: could the management of the power companies enter the market for mergers with hidden values, making it very profitable for new owners to gain access to the established companies by taking over 'societal' capital hitherto bound in the power sector as a common good?

Technological development and regulatory regimes

Another of the responses to the emerging energy debate came from the Danish Society of Engineers (IDA - *Ingeniørforeningen i Danmark*). The society constituted a working group to focus on those aspects of the reform that would influence technological innovation and environmental protection. The question was asked, how the reform would influence the successful Danish policies for developing technologies based on renewable energy sources and still maintaining a highly efficient utilisation of the traditional large-scale power utilities. The perspective was also to identify how future regulatory activities could be installed to maintain environmental and technological priorities.

By developing three future scenarios[10] with structure of ownership and markets for power and mergers involving all vertical levels from the power producing utilities, over the networks and distributors, to the consumers of power, the Danish Society of Engineers was able to highlight the potential impact resulting in these different scenarios. In the final report (IDA, 1998), it was concluded that only one of the scenarios, avoiding vertical integration between the utilities and distributors, would allow for both future promotion of technology development and a working regulatory system. Especially the identification of objects of regulation in terms of companies, technologies, and environmental measures created a problem for establishing a market-based competition and simultaneously securing established and future environmental policy priorities. If distributors were separated from the power companies and acted as agents for the consumers, they would have incentives to supply the needed test sites and an interest in investing capital in the risky new technologies and in further developing a demand-side management for power reductions and efficiency. In the other scenarios, vertical integration between power utilities and the distributors especially turned out to be the weak point in the liberalisation strategy. By taking over distributors and power brokers, the utilities could gain a market access that easily could be translated into local market dominance and limiting the impact of regulation.

Contrary to the common discussion that took place in the newspapers, the report considered price competition and even lower prices on consumer markets highly unlikely, and it was argued that continued focus on this feature of the reform was misleading. Even though the Danish power sector had produced some over-capacity, it had still maintained very low cost prices for power, both by combining power production and heating and by establishing a closely monitored system of power utilities. In a roundtable discussion organised by the Danish Society of Engineers, the actors common view supported that it was not likely that consumers in Denmark could expect lower prices. Traditional regulated monopoly had been shown to be quite efficient in this respect. This also was the case in other countries liberalising the power sector (Russell & Bunting, 1998). The more likely impact of liberalisation, when profits and new taxes were paid, would be higher consumer prices, lowered prices for large power users in industry, and first and foremost a restructuring of ownership and the activities of power companies as a result of mergers and acquisitions.

A first sketch

The process seemed to be reaching a conclusion after about one year, when the Energy Agency, under the guidance of the Minister of Energy and Environment, drafted a proposal for the architecture of the new power market in Denmark. The model became public during the summer of 1998 when a 13-page memorandum from the Energy Agency was circulated and discussed as the unofficial strategy of the ministry. At this point, it was possible to recognise elements of the ideas and notions argued by the Danish Engineers Society, energy researchers, and grassroots. And the first response was to see this as a victory for the critical accounts of the liberalisation policy.

The Minister of Energy and Environment, although, still under pressure from the power utilities and industry, distanced himself from the memorandum and referred to it as some unfinished thoughts. It never became the official proposal, but it did provide a hint of the framework within which the legislation of the power reform was to be unfolded.

Investment capital and freedom of action for utilities

Traditional actors within the arena, like the power suppliers or the distributors, were much more reserved, since for many years they had governed an energy system, which they interpreted as successful and functioning well. The existing energy system based on long investment horizons, monopoly, continued technological innovations, and even adjusted towards renewable energy, was a system in balance without considerable lack of optimisation.

But as the reform was inevitable the future role and room for action assigned to the power companies entered the agenda of the utilities, and in combination with this, how the ownership of these companies would develop. The companies needed capital in order to finance long-term strategies to expand their activities, market their product, and renew and eventually extend their power production capacity. Without investment capital, it would be difficult to operate as a competitive agent in a market of scale-intensive providers competing on price, and to access long-term delivery contracts across national borders. The competition was going to be strongly enforced by the large companies in the close vicinity of Denmark. Especially the German power provider, Preussen Elektra, and the Swedish-based but partly German-owned power companies, Sydkraft and Vattenfall, were expected to enter the Danish market. But also other companies in energy supply, such as Royal Dutch Shell and Norsk Hydro, could eventually enter the market as power providers.

When the discussion about the reform was reaching its climax, representatives of the power companies formed an alliance with both the industrialists and the metal workers union fighting for more freedom for the utilities as independent actors on the future arena. Assisted by a range of reports from international business consultant firms, banks, the Ministry of Trade and Industry, and competition authorities, they took over the arguments about price reductions based on expected efficiency gains, even though the price reduction arguments had proven to be rather misleading. This led to a renewed entrance into the arena of one of the earlier dominant actors, but re-enacting the role of the utilities as commercial entities in an alliance with industry.

In accordance with their public statements, they urged at the same time that a government sponsored commission be set up to negotiate restructuring the market, and they accused the Ministry of Energy of not taking their views into account. They also tried to argue for the exclusion of grassroots, consumers, and environmental movements from the commission. But the Ministry of Energy rejected this, claiming that no one should be excluded from the debate by putting it behind closed doors, as would be the case in a commission.

Economic balance and state finances

A rewritten proposal was first presented officially in late 1998 and the political parties in the parliament invited to negotiate the legislation. The negotiation was situated in the Energy Policy Committee in the parliament, which during the negotiations supplemented their knowledge of the consequences of different political choices and institutional set-ups by posing a large number of questions to the Energy Agency. The questions resulted in approximately 200 small memoranda to the committee, making it rather difficult for non-professional actors to keep up with the reform process. A breakthrough was reached, when the committee pinpointed that it would be possible for the Finance Ministry to gain some income to cover the fiscal deficit by capitalising the power sector. As with the re-entrance of the power utilities in a new role mentioned earlier, the state fiscal interests were introduced into the arena and completely reshaped the discussion about the role of ownership to the existing power infrastructure. From being a strategic infrastructure and resource, power production became an object of taxation making tax revenues an obligatory passage point in any further reform initiative.

This proposal was put on the agenda, despite the fact that the utilities were not state-owned but established as non-commercial companies governed by local communities or by local consumers. However, this intervention succeeded in enrolling the Minister of Finance and managed to shift the focus of the public debate. Thus, implementation was now also seen as a fiscal policy problem. It was even argued that the state would have to gain access to the hidden values of the power companies in order to avoid an unintended overheating of the economy, which could result if new shareholders in the power companies sold their shares and spent the revenues for private consumption.

THE ACTUAL REFORM: FINANCING STATE DEFICITS RE-CONFIGURES THE ARENA

The final legislation for the power reform was passed in May 1999 after a hectic negotiation process during the winter. It ended with a compromise between the government and the main opposition parties. The legislation combined a rapid opening of the power market for competition, elements of consumer protection and local influence, and the possibility to enforce high environmental standards as well as to deliver a fiscal contribution to state budgets, both in the short and in the long perspective. Fiscal government interests played an important part in the reform and determined indirectly how other policy goals were institutionalised once the path of liberalisation was finally sketched. The elements of the power reform were approved by the European Commission in the Autumn 2000.

To pay a tribute to the Danish budget deficit, a core element of the reform was that the sector in years to come would have to pay a couple of billion Danish Crowns every year to the Finance Ministry. This has been implemented by imposing a new tariff on distribution. Until recently, the companies did not pay taxes due to the non-profit nature of the sector. This has now come to an end. Also, subsidies for renewable energy sources have been rejected. In the future, trading PSO-power and eventually green certificates through the marketplace will finance renewable energy sources. In essence, the consumers will pay for the development and maintenance of renewable energy facilities. In the end, this may result in even higher power prices than before. And the irony may be that the customers will have to pay for the capital costs for the existing power installations for a second time.

The political agreement about the power reform states that the traditional Danish power companies will be allowed to gain and accumulate profit as private companies, abolishing the former 'non-profit' and 'only-paying-the-cost' principles. Consequently, they will also have to accept the risk of eventual stranded costs of taking over the responsibility for the existing facilities. The opening balance of the companies would be a question for further negotiations, when the reform is implemented. This also includes the valuation of company assets and the limitations on loans and sale of the facilities. In fact, this has turned out to be one of the conflict areas of the reform, and has led to further negotiations and recently also to a revision of the law[11] to allow for capital transfer from the network operating the companies to the utilities. This move has been taken after negotiations between the Finance Ministry and the Ministry of Energy and Environment to compensate for the power companies' lack of accumulated investment capital in the past, due to their non-profit status.

With regard to the opening of the market, the reform moves even faster than required by the directive, giving consumers access to choose between power suppliers. By April 2000, consumers purchasing more than 10 GWh, and by 1 January 2001, consumers demanding more than 1 GWh a year will obtain access to the marketplace. Finally, by 1 January 2003, every consumer will be able to choose between the different suppliers.

Parallel to this, access to the distribution network will be regulated by fixed prices and an obligation for transmission companies to allow for third-party power transmission. This has been established in response to the opening for foreign companies to gain access to distribution companies. By giving almost every consumer access to buy power on the market, and at the same time securing access to the distribution network and securing fixed prices for third-party access, the interest of utilities in purchasing distribution companies in order to obtain a *de facto* regional monopoly is limited. Transmission tariffs will not be subject to negotiations between power providers and the consumer. This is in contrast to e.g. Germany, where third-party access is subject to negotiation between the parties.

The Danish system, which is similar to systems operating in the other Nordic countries and in UK, is supposed to facilitate fair competition and transparency in price setting (Olsen, 1999), since the task of transmission companies is defined as a core element of the market framework. It also creates space and opportunities for setting up institutions and agents in the marketplace to further the exchange of power and price setting, depending on actual supply and demand. The Nordic Power Pool Company, based in Norway but with a new branch established in Denmark, is facilitating such a marketplace. It quotes the current day-to-day price of power in the Nordic countries and facilitates the trade and exchange of power between the suppliers and the distribution companies in the Nordic countries.

The legislation continued the special Danish tradition of public influence in the power sector. Local communities, consumer societies, and a couple of limited companies still own the power sector. And it is explicitly stated that in order to obtain a concession to operate a local distribution network, the company asking for the concession has to install a majority of consumers on its board. On the boards of companies accepting the mentioned provider obligation, at least one-third of the members must represent consumer interests.

A representative of the state agencies (the Energy Agency or the Competition Agency) will have a seat on the boards of the system operators, but the network companies will keep the ownership. Furthermore, the legislation gives the state an option to buy the network at a fixed price if the system operators were to be sold. This construction assures that the system operators will act as independent bodies in the marketplace, monitored and regulated by the public agencies.

In line with the directive, the legislation addresses the question of PSO by extending the rules of the temporary legislation in law 486 of 1996 to the new arrangement. The system operators and the network companies are obliged to purchase the precedent PSO-power, e.g. from renewable energy sources, waste, and biological fuels, and from local, gas-fuelled co-generation plants. Prices are calculated on the basis of the cost to build, operate, and transfer the produced power to the distribution network. The actual prices for the PSO-power types are higher than the price from the large co-generation plants and the quoted day-to-day prices of the Nordic Power Pool. Consumers pay an extra tariff to the net-operators to cover the higher PSO-price, and they distribute the money to the PSO-suppliers.

The political agreement behind the power reform also specifies the demand, requirements, and conditions for building new energy facilities utilising renewable energy sources, dropping old facilities, and building offshore windmills. It has also been agreed among the parties behind the legislation that PSO-power from the year 2002 will also be exchanged within a separate market framework, called a 'green market' for renewable energy sources. According to this scenario, suppliers will receive green certificates according to the amount of power produced. And consumers will be obliged to buy a certain quota of green certificates when they purchase their power. However, this market has not been institutionalised in any form but was passed as a 'black box' within the legislation (Olsen, 1999). The delegation of environmental policy expectations to this new market institution has therefore temporarily excluded concerns for continued development of environmentally friendly energy sources from the debate, since no actors are able to predict the impact of this future 'green market'.

The still open-ended character of the reform has been demonstrated by the number of mandates that the Ministry of Energy has to fill with ministerial orders and by the already introduced revisions of the law. These deal with the capital funds of the utilities and the uneven share of installed production capacity based on renewable sources in the two power network regions of Denmark.[12] Concerning the implementation of the legislation, concessions for the network operators, the transmission companies, the provider-obligation companies, and the power producers will be given at the beginning of the year 2001. Still pending are the details for the price control of power to customers being serviced by the provider-obligation companies, the calculations of costs of production and consequently prices for power from the installed wind turbines and other installations based on renewable sources, and the construction of the 'green market'.

IMPLICATIONS FOR MARKETS AND OBJECTS OF REGULATION

In this section, the implications of the power reform in its present form will be discussed. As shown, the reform turned out to be a compromise that balanced the interests of a number of different actors and left a rather long list of unresolved problems to be negotiated in the future.

Market constructions and company influence

As already stated, it is rather unlikely that the liberalised power sector, with increased competition among power providers, will result in lower prices for consumers. A few major power-consuming industries and a potentially growing number of companies that choose to produce power in-house, in combination with other processes, may gain price advantages, but the majority of consumers and smaller industries will not feel the change. While power prices may not be the main impact of the reform, the access to consumers, and size and control of capacity may play an important role in the responses of companies to the liberalisation. Important new markets for acquisitions and mergers between power utilities, distributors and brokers will be constructed. This can already be seen in the mergers between large power utilities in each of the two regions in Denmark. And further national and international concentration can be expected.

However, the Danish regulations have limited the possibility for power utilities and international power conglomerates to completely control distributors and the transmission network. The demand for local influence on boards of distributor companies is supposed to work as an indirect limitation to large power conglomerates control of the access to consumers. The important control of vertical integration that has proven difficult for policy and competition agencies has been delegated to the local actors.

Objects of regulation

The new regulation of the power sector is based on the merger of the two existing regulating bodies, creating a new path to be developed in the Danish administrative tradition. The Competition Agency and the Power Price Board that have traditionally regulated prices and protected consumer interests are merged with the Energy Agency. The Energy Agency has been concerned with approving new facilities, giving licenses, managing R&D and subsidy budgets, facilitating energy saving and new energy taxation. The new authority – the Energy Inspectorate – will maintain some of the earlier tasks in their translated form, together with new obligations deriving from the market regime. The Energy Inspectorate will continue monitoring and regulating monopoly activities and monopoly prices for third-party access to the network, monitor system services and provider- obligation companies, and eventually also include regulatory measures concerning the heat and gas sector. Especially the grey zones between competition and monopoly activities will be monitored, in principle to counteract behaviour that twists competition.

Another new feature of the regulating Energy Inspectorate is, that it will be governed by a board composed of independent experts instead of actors from the power sector itself, and that it will become independent of the Ministry of Energy. This moves the centre of regulation from the ministerial system based on negotiations between involved actors, to a new expert-driven body, independent of government. This translation is not only a shift in the institutional frame for the interaction between companies, customers and regulators, but also a shift in the objects of regulation, from being defined as an integral part of energy planning to becoming a rule-driven expert system focusing on specific features of the energy system. In this process, cost structures are supposed to become more visible and consequently become objects for regulation and deliver the accounts that make monitoring by this expert regime possible.

Objects of policy in the environmental and technological debates now have to be translated into quite different measures, whereby the policy objectives are delegated to market forces and price structures. What was negotiated as explicit technological and environmental options, will through the translation into economic measures become open to different interpretations and thereby become objects for other actors' interests.

As a consequence of the restructured regulatory regime, the regulation of PSO, the creation and monitoring of the 'green' power market, and the support for specific technologies that secure the use of renewable energy, sources will be more difficult in the future. The liberalisation of the power sector is not in this sense a neutral change that makes the sector more dynamic and in all cases better able to adapt to future needs. It shifts focus both at the strategic level and for the efficiency and objectives of regulation. From being a system of subsidies for new energy installations financed by government, the new regime demands continuous negotiation of the levels and redistribution mechanisms for PSO-based fees and later for the creation and circulation of 'green certificates'.

IMPACTS ON INNOVATION AND TECHNOLOGY USE

The shift from the traditional national system of regulation to a new market-based energy supply system also leads to a shift in priorities for and the focus of innovative activities. The liberalised power markets in Europe and in Denmark will still support innovations, but they will be of a different kind, and the actors involved may change.

The first observation to make is that almost no traditional large-scale power utilities have prioritised renewable energy technologies by themselves. The development of the Danish cost-effective wind turbine for power production has taken about 15-20 years of experimentation and continued technological innovation. This has been made possible through energy supply policies favouring power produced by wind turbines but still using market-demand means to put pressure on the wind turbine industry to enhance the product. These policies have included protection of new markets and 'infant industry' support. Such long-term investments motivated by environmental concerns were not supported by the traditional regime of the power utilities in Denmark. Although they also made a long-term investment commitment, the focus instead was on large-scale power generators, capacity growth, and low prices. The creation of a separate development arena for renewable energy sources was inevitable in order to achieve the energy policy goals.

Another observation is the lack of commitment by the new commercialised private utilities and distributors to fund innovation programmes for technology improvements. The reshaped power suppliers have already at this stage hinted that they will not fund public or semi-public R&D. Instead, as they eventually enter into cross-border mergers or international strategic alliances, they will increasingly become part of an international energy innovation system. This would transform the Danish system of innovation of capital goods for the energy sector, such as pipelines, small-scale co-generation systems, NO_X-utilities, etc., and it will end up being disconnected from Danish energy policies focusing on new small-scale technologies. This shows the impact of industrial structures and integration on technology priorities, both concerning investments and the character of the technology.

From the perspective of commercialised utilities, and taking into account the competitive character of power markets in the future, investments in power production equipment will search for existing, well established, and price-competitive technologies with a shorter turn around period than the hitherto dominating steam turbine plants. Especially the competitive character of the market and the pressure for CO_2-optimised utilities will make long-term investments more risky. This tendency can be seen among a number of utilities in Europe now prioritising combined cycle gas-turbine technologies, eventually including co-generation. This technology has shorter life and pay back times than the traditional power generators (Islas, 1997). So liberalisation will affect technology priorities and imply a shift in the planning and investment horizons favouring proven established technologies.

For a period of time, existing energy technologies are 'frozen' at their actual historical level of development, and further developments will have to be funded by growth in markets outside Denmark. In the case of wind turbines, this is already the case and does not stop further innovations of this energy technology. Other energy technologies still not developed will have to be identified and be given support by special funds and 'green certificates'. Thus, they will also remain economically vulnerable to minor changes in the 'green market' and be exposed to unstable investments and changing project support. The argument for PSO-policies to be able to cover the need for 'green power' may be right in a stationary situation where the alternatives are known and fully developed. However, except for wind turbines, most alternative energy technologies are still under construction and will need further innovations to become marketable.

Consequently, the new regulatory regime in the power sector will hamper the development of technologies based on renewable energy sources, and a dedicated policy response, including government funding, is needed in order to keep alive the innovation system for energy technologies based on renewable energy sources. This implies the creation of a new development arena for 'green technologies' to be nurtured by a separate institutional framework. In this respect, liberalisation has contributed to a further split between already diverging paths of technological development.

TECHNOLOGY STUDIES AND POLICY PROCESSES

Technology studies have dealt with the processes of innovation and co-production of technologies and modern society. Policy processes have also been part of these studies, but mostly just so far as they influence the spheres of technology construction and scientific institutions and not as separate networks. Only sometimes (and not in any continuous and stable way) have policy processes addressed technological issues or even taken technology seriously. This, together with the analytical focus of technology studies, has limited its ability to give advice about technology policy and priorities for investments etc. In this chapter, an attempt has been made to bridge technology and policy studies by identifying a common framework of actor-networks staging, in this case, the arenas for power policy and development.

In the power arena, the object of policy has been shifting, as have the roles delegated to technology, environmental concerns, markets, and regulatory regimes. Questions can be raised concerning the result of either delegations made to specific energy technologies to produce an environmental result, or to similar delegations made to market structures and regulatory regimes. All of these constructions produce rather closed actor-worlds with a focus on certain types of expertise. This was the case with the engineering-based actor-world in the traditional power utilities and will be the case in the new regulatory bodies to monitor competition and prices that are now separated from the companies' strategic decisions and investment priorities, creating new actor-worlds to set the stage for the future. Not only is the object of regulation translated, including changes in the network of actors, but also the types of competence involved and the required repertoire of actions are changed, involving different areas of knowledge and new actors.

One major challenge has been the rather strong preoccupation of policy processes with economic rhetoric and with the delegation of policy measures to institutional constructions. This is partly 'black-boxing' or simplifying the role of technology and environment as both actors and mediators in the process. This shows the importance in policy processes of enrolment strategies as an integral part of the boundary work that, if successful, will result in changes in the dominant rhetoric and make certain elements of this the defining obligatory passage points. This is also the case for new energy technologies, which will now have to be translated into price-performance and competition measures to gain access to the energy markets.

Although we have been dealing with a major transformation of the power sector, elements can be identified that show both continuity in the Danish power arena as well as re-conceptualisation. Especially the environmental priorities and the local ownership structure are examples of the continuity that has been integrated into the new power arena, while the restructuring of companies and rules for accounting and regulation have undergone a re-conceptualisation. From having been defined in terms of efficiency, they are now to be defined in terms of market competition and price relations. In the different stages of the policy processes studied, unfolding on the power arena, different types of stabilisation become apparent. In certain stages, temporary stabilisation seems to be dominated by rather symbolic constructions, while in others, the stabilised actor-networks have certain artefacts as their rather material core. The first is exemplified by the focus on competition efficiency and maintaining the idea of progress and continued adjustment due to economic processes. The second is exemplified by the large traditional technical system of power networks that emphasises the functioning of power-provider technologies and environmental priorities delegated to specific technical solutions such as wind turbines.

We have in this chapter investigated a possible integration of technology studies into policy analysis. This has been done by taking technology and environmental concerns explicitly into account in the analysis of changes in the regulatory regime for the power sector and by identifying consequences of liberalisation for industrial structures and markets. The importance of detailing the technological and environmental priorities implied by policy has been illustrated, leading to a critique of the role of the frequent simplifications and exclusions resulting from the rather dominant policy rhetoric of the actor-world inspired by economic theory. This has resulted in an exclusion of dialogues about the choice of technology and ways to satisfy environmental priorities.

NOTES

1 The empirical material for this study has been collected partly in connection with a project in the Danish Engineers Society on the liberalisation of the power sector in Denmark, resulting in the publication 'Liberaliseringen af elsektoren – erhvervs- og miljøpolitiske

konsekvenser' (IDA, 1998). The authors have continued the research, following the transformation process up to its contemporary conclusion. The empirical material consists mainly of the policy papers and documents written by the involved actors and some interviews. Due to its overwhelming magnitude, this material, is not explicitly cited in the text.

2 Including social actors, institutions, artifacts, visions, and theoretical objects.

3 A quite similar framework to the one described with the notions of 'arena' and 'actor-worlds' can be found in the notion of 'social worlds / arenas' as introduced in studies of professions in the sociology of Anselm Strauss and others (Clarke, 1991). When we were developing the arena framework introduced above, this strand of theory was not known to us, but it shows a number of analogies, both in scope and theoretical framework, which support the approach and inspire to further development.

4 The sector has for historical and geographical reasons been divided into two parallel regional structures, institutionalised in two different corporations (ELSAM in Jutland and Funen, ELKRAFT on Zealand). Both corporations were created by the co-operation and mergers of a number of local power utilities and distributors during the last four decades and have developed into more and more internally linked corporations with respect to ownership, trading relations, distribution, co-operation and regulation. The organisational co-ordination between the two bodies has been supported by their membership in the Danish Power Suppliers Association (DEF), which acts as a political representative on behalf of the two corporations, the linked power utilities, and the distribution companies.

5 The oil crisis also prompted the Danish oil and gas sector to explore more intensively and to extract more oil and gas from the North Sea. This might have influenced the power production sector, as the gas potentially could have substituted coal and oil as fuel for power production. But instead, the gas arena never closely influenced the power arena, since the government decided that gas should primarily be utilised as a substitute for oil for heating and partly for electric power for cooking. While gas as fuel for combined cycle co-generation plants has entered the power arena in recent years, furthered by a growing interest in smaller power generation plants, it did not re-structure the power arena before the influence of liberalisation took off.

6 For further details on this controversy, see e.g. Jørgensen and Karnøe (1995).

7 Although the providers of obligations and environmental protection measures were not in the forefront when negotiations about the directive started, they were added during the process due to changing priorities in the EU. The directive itself thereby represents a move from the early phases of the 'inner market' policies primarily focused on competition and market integration to the broader scope of EU policy.

8 The original ambition of the Ministry of Energy was to include power, heat and gas in one reform, thus creating only a single energy sector governed by one law. Partly due to the forthcoming gas directive, which copies liberalisation from the power sector, and partly because of political considerations and the different interests of the actors involved, the focus was concentrated on the power sector but included co-generation elements from the heating sector.

9 To monitor the implementation of EU directives, national governments have to notify the Commission about new legislation and must await an approval before it is considered legal in EU terms. In terms of fair competition, state subsidies, public procurement measures, and PSO, the European Commission apparently accepted the Danish interpretation, which underlined the Commission's acceptance of adjusting liberalisation to national policy goals, especially concerning environment and other nationally significant institutional or natural circumstances.

10 For a further elaboration on the use of sociotechnical scenarios see Geels, chapter 13, in this volume.

11 In December 1999, the law was revised to open for transfer of investment capital funds from the distribution companies to the utilities to prepare them for the upcoming competition from foreign companies and the need for renewing the power production facilities.

12 While the first discussion mentioned earlier was already controversial when the law was passed, the revision of May 2000 shows how rather complicated the PSO-method of calculating and distributing fees is. The revision also opens for an exemption from the PSO-fees for companies establishing 'in-house' power production facilities, again due to the rather complicated regulation of the power market.

BIBLIOGRAPHY

Benson, J. K. (1981), *Networks and policy sectors: A framework for extending interoganisational analysis,* Missuouri.

Callon, M. (1986), 'The Sociology of an Actor-Network: The Case of the Electric Vehicle,' in Michel Callon, John Law & Arie Rip (eds), *Mapping the Dynamics of Science and Technology*, London: Macmillan.

Clarke, A. E. (1991), 'Social Worlds / Arenas Theory as Organisational Theory,' Chapter 9 in D.R. Maines (ed.), *Social Organisation and Social Process: Essays in honour of Anselm Strauss,* New York: de Gruyter, pp. 119-158.

Collier, U. (1998), 'Liberalisation in the energy sector – Environmental threat or opportunity?' Chapter 6, in U. Collier (ed.), *Deregulation in the European Union – Environmental Perspectives,* Routledge, pp. 93-113.

Fink, H. (1996), 'Arenabegreber: mellem ørkensand og ørkensand' (Notions of Arenas: between Desert Sand and Desert Sand), in Hans Fink (ed.), *Arenaer: om politik og iscenesættelse* (Arenas: of Politics and Staging), Aarhus: Aarhus Universitetsforlag, pp. 7-22.

Hirsh, R. F. & Serchuk, A. H. (1996), 'Momentum Shifts in the American Electricity Utility System: Catastrophic Change – or No Change at All?' in *Technology and Culture,* **37**, no.2, pp. 280-311.

Hvelplund, F. & Lund, F. (1995), *Demokrati og forandring: Energihandlingsplan 96* (Democracy and Change: The 96 Energy Action Plan), Aalborg: Aalborg Universitets Forlag, 514 pages.

IDA (1998), *Liberalisering af elsektoren - erhvervs- og miljøpolitiske* (Liberalising the Power Sector - Consequences for Industrial and Environmental Policies), Copenhagen: Danish Engineers Society, 49 pages.

Islas, J. (1997), 'Getting round the lock-in in electricity generating systems: the example of the gas turbine,' *Research Policy*, **.26**, pp. 49-66.

Jørgensen, U. & Karnøe, P. (1995), 'The Danish Wind-Turbine Story: Technical Solutions to Political Visions?' Chapter 4, in A. Rip, T. J. Misa & J. Schot (eds), *Managing Technology in Society - The Approach of Constructive Technology Management,* Pinter Publishers, pp. 57-82.

Jørgensen, U. & Karnøe, P. (1996), *Dansk vindmølleindustris internationale position og udviklingsbetingelser* (Danish Wind Turbine Industries' International Position and Conditions for Development), Copenhagen: AKF report, 230 pages.

Jørgensen, U. & Sørensen, O. (1999), 'Arenas of Development - A Space Populated by Actor-worlds, Artefacts, and Surprises' *Technology Analysis & Strategic Management,* **11**, (3), pp. 409-429, reprinted with revisions as chapter 7 in this volume.

Konkurrence i energisektoren (1998), (Competition in the Energy Sector), Copenhagen: Konkurrencestyrelsen, 195 pages.

Latour, B. (1988), 'The Prince for Machine as well as Machinations'. Chapter 2 in B. Elliott (ed.), *Technology and Social Progress*, Edinburgh: Edinburgh University Press, pp. 20 – 43.

Olsen, O. J. (1999), 'Den danske elreform' (The Danish Power Reform), *Samfundsøkonomen*, no.4, pp.5-9.

Owen, G. (1999), *Public purpose or private benefit? The politics of energy conservation*, Manchester: Manchester University Press, 233 pages.

Petersen, F. (1994), 'Atomalder uden kernekraft' (Atomic Age without Nuclear Power), in Hans Buhl & Henry Nielsen (eds), *Made in Denmark* (in Danish), Klim, pp. 197-215.

Rhodes, R. A. W. & Marsh, D. (1992), 'New directions in the study of policy networks,' in *European Journal of Political Research*, **21**, (1-2), Kluwer Academic Publishers, pp. 181-205.

Russell, S. & Bunting, A. (1998), 'Privatisation, Electricity Markets and New Energy Technologies,' in *Technology in the Making: Shaping Artefacts, Building Australian Society.*

Strunge, L. (1995), *På vej mod en positiv industripolitik i EU: Konstitution af rationaler, institutioner og strukturer,* (Towards a Positive Industrial Policy in the EU: The Constitution of Rationales, Institutions, and Structures), Roskilde: Roskilde University, 312 pages.

12. The Political Control of Large Socio-technical Systems: New Concepts and Empirical Applications from a Multi-disciplinary Perspective[1]

K. Matthias Weber

INTRODUCTION

The starting point for this analysis is the question how society, and government in particular, can deal with technological changes and challenges in increasingly complex large socio-technical systems (LSTS). The term 'socio-technical' has been chosen to highlight that new technologies cannot be analysed in isolation from their social context. 'Large' is a distinctive characteristic of technological systems such as energy supply, transport, information and telecommunication.

One of the tasks of governments is to help shape new technologies in a sustainable way, and to smooth problems and conflicts they may bring about. This is a particularly difficult task in the case of large socio-technical systems due to their high degree of complexity and interdependence. However, while previous research on such systems has often emphasised their rigidity and stability, this paper looks at the possibilities to transform them in spite of such constraints.

The reasons for investigating large socio-technical systems thus go beyond purely scientific interest. The aim is to provide a scientific foundation to inform decision-making processes. In other words, the question is if, how, and by what means governments can influence the process of change of large socio-technical systems. In practice, a broad range of actions can be and are taken by governments in order to steer these large systems, but only limited knowledge is available on what the actual impacts and side-effects of these measures will be. There is clearly scope for improving our understanding of policy actions and of their impacts on system change in particular.

Real-world LSTS are shaped by a great diversity of factors, operating in different realms and at different levels. Moreover, policy actions are just one among other classes of factors, and they have direct as well as indirect impacts, both intended and unintended. Whereas scientific research often emphasises particular types of factors, it is necessary from a policy-oriented perspective to take a more holistic approach and go beyond the narrowly defined views of individual disciplines. What is needed to underpin decision-making is an integrated and inherently multi-disciplinary research approach which can provide a useful basis for analysing change in large socio-technical systems and for informing strategic decision-making. It should enable exploration of the impacts of current policy decisions within the context of different future scenarios, and thus facilitate constructive and forward-looking policy analysis.

In other words, beyond the research interest to integrate scientific perspectives, the main objective is to make social science theory constructive for policy purposes. Some first steps in this direction are made. A short overview is given of current research streams dealing with the emergence and control of new technologies. An integrated research perspective is presented, formulated in a systems language ('The PET systems approach'). Four key aspects of the PET-perspective are then selected and subsequently underpinned by empirical data from two case examples: the innovation diffusion of combined heat and power technology, and the emergence of new urban mobility systems. This analysis let us formulate a number of policy conclusions, both case-specific and generic, allowing us to assess the applicability of the PET systems approach as a tool to support policy-making.

CURRENT PERSPECTIVES ON TECHNOLOGICAL CHANGE FROM DIFFERENT DISCIPLINES

Recent research on the process of technological change from many scientific disciplines points to the great variety of determining factors and impacts that ought to be taken into account by decision-makers.

Economics has a long tradition of dealing with technological change. Research in the neo-classical realm has moved from considering innovation as an exogenous force towards endogenising it into models of growth and structural change (Romer 1990, Helpman 1992). However, technology has remained a rather abstract element, taking little account of the specificities of technology and the differences between phases of the process of technological change. Since the early 1980s, evolutionary economics has given stronger emphasis to the analysis of technological characteristics for processes of economic change, and the specific development dynamics they imply. Concepts such as technological trajectories, technological regimes and regime shifts (Dosi 1982, Kemp et al. 1994), while remaining at a fairly generic level, offered new insights into policy formulation. More recently, analyses of national, regional and sectoral innovation systems attempt to integrate a broad range of aspects relevant to innovation processes (Lundvall 1992, Nelson 1993, Edquist 1997, Braczyk *et al.* 1998). They consider prominently macro-level institutional elements, but also take into account micro-level mechanisms, such as patterns of collaboration, learning and network development. They do not really open up the black box of technology, nor consider the full range of relevant social processes underlying technological change, but interpret innovation and technological change within a wider systemic context.

These latter two shortcomings are prominently addressed in the sociology of science and technology, where typically a more micro-oriented perspective has been adopted, based frequently on historical case studies. The evolution of networks and the shaping of technologies by actors and entrepreneurs have been one key research interest in this area (Bijker 1987, Callon 1992). Other lines of investigation have been dealing with belief systems and mental frameworks of actors, and with the analysis of social institutions (c.f. Dierkes & Hoffmann 1992, Jasanoff *et al.* 1995). Specifically on large technical systems, several mainly historical studies have been carried out, concentrating first on the emergence and later on the transformation of such systems (Hughes & Mayntz 1988, Summerton 1994). Following a systems approach, emphasis is put on cultural aspects of system transformation, on the role of key actors (e.g. system builders) and on the reconfiguration of networks of actors. The technical dimension (e.g. in terms of infrastructure interdependencies) is also taken into account. However, the conceptual tools provided by the work on large technical systems have not yet been applied for systematic comparative research.

With respect to the process of change, several concepts have been coined to capture certain aspects of dynamism (e.g. 'alignment', 'closure', etc.), but mainly for descriptive rather than explanatory purposes. Only in recent years, and in parallel with evolutionary economics, have notions of self-organisation and evolution been transferred to the social sciences in order to provide an underlying rationale of how socio-technical change comes about.[2]

Similar interests have emerged within the realm of political sciences. A move has been made away from linear stimulus-response models of policy measures impinging directly on technology development (e.g. by means of regulation or technology policy), towards more comprehensive approaches. The impact of political structures and processes on the outcome of technological innovation has been recognised and is reflected in empirical studies as well as in theoretical approaches, using either a systems or a networks metaphor (Mayntz & Scharpf 1995, Görlitz 1995, Busshoff 1992). Self-organisational or self-referential models of political control address how the political process itself is intertwined with the processes of technological and social change (Görlitz & Druwe 1990). These insights have been used to inform and underpin policy strategies to foster change in areas such as sustainable agriculture or technology transfer (Görlitz 1994).

A detailed understanding of how the technology in question operates, how it is embedded in a wider technical system, and what technical constraints are imposed on decision options should be an important pre-requisite of any sociological, political or economic analysis but in practice this has frequently been neglected. Attempts have been made to bring the technology element closer to the policy-making process by the first generation of Technology Assessment (TA) exercises. This early type of TA sought to address the information deficit in policy-making related to new technologies, but it did not explore in depth the social processes associated with their emergence.

A crucial problem in policy-making in relation to technology is the need to look ahead. Prospective analyses of various kinds have thus been requested. Traditional TA was one response to deal with the information problem, as well as quantitative systems analyses and simulations (particularly in the field of energy and transport policy), and early Delphi-type technology foresight exercises (which were restricted to 'impact analysis' and the elicitation of future expectations regarding the role of emerging technologies in society).

Another inroad focuses on how this information can be effectively introduced into decision-making and how different actors and institutions intervene in this process. By their involvement into the process of scenario-development, the positions and attitudes of actors are endogenised. This approach has been pursued at different levels, ranging from firm strategies (Godet 1993), via participatory mediation processes (Geurts & Mayer 1996), to the second generation of foresight exercises at national and regional level.[3]

Technological and social analyses of technological change are fused in Constructive Technology Assessment (CTA). Based on processes of societal dialogue, the different relevant aspects and dimensions of a technology are considered through the participation of different actor and stakeholder groups. In general, CTA targets at a more active involvement of researchers in the decision-making process, e.g. as mediators in societal discourses (Rip, Misa & Schot 1995). This approach reflects the desire to integrate technological expert knowledge as well as different social interests into a continuous learning and negotiation process which can be accepted by all participants, and thus underpin political decisions. Strategic Niche Management (SNM) goes even a step further by trying to inform and guide the networking processes that need to accompany the uptake of new sustainable technologies, and thus bring them from an experimental stage to wider diffusion (Kemp *et al.* 1998, Weber *et al.* 1999)

Scenario development can play an important role in these social strategies as a tool to inform social interaction, to make alternative pathways explicit, and to explore options for policy action. Until now, scenario-development exercises have been mainly process-based; social theory has not been sufficiently comprehensive to deal with the type of holistic and strategic questions policy-makers are confronted with. The theoretical and conceptual knowledge accumulated has remained untapped, or only been fed implicitly into these scenario exercises through the mental frameworks of the experts involved.

The approach of socio-technical scenarios combines the future possibilities of scientific-technological advancement with those of their societal embedding and entrenchment (Geels chapter 13, in this volume, Weber 1999). By trying to look ahead, this line of work is certainly an unorthodox methodological novelty in the sociology of science and technology. The links with established social science research approaches could be improved by establishing a comprehensive and explicit theoretical foundation of potential driving forces and mechanisms underlying socio-technical change. It would also be important to provide a broad range of aspects that could potentially matter for developing a scenario to make the process of constructing scenarios more straightforward and transferable. This deficit is addressed by the PET system approach that will be introduced in the next section. By integrating the different complementary elements of theories of social and technological change into a framework and methodology for scenario development, it aims to provide an explicit foundation for forward-looking policy analysis and decision-making, but one which is still built on a sound theoretical foundation.

THE PET SYSTEM APPROACH: A MULTI-DISCIPLINARY CONCEPTUAL FRAMEWORK

While all the aforementioned perspectives provide important insights into the process of technological change, there is no comprehensive framework available that would provide a foundation for the analysis of means and strategies of political control. The PET system approach represents an attempt to integrate existing lines of theory into such a framework for constructive, forward-looking and problem-oriented policy research. PET stands here for Politics, Economics and Technology, to emphasise three of the key realms of society considered (for more details, see Weber 1998, 1999). The systems language has been chosen to make the complexity and inter-relatedness of the subject matter explicit. This is also why this perspective is particularly suited for the analysis of large socio-technical systems (where it moreover helps lay out requirements for comparative and exploratory research).

Overview and delimitation of the PET-system

One of the critical questions of any systems analysis is that of defining and delimiting the system under study. In fact, there are no clearly delimited socio-technical systems in reality. A system analysis is just a highly explicit method of delimiting and studying a selected segment of socio-technical reality in order to keep its description manageable. A notion of open systems is needed here, to highlight the entrenchment of the object of study in its wider social context.

In practice, it can be quite difficult to justify the delimitation of such a specific segment. A useful delimitation ought to minimise feedbacks from the system to its environment in order to simplify the analysis and assign the system a distinctive identity. If later empirical analysis reveals that the system has in fact a significant influence on the environment, then the definition of the system boundaries should be reconsidered. Relevant changes in the systems environment are still taken into account as exogenous factors. System delimitations therefore need to be justified individually for each empirical study. While in some cases, it may be appropriate to define a national system, in others a regional or sectoral delimitation may be better. Due to changing patterns of interaction, a system's boundaries, like its internal structure, can also change over time. Reconfigurations, decomposition or integration with formerly exogenous elements can occur.

Keeping these considerations in mind, the PET system should be interpreted as a guiding pattern to structure the system description of a particular case under study. As depicted schematically in Figure 12.1, the PET system comprises five analytical levels, which cover both the micro-perspective of actors and interactions, and the macro-perspective of structural transformations. It puts technology in the centre of a network of social relationships, to reflect the interdependence between technology and social elements of the system. In addition, the whole system is embedded in its environment.

Figure 12.1 The basic elements of the PET system perspective

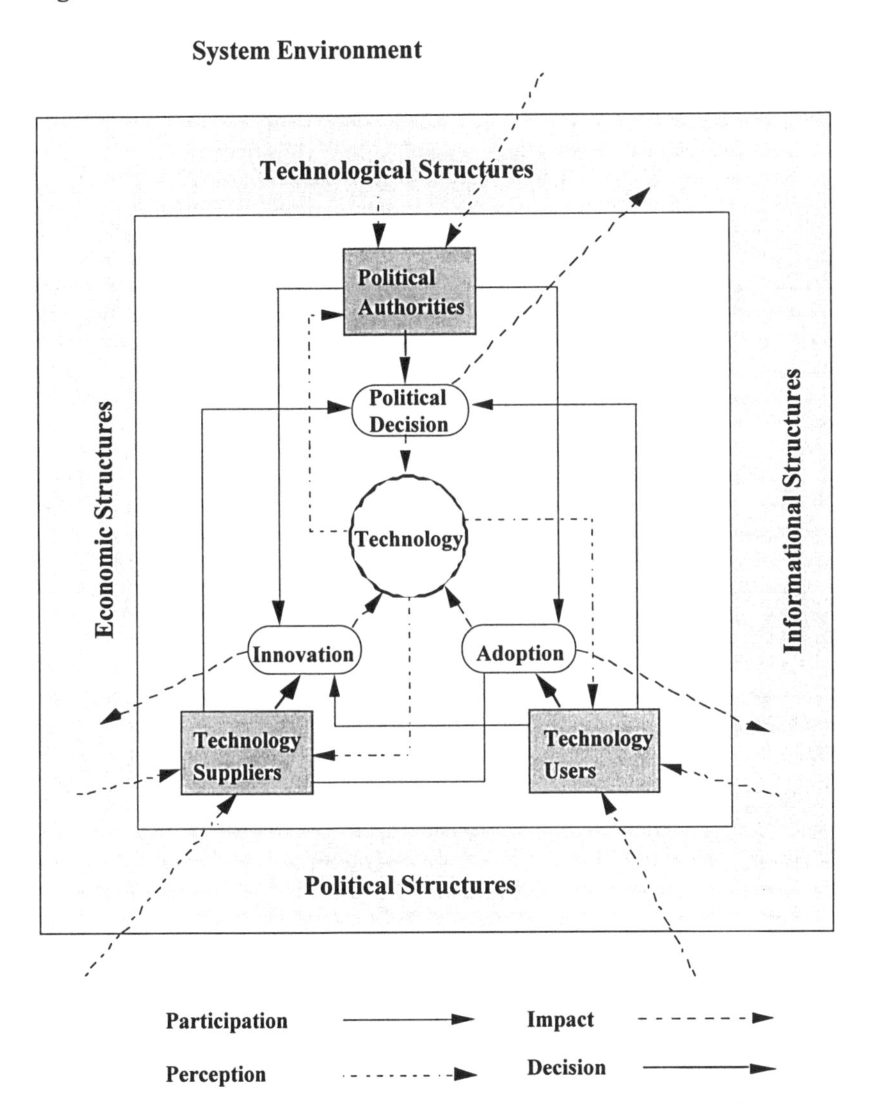

The five levels of analysis of the PET-system are thus as follows:

1. The *technology in question and its alternatives,* which are subject to changes, transformations and improvements (i.e. they are still 'malleable'), or even confronted with the emergence of substantially new niche solutions.

2. *Actors* and their motivations, interests and possibilities (mainly users and suppliers of technology, and political authorities, but also intermediary organisations and interest groups).
3. Their *interactions:* that determine the generation and selection of technologies and of technology-related strategies, captured here mainly in terms of innovation, adoption and political decisions.
4. *System structures:* which provide the evolving settings and institutions inside which interactions take place and which in turn are the aggregate outcomes of interactions over long periods of time.
5. The *exogenous system environment*, here conceptualised as being unaffected by the interactions in the system. The system as a whole adapts to the changes taking place in the system environment.

Technology - the object of investigation

The analytical separation of technological elements from social ones may not be shared by many social scientists, but is essential to have a good understanding of the technical and scientific aspects of a specific technology in order to be able to deal with its social and organisational aspects. However, this does not question the fact that social and technological aspects are deeply integrated.

Apart from the simplest examples, almost all technologies are constituted of several technical components, hence representing a system in a purely technical[4] sense. Large socio-technical systems are so pervasive in character that they comprise entire countries (electricity supply) or even the world (telecommunication). If the term *technology* is used in relation to large socio-technical systems (such as energy supply) it refers to a specific sub-system (such as a CHP plant) whereas the systemic technological context of this subsystem is captured by the notion of *technological structure.* Obviously, both these levels are interconnected.

A technology represents more than the technical hardware of a system. It comprises also the entire body of knowledge (codified and tacit) and organisational practices necessary to make it operate. This 'societal knowledge base' with respect to a technology can be captured in a simplified way by a number of main assessment dimensions which reflect those aspects, characteristics and impacts of the niche technology which are perceived as relevant by the different actors and stakeholders. These dimensions can be of a technical and economic nature, but also refer to a technology's organisational implications, its longer-term impacts on the structure of power relations, or its environmental impact. They comprise also the different kinds of long-term 'visions' that actors may have. Consequently, to characterise a technology and its emergence, a multi-dimensional and multi-actor framework is needed.[5]

Thus, whereas a technology may be on the one hand subjected to the limitations of what is physically feasible, the remaining set of options is further constrained by the technological, social, economic and political context. This context is composed first of all of different types of actors who assess different aspects of a technology in different ways, in line with their respective interests. Secondly, there are several structural settings in place which constrain both actors as well as technology.

Structures of the PET-system[6]

Technology develops within a socio-economic and technological environment, which can be structured along some main categories. These structures are not regarded as fixed, but as being in a permanent process of change, mostly slow and incremental, but on occasions also fast and revolutionary. Being the result of a historical process, these structures could be regarded as the accumulated result of decisions made by the actors in the course of time. They play an important role in framing and constraining individual decisions, but in turn they are also shaped by individual decisions. In other words, we get feedback between the macro-level of structures and the micro-level of actors.

1. *Economic structures* are a classical element of analysis, referring to the types and sizes of companies in an industry, its organisation (e.g. monopolised/liberalised, private/public ownership), and the structures of its main market, on both technology-supply side and technology-user side. I also include what is usually defined as the regulatory regime for a technology.
2. *Structures of the policy-making process* determine to a large extent the outcome of the negotiation processes leading to the formulation and implementation of public policies and regulatory measures. Especially the interfaces between government and industry 'shape' to a considerable extent the political decisions with respect to emerging technologies.
3. *Structures for the generation and supply of (technological) information* have been recognised as increasingly important. They comprise elements such as the organisational patterns of R&D-activities in a country or an industry, the science and technology systems and associated policies, channels of information dissemination and institutions like the patent system.
4. *Technical structures* are defined as the physical settings as part of which the diffusion of a new technology takes place, for example technical inter-dependencies and constraints, or the structure of the existing capital stock.

Interaction arenas and decision-making processes

Whereas structural transformations represent the changes taking place at the macro-level of the system, they build on the underlying processes at the micro-level. On the one hand, accumulated individual decisions, once taken, shape and determine how the structural settings change (dashed arrows in Figure 12.1). For example, a decision to adopt a new technology has an incremental effect on the technological structures of a country. Political decisions can even transform economic structures quite abruptly. On the other hand, it is evident that the actual evolution of these structures (e.g. the perception of structural changes at the micro-level [dashed-dotted arrows]) shape and affect the positions of the different actors and stakeholders with respect to new technological options.

'Decisions' are distinguished from the 'interaction arenas' in which these decisions are prepared through negotiation or bargaining processes.[7] Although e.g. the decision to adopt a technology is finally taken by an individual company, it is the result of interactions with others, e.g. with technology suppliers, regulatory bodies, or consultants. A decision (bold arrows) is the final result of such a negotiation process in conjunction with the interests, goals and expectations of the decision-taking actor. These negotiation and decision-making processes take place within the limits of certain rules and institutional settings. Being clearly a simplification of reality, three main types of decisions and of related interaction arenas are distinguished as shaping the innovation and diffusion of new technologies:[8]

1. Innovation decisions, i.e. the decision of a supplier to develop and market a new technology.
2. Adoption decisions made by technology users, after having explored alternative options, and the different requirements to be fulfilled.
3. Political decisions taken by political authorities like governments, e.g. the introduction of new specific regulations, or the reorganisation of an industry as an example of a structural measure.

Political choices regarding new control measures are analysed in a similar way as technological choices, namely as the result of a decision-making process, involving not only government bodies but also representatives from industry or from other relevant social groups. The innovation and adoption of new technologies, and the decisions of political control are finally determined by overlapping groups of actors involved, but in different arenas of interaction, each with its specific mechanisms of selection and variation.

Furthermore, different means can be distinguished by which interactions can take place. Four different types of means are proposed, namely *financial exchange* expressed in monetary terms, *execution of power* which usually takes the form of legal acts, *information exchange*, and *transfer of artefacts* such as technical hardware.[9]

The organisational settings in which these means are used can be described in terms of markets, hierarchies or networks. They exist at different levels and in different realms of the system. Administrative hierarchies are responsible for the implementation of political decision, but consultation processes in the networks of interest groups also play an important role in the preparation of political decisions. Technological innovations take often place in network-type settings (Callon 1992, Lundvall 1988) involving suppliers and users of technology, universities, and research centres, but also intermediary agencies, interest groups and political institutions such as ministries. These *inter-institutional networks* pervade the system and are responsible for shaping a conducive or detrimental 'climate' for a new technology.

Actors: Objectives, perceptions, and expectations

Negotiation processes precede the final decisions regarding a new technology, but apart from these negotiations, it needs to be explained by which general motivations and processes an actor – be it an individual, a private company or a public body – takes a decision. Essentially, the same model of the decision-making process is assumed for the three main types of actors considered (users and suppliers of technology, and political authorities).

Here I utilise the actor model suggested by Nelson and Winter (1982) which, in contrast to the traditional model of maximising agents, is based on notions of routines and satisficing behaviour under uncertainty. It is complemented by the concept of self-reference of decision-making as introduced by Maturana and Varela (1980). Based on perceived, but limited and filtered information about developments in available and emerging technical options,[10] an actor (user, supplier, political authority) who is confronted with the necessity and the possibility to spend limited resources on different types of activities, interacts and negotiates with other actors. This serves the purpose of reducing the uncertainty about the presumably best possible decision to be made. In the course of this negotiation process the four different types of means are used by the actors involved in order to affect the final decision so that they match their own perceived interests as well as possible. These interests can be seen as a combination of 'external' interests[11] (i.e. in the sense of economic or political success) and of 'internal' interests (reflecting the self-referential aspects of decision-making).

Self-reference is a particularly useful concept in the context of perception and filtering of information, because it addresses the constraining role of expectations within established mental frameworks of actors. These expectations contribute to reducing uncertainty and thus to the stabilisation of the decision process by excluding non-conventional or unexpected options from the actor's mental universe.

System dynamics

The overall description of technological change in terms of structures, actors and technological characteristics, as schematically sketched in Figure 12.1, is useful for structuring empirical information. However, the PET-terminology should also help to understand better the dynamics of technological change in LSTS and its underlying mechanisms. Basically, an evolutionary argumentation is going to be applied to describe the dynamics of the PET-system, but complemented by a number of additional elements.

A first important mechanism driving change in the PET-system consists of *external pressures* which may be exerted by constraints and requirements imposed by the system environment, such as societal problems, resource scarcity, major political changes or scientific discoveries from outside the technological realm under study.

Within the PET-system, the main driving forces of *change and variation* are on the one hand rooted in the auto-dynamics of technology, opening up new opportunities for future development paths. On the other hand, different types of needs – technical, economic and social – can induce efforts to generate variety and innovation. Such needs can be driven by exogenous developments (see before), but can also be rooted in the interests, problems and objectives of the actors in the system. There are first of all technological responses to such pressures and variations in terms of improvements along the different assessment dimensions, but such responses can also be seen in changes of political or economic structures.[12] Uncertainty adds to the variation in the search for new solutions because *ex ante* different paths can be worth exploring. Finally, the diversity of knowledge bases, skills and organisational features of novelty-generating actors contributes to variety.

Already these variety-creating mechanisms embody elements of selection. Constraints are imposed by dominant mental frameworks and designs, which are shared among the community of researchers, engineers and decision-makers (Tushman & Anderson 1986). Moreover, expectations concerning the operation of the different decision arenas in agreement with their respective internal logic represent a pre-selective force. Similarly the perception of structural interdependencies in the system, i.e. the expected constraints imposed by existing technological, economic, informational or political structures, needs to be mentioned. In other words, the 'selective force of anticipated selection mechanisms' contributes to the *ex ante* adjustment to certain technological trajectories, even before any effective decision is taken.

One can thus distinguish three key mechanisms of *technology selection*: the constraining force of mental frameworks and expectations on innovation, the selection of technologies by users, and the selection by the regulatory and political context.[13] These are further embedded in the structural evolution of the PET-system, which, in turn, is the aggregate result of the actors' decisions on technology-related issues. Any decision of an actor has implications not only for the technology in question, but also for system structures. Adoption decisions directly change the technological structures; the reorganisation of an industry changes the conditions for setting up certain technologies. On the other hand, the structures as perceived by the actors determine to a large extent their positions and decisions with regard to a specific technology. In other words, system structures represent an important constraining element of a technology's selection environment, but they are interrelated and co-evolving with the technology itself and the actors' positions.

Hence, the notion of selection environment as applied here is much broader and at the same time more differentiated than the initial definition used in economics, which was mainly confined to adoption decisions in a competitive market. In the PET system approach, all types of actors participate in the three arenas of interaction (innovation, adoption, and policy-making) and contribute to technology selection, using a variety of interaction means (information, money, legal measures, and artefacts).

Obviously, for both variation and selection, the prior history and development path of a system has an important role to play in shaping structural settings, mental frameworks, rules and institutions.

System change can be delayed by a number of *stabilising mechanisms*. In addition to dominant mental frameworks and shared expectations, the interdependence between the elements of the system represents a major barrier to the inducement of change. Such interdependencies can exist between all types of system structures. For example, technological structures of a large socio-technical system such as energy supply tend to reflect the decision structures in the political realm. Other important mechanisms which delay change are network externalities (e.g. between technologies and different types of structural settings), the longevity of technical hardware (e.g. vintage effect), complementarities between technologies (e.g. economies of scale and scope), learning and epidemic effects. In order for a new technology to diffuse successfully, it has to 'match' with the rest of the system which itself is changing – not only in the technical sense, but also with regard to organisational and cognitive aspects.

This interpretation implies a notion of a smooth co-evolution, rather than that of discontinuous jumps and technological revolutions. In large interrelated socio-technical systems it is very difficult to introduce fundamental changes due to structural resistances. However, past histories of such systems show that once the right matches are established, a fast process of change with pervasive impacts can happen. The underlying reasons for initiating such transformations can be exogenous to the system, as well as coming from within the system.[14] Changing the dominant paradigm in terms of structural settings (technological, economic, informational and political), mental frameworks and related sets of interests can be regarded as a process that requires a very high activation impulse. In other words, it may be quite difficult to trigger self-reinforcing processes to establish a new and consistent evolutionary, or even revolutionary, pathway. The existence of such *reinforcing mechanisms* is crucial to enable fast and radical processes of change that establish a new technology throughout the system. For example, successful demonstration or pilot projects often have a multiplier effect that is critical to the diffusion of a new technology. Similarly, the propagation of the results of learning process in local niches is important to make a new technology fully competitive within the dominant system. Also the creation of high expectations can be important to overcome 'lock-in' (Arthur 1988) to the dominant system or technology trajectory. However, too high expectations can also be counter-productive if they are not met and thus discredit the new technology.

In order for those reinforcement mechanisms to be fully effective, the right timing of stimulation and policy measures is crucial. Such windows of opportunity can arise when a government introduces a major structural or regulatory reform of an industry, which also offers an opportunity to stimulate a radical departure from the existing technological structure.

Thus policy has several possible avenues for inducing change even in highly complex and interdependent systems. However, their effectiveness is very much a matter of using reinforcing effects, and of the appropriate timing and fine-tuning of individual measures. Structural shocks (e.g. a far-reaching regulatory reform) can have significant impacts, but so does the careful management of bottom-up learning processes, supported by targeted technology-push measures. The PET system perspective does not suggest straightforward technology policy strategies, but offers a systematic and comprehensive pattern to guide a differentiated empirical analysis.

SELECTED KEY ASPECTS UNDER STUDY

The PET systems approach can be translated into a research methodology to elaborate future scenarios on which to build a policy analysis (Weber 1999). Though a comprehensive system analysis of the processes and scenarios of technological change for the two cases of energy supply and urban transport would go beyond the scope of this contribution, four key aspects of the PET approach are explored. They refer to the structural and the behavioural levels of the system, the dynamics of change, and the policy strategies pursued. I explore the validity and usefulness of the PET-approach, focusing on these four aspects, complemented by some background information on each of the two technologies and by a brief exploration of future perspectives.

Interdependence and co-evolution of system structures

The PET system approach argues that the different types of structures represent the intra-systemic context of the social processes by which a new technology emerges and becomes established. In phases of slow and incremental change, these structures represent a fairly stable context for the technology and social interactions. If, on the contrary, major behavioural or technological changes lead to severe mismatches with the structural context, adjustment processes between all three levels are induced as a consequence of the interdependence of actors, technologies and structures. This interdependence implies also that a technological innovation or any other significant decision cannot be compensated for simply by an adjustment of one structural aspect only but will in most cases require a co-evolutionary transformation of the entire structural context. Similarly, a major change or reform of one structural component will induce further co-evolutionary structural adjustments.

The diversity of perceptions and decisions under uncertainty

Such structural co-evolution and related structural changes are based on decisions at the micro level. The perception of structural changes by decision-making actors is the key mechanism for interconnecting structural changes with each other. Obviously, a decision does not only depend on the actors' perception of their environment (which includes the system structures), but also by their mental framework and specific problem situation. In line with the self-referential decision-making model, a great diversity of actor positions and perceptions can be expected, as well as the existence of prejudices and non-rational decisions, and a great deal of uncertainty. Different types of actors obviously perceive their environment in different ways, but also differences among actors who are in a similar situation are thus to be expected. Co-operation and networking contribute to reducing this uncertainty, to learning about other actors' positions, and thus to limiting the diversity of actor positions.

Dynamics of change and the role of reinforcement mechanisms

In the PET systems approach several mechanisms have been indicated as crucial for explaining the dynamics of change: that generate variety, select alternatives at different levels, and reinforce or delay the process of change. In the case studies, it should be possible to identify these empirically. Of particular interest is the argument that these reinforcing and delaying mechanisms are effective levers of change and thus good entry points for policy measures. Given the dynamic and non-linear character of such reinforcements and delays, the timing of policy measures can also be expected to be crucial for their success.

The effectiveness of combined policy strategies

Several possible intervention points for policy are suggested by the PET system approach: structural reforms, bottom-up technology-oriented measures, support for networking and the intelligent use of dynamic reinforcement mechanisms. The question which of these are likely to be particularly effective cannot be answered in general. The case studies should allow examination of the operation of each of these policy avenues, and point to some initial lessons. In line with the arguments regarding co-evolution and interdependence, it is expected that a combination of measures is more effective than single structural, behavioural or technological initiatives.

COMBINED HEAT AND POWER FOR ENERGY SUPPLY[15]

The first LSTS under study is energy supply, and the specific technology is co-generation (also called combined heat and power, CHP). The interesting feature of this case is that the energy supply systems in many European countries are currently undergoing a major reconfiguration, which open up opportunities for changing also the respective technological and systemic trajectories.

Background

Co-generation is the combined production of electricity and useful forms of heat in the same process. It is thus part of both the systems of power and of heat supply. Three main types of applications can be distinguished. Large-scale district heating systems are usually based on steam engines, and since the 1980s increasingly also on gas turbine or combined cycle plants. Industrial CHP applications tend to be somewhat smaller in size. Steam engine plants have been widely replaced by gas turbines in recent years. Small-scale CHP is a quite new development for which mainly reciprocating engines are used.

From a political perspective, the key issue in the current debates on energy supply is liberalisation and privatisation, a development, which is about to alter significantly the structure and organisation of the large socio-technical system of energy supply. This change of the regulatory regime represents a bifurcation point for the future development of energy supply. Depending on the specification of the liberalisation policy and on the implementation of complementary technology policy measures, several different scenarios seem possible.

With regard to CHP this raises the issue of the appropriate strategies to promote and advance the diffusion of CHP, a technology that is widely recognised to offer major environmental benefits. These 'appropriate' strategies depend on the formulation of energy supply framework to be introduced.

The PET-systems approach points to the aspects that ought to be taken into account in such an analysis, as well as to possible inroads of how the promotion of CHP in a liberalised framework could be addressed.

The analysis in the following section will pay particular attention to requirements of CHP with regard to such a liberalised framework. It is based essentially on the cases of the UK, Germany and the Netherlands, but complemented by information from other European countries.

Interdependence and co-evolution of system structures

Liberalisation, deregulation and privatisation have started to change the structures of energy supply. While originally an economic transformation, these policy initiatives are also altering power relations in the political realm and the roles which some of the actors play with regard to knowledge generation and supply. For example, public utilities were quite actively involved in R&D activities on new energy technologies in the past, but once they are operating in a fully competitive environment they are more reluctant to do so. Finally, the restructuring of the industry has implications for the structure of power generation which, under a liberalised framework, can be operated by public utility companies as well as by private operators.

Traditionally, power supply in most European countries was conducted by centralised monopolies, exploiting economies of scale and constructing large power-only plants. CHP, which requires the local consumption of heat, does not fit into such a centralised structure except under very specific circumstances. Such circumstances are offered by local monopoly suppliers of heat and power, which could exploit the economies of scope offered by CHP, e.g. in many German cities.

Liberalisation seems to have improved the conditions for CHP in legal terms. Private companies can far more easily set up local generation and supply; formal and administrative barriers to access to the generation and supply business have been abolished. On the other hand several structural conditions inhibit the widespread introduction of CHP. The established public utilities have maintained a very strong position, and until recently have fought to preserve a centralised supply structure. Only when they are able to exploit the benefits of CHP, e.g. by moving also into the heat supply business, do they show a growing willingness to support the new technology.

The most powerful tool of the former monopolies, which continued to operate the main electricity grid, turned out to be the control of the conditions for grid connection. Initially, it revealed to be quite difficult to control the technical connection of small CHP units, which did not deliver a fully predictable power supply to the grid. The economic terms of exchange were comparatively unfavourable for CHP. As a consequence, CHP is still most appropriate in cases where a significant share of the power is consumed on site by the generating company, and not exported to the grid. Improvements have been made in the meantime that solved the technical side of the problem, but not necessarily the economic dispute. It also took quite some time until the political decision structures changed in a way which gave the proponents of CHP access to the decision-making process and enabled them to lobby for it. In fact, in most countries established coalitions of energy supply companies contributed to the preservation of the old structures.

Liberalisation can have double-edged implications for CHP. In countries where already powerful local actors were in place and supporting the operation of CHP, its operation tends to be threatened by a fully competitive framework. Although being economically and environmentally beneficial in the long run, CHP requires comparatively high initial investments. A very competitive economic environment forces the actors to respect shorter-term economic requirements.

A major impediment for CHP has also been the lack of knowledge and information about CHP. Only once sufficient successful applications were known the reluctance and prejudices with respect to the technology disappeared. However, this required adequate information dissemination structures to be in place, a condition that was not met until very recently in many countries.

This short overview shows that in fact several structural adjustments were necessary (technological, economic, political, and informational) to enable a wider diffusion of CHP. The initial change of the economic structures, i.e.

the introduction of a new liberalised regulatory framework, was just the triggering event for farther reaching changes.

The diversity of perceptions and decisions under uncertainty

The liberalisation process brought about many changes and uncertainties for the actors involved in energy supply, in terms of both actors entering or leaving the scene and their changing portfolios of activities. Technology suppliers moved into the energy service supply business, energy supply utilities reinforced their activities in the generation business, political authorities started to replace an interventionist approach for a more hands-off philosophy, transferring many responsibilities to regulatory and industrial bodies. Industrial associations and other interest groups had to reorient their strategies in response to the less prominent role that government was willing to play in the energy supply field. New private companies were founded to provide sophisticated energy service solutions to end-users. These new companies became soon aware of the particular opportunities offered by operating CHP plants, often to the irritation of the established large, formerly public, companies.

However, liberalisation meant first of all a higher level of uncertainty for most actors, and thus high risks for new companies that intended to get involved with a new technology such as CHP. It was highly unclear whether the new framework, once fully established, would be conducive or detrimental to CHP. The grid connection conditions and export tariffs were critical issues, adding to the technological risks at play anyway. The main advantages of CHP, namely its environmental benefits, have never been fully translated into economic terms and legal provisions. However, in many countries voluntary or compulsory agreements between auto-generators and utility companies on export tariffs have been achieved and reduced the prevailing uncertainty, but only after long and cumbersome negotiations.

Finally, neither the regulatory framework nor the mental frameworks of decision-makers in companies and public authorities did change as fast as the technology. Prejudices against CHP were quite common, especially in countries, which had never built up a major CHP capacity in their history.

But even under these conditions of uncertainty, the opportunities for establishing CHP could be exploited by companies that were willing to introduce a novel approach to energy service supply in close co-operation with their customers. The mutual learning processes between new innovative technology and service providers on the one hand and energy end-users on the other have emerged as crucial to the success especially of small-scale CHP.

The behaviour that could be observed during this uncertain liberalisation period shows that many barriers to a wider uptake of the technology reside in the difficulties of changing the mental frameworks of actors, their risk-averse behaviour, and routinised internal decision procedures. They reflect the difficulties the actors faced in coping with the new structural conditions.

Dynamics of change and the role of reinforcement mechanisms

In spite of these adjustment problems, CHP has been quite successful in a number of European countries. Several mechanisms operated in favour of CHP. Success stories contributed to destroying prejudices against the technology, and convinced further users to adopt it. Learning from initial applications has played an important role for customising CHP to the needs of specific user groups and thus to create niches. In some countries, the standardisation of small 'packaged' CHP plants or of the respective design procedures simplified technology and reduced costs. The multiplier effect that can be achieved by an active information dissemination policy was important in all countries investigated, though carried out in some cases by public agencies, and in others by industrial organisations.

The existence or creation of an industrial support organisation was crucial for the promotion of CHP and for bringing the different interested parties together. At the political level, the support lent to CHP by key organisations and parliamentary committees reinforced the role and visibility of CHP on the political agendas.

In spite of these positive impacts of liberalisation on the uptake of CHP, one ought to be careful with a general judgement. Depending on the details of implementation and the specific situation in a country, liberalisation and competition can have a number of detrimental impacts on CHP as well. A tendency to short-termism in energy-related investment could be observed in several countries that counteracted the removal of other barriers. Moreover, the economies of scope which local utility companies could exploit by means of CHP were called into question by the introduction of competition between the different energy supply chains (especially between heat and gas). These examples show that often apparently minor mechanisms have been important for giving CHP the necessary impulses, or for preventing its wider diffusion.

The effectiveness of combined policy strategies

While in general the 'top down' introduction of a liberalised framework seems to have opened up opportunities for CHP, its impact depended ultimately on the fine-tuning of the detailed regulation in the liberalised framework and on additional supportive 'bottom-up' technology policy measures. The British example shows that without particular attention to the requirements of CHP, a new framework can have a very limited effect only.

The case-studies also show that the system structures need to be compatible with CHP if the technology is to diffuse. This means that certain monopolistic elements, especially the existence of horizontally integrated local supply companies, whether public or private, are important for CHP. Consequently, enforced liberalisation is not necessarily the right way for all countries, but can be detrimental for CHP, especially if it leads to the breaking-up of organisational structures that are conducive to its diffusion. In fact, it is hard to imagine a similarly top-down policy as British liberalisation being implemented in the Netherlands or Germany.

With respect to the reconfiguration of the energy supply system towards a more decentralised system based on a significant share of CHP, the main difficulty is the strategic management of the co-evolution of the different elements of the system. It should be done in a way that maintains existing conducive elements while new conducive elements of the liberalised framework are put in place. Both technology-oriented measures and 'bottom-up' networking policies appear to be necessary to complement 'top-down' structural reforms and to achieve a smooth transition to a new, decentralised and more sustainable system. The Dutch example comes in many respects quite close to such a combined strategy, as reflected also in the fast diffusion of CHP.

Exploration and policy implications

When looking forward, the fate of CHP after the transition from a monopolistic to a liberalised system – a major structural change of the system of energy supply – is still unclear, because many of the mechanisms under the new framework have not yet been settled in most countries. Out of the multitude of alternative development paths, three possible scenarios can be constructed, even if the transition paths for individual countries may differ considerably and depend on issues of political, economic and innovation culture:

1. Moderate liberalisation in line with the requirements of CHP: would enable the technology to diffuse further at a moderate pace;
2. Full application of liberal principles: would probably lead to a quite fast uptake in those countries where its diffusion has been blocked until now, though clearly limited by an upper threshold level. In countries where CHP is already well established, full liberalisation could endanger the operation of many schemes;
3. Political enforcement of CHP: could reinforce the diffusion process, but might also have negative economic side-effects.

NEW URBAN MOBILITY SYSTEMS[16]

The second LSTS addressed is urban transport: specifically new urban mobility systems (NUMS). Recently, several demonstration projects have been carried out to refine possible design configurations. It is still unclear how fast NUMS will be introduced, and what consequences this will have. Several elements are already available, but they have not yet been fully integrated into a competitive concept. The uptake of NUMS will depend to a large extent on a transformation of the dominant organisational, legal and economic conditions for mobility provision, and also substantially on changes in mobility behaviour of urban citizens.

Background

In spite of the fact that new urban mobility systems are still being defined as a transport concept, they can nevertheless be characterised by the following elements:

1. the use of new types of vehicles as part of a mobility service package;
2. seamless inter-modal integration of individually driven vehicles with traditional large-scale public transport systems;
3. the application of highly developed travel information systems.

'Individualised public transport' is one of the labels that has been coined to describe recent experiments such as the French Tulip and Praxitele systems. They are based on rental schemes for small electric vehicles, with smart card access and payment.[17] Vehicle parks at strategic locations in a city are the hire and return points for these vehicles, e.g. in the vicinity of metro or railway stations.

Professional car-sharing schemes, pioneered in Germany and Switzerland, have also evolved into new approaches to urban mobility provision (Knie & Petersen 1999, Harms & Truffer 1998).

Apart from the integration of different mobility forms in a service approach, new urban mobility systems are expected to complement the wider use of 'soft' transport modes such as walking or cycling, and to bridge the gap between public and individual transport. As they also aim for a more conscious use of transport services they contribute to a reduction of mobility needs. In this sense, the technological component is complemented by a strong social component, because new mobility systems imply changes in the mobility behaviour of citizens.

The politically most pressing question with respect to urban transport is how to promote a reorientation of mobility and transport services towards a more efficient, market-based provision on the one hand and a reduction of congestion and emissions on the other. Such a reorientation certainly requires the introduction of new technological elements (telematics, new vehicles, information systems) but it has also significant implications with respect to economic and regulatory frameworks (e.g. cost attribution, liberalisation, access restrictions, land use planning), the organisation of transport services (e.g. co-ordination of transport modes), and user behaviour (e.g. mobility needs and demand). New urban mobility systems are regarded as one of the options to alleviate the congestion and emissions problems, but they are only viable within particular organisational settings which, currently can only be realised for niche applications.

While being in principle feasible, NUMS are still in an infant state and cannot compete on equal terms within the dominant regime of urban transport provision, for both economic and social reasons. It still needs to be stabilised and developed together with the respective user requirements. This implies that a strategy for promoting new urban mobility systems should comprise a whole set of complementary measures in order to improve the technology, to transform the dominant urban transport regime, to change mobility behaviour and to introduce new types of mobility policies.

A key issue is therefore not just to develop a strategy to promote and advance the diffusion of new urban mobility concepts, but also how this strategy would fit into the context of different future scenarios of urban mobility in technological, economic, regulatory, organisational, social, and political terms.

Empirical material on this example has been drawn from cities in a number of European countries. It needs to be emphasised that the process of transformation of NUMS is far less advanced than in the case of CHP and energy supply. Only a few examples exist in which one could speak of an onset of structural change in the supply of urban mobility services; most cities and schemes have not yet gone beyond the experimental or demonstration stage. Therefore, the discussion of the different aspects of the case will have to remain exploratory in character.

Interdependence and co-evolution of system structures

Several structural implications of the establishment of new urban mobility systems can already be identified on the basis of the first experiments and demonstration projects carried out in the last years. For example, the established infrastructures might have to be changed or complemented in order to enable the recharging of electric vehicles, one of the potential key elements of NUMS. However, the most important technological challenge is the establishment of an information system to improve the co-ordination of all transport services in an urban area. Another structural matter concerns the spatial structure of cities and their hostility to the use of 'softer' transport solutions. Spatial and infrastructure planning has until now aimed predominantly at expanding transport, rather than at reducing mobility needs.

In terms of economic structures, a new legal and regulatory framework would be necessary to facilitate the integration of modes and services. The organisation of the transport service industry is very much oriented towards single modes, with limited co-ordination between them. It can be questioned whether the current trend towards increasingly competitive transport markets (especially for buses and railways) will be conducive to such an integration of services, which is a precondition for new urban mobility systems.

In addition, the lack of economic incentives for transport users to use means with lower social costs are an important impediment to their wider use. Transparency of costs is therefore an important element of an economic and regulatory structure by which new urban mobility concepts would be favoured.

On the technology supply side, the organisation and structure of the manufacturing industry is of relevance because the dominance of large companies makes it difficult to move towards technological solutions which would require a departure from the currently dominant technology.

The scope to introduce new structural and regulatory solutions depends also on the political structures. As with energy supply, it will be difficult to get support for initiating structural changes. Political support for experimenting with NUMS would also require a political culture that is conducive to experimental approaches and not geared towards the preservation of the *status quo*. Finally, the competencies of local authorities are important for urban mobility issues, because they need to drive the introduction of new solutions.

Structural problems in accessing information and disseminating experiences have further inhibited the acceptance of new mobility concepts. Information provision needs to be supported by appropriate organisational structures for collecting and disseminating it. For instance, the co-ordination of timetables has created many difficulties in the liberalised rail and bus market.

The diversity of perceptions and decisions under uncertainty

The first demonstration projects with new mobility concepts have shown how difficult it is to co-ordinate the activities of many different relevant actors and gain their support for such new concepts and grapple with user needs, manufacturers' and operators' interests, and the priorities of government (Weber *et al.* 1999). The development of support networks is therefore a crucial task to prepare demonstration projects, especially if they are meant to be just the first stage of a strategy to change the perceptions and positions of the key actors, and thus transform urban mobility.

Manufacturers may on the one hand have an interest in new technology options, which would open up new markets, but they are also often conservative about supporting alternatives that might threaten their established products. The major R&D efforts go into the improvement of traditional vehicles in order to improve e.g. environmental or safety characteristics. The partial support by the French manufacturers for innovative and radically new concepts such as Tulip or Praxitele is rather an exception. As shown by the Swiss car-sharing case, but also by the origins of the Smart car, newcomers may be needed to introduce conceptually innovative solutions.

Transport operators play a crucial role for new urban mobility concepts because they need to provide good interfaces with the established public transport modes. In terms of their approach to mobility provision, they tend to focus on their established mode only (bus, rail) rather than thinking in terms of integrated mobility chains that require co-operation with other providers.

National authorities (including planning authorities) have an important role to play to set the framework conditions that give an orientation for future technological developments in urban mobility provision. Moreover, until new urban mobility systems are sufficiently advanced to compete in a market-based system and induce a transformation of the dominant practices in the LSTS of urban transport, they need some kind of protection to enable learning and improvement processes.

Industrial associations are strong mobilising organisations, which not only shape public opinion, but also express industrial interests and provide information about new concepts. Citizens' organisation and associations of promoters of new urban mobility systems represent platforms for exchange, but they have also been important for launching new concepts in the public and political debates.

End-users will ultimately have to accept and use new mobility concepts (to a far greater extent than in the case of energy). New mobility systems need to fit the lifestyles at least of selected 'pioneer users' in order to pave the way for making new options acceptable and to change conventional ways of thinking about mobility (Hunecke & Sibum 1997). For example, Swiss pioneer users showed that car-sharing, which at a first glance seems to imply a reduction of mobility, in fact improved their perceived quality of life (Harms & Truffer 1998).

Dominant mental frameworks and future expectations affect the acceptance of new concepts. Currently, the dominant, individual private car-based mobility system is so strong that other options can hardly be imagined.

Moreover, as NUMS are still an open concept, it is also more difficult to get active support than for a quite well-defined technology such as CHP with its well-known environmental benefits.

These examples show that established ways of thinking represent a crucial barrier to the introduction of new urban mobility systems. However targeted networking initiatives, the involvement of a wide range of actors in the process of shaping the technology and the lead function of pioneer users can help change future visions and expectations, reduce uncertainty and facilitate complementary actions by the different actors.

Dynamics of change and the role of reinforcement mechanisms

The dynamics of change in the case of new urban mobility systems needs to be seen in its technological, structural and behavioural dimensions. Without significant structural changes, it will not be possible to achieve a widespread use of new urban mobility systems. Neither can it be achieved if the individual patterns of mobility behaviour remain as they are. There are clearly several delaying mechanisms at work, which prevent them from being introduced, e.g. the lack of inter-modal integration of mobility services. Another impediment for NUMS are the economies of scale and scope that have been achieved in the mass production of conventional vehicles, and the complementarity of present infrastructures and vehicles.

Until now the emergence of new urban mobility systems has been determined by the joint operation of five main forces. Technology-wise, new options have become available, but they are still in a phase of stabilisation. The problem pressure in urban areas has certainly been one of the main driving forces behind the search for new urban mobility systems. Uncertainty about future mobility needs of citizens, and the interests of manufacturers and operators have also been important determinants. Finally, a multitude of structural constraints of the LSTS shaped the path that the technology has taken until now.

On the other hand, the introduction of new urban mobility concepts benefits from a number of reinforcement mechanisms. Good and successful examples have changed attitudes, and also contributed to learning processes for improving the new concepts technically and organisationally. The stimulation of a certain degree of competition, both among various new alternatives, and between alternative and established technologies, has been a powerful element for improving and refining NUMS. A difficult balance needs to be struck between competition and protection in order to keep on the one hand a certain pressure for innovation, while on the other hand protecting experimental NUMS. The establishment of shared guiding visions among those actors involved or at least interested in a new concept can help a lot to align and co-ordinate their actions implicitly.

The next and critical step towards a reconfiguration of the LSTS of urban transport will be to go beyond isolated cases and to establish new urban mobility systems more widely.[18] At that stage, issues of standardisation of the design and organisation of such systems will become critical, for example to make the systems inter-operable across different cities and to convince other than the pioneer users of the benefits of NUMS.

The effectiveness of combined policy strategies

With respect to a more sustainable reconfiguration of urban mobility provision, it seems to be unlikely that the liberalisation of transport services and the introduction of a more transparent market-based pricing approach alone would be sufficient to stimulate the introduction and uptake of new urban mobility systems. Further mechanisms are needed, especially a better co-ordination and co-operation of different modes as part of a NUMS-based concept, but also framing planning policies and environmental regulations that favour more sustainable mobility service concepts.

Beyond these structural policy elements, a niche development strategy for introducing new urban mobility systems would require technological improvements and a learning process about how to adjust the possibilities of NUMS to the needs of more than just a few pioneer citizens. This implies also helping citizens learn about alternative ways of satisfying their mobility needs. Similar changes in attitude would be necessary on the part of established transport operators and industrial manufacturers who, for the moment, show only a very limited interest in NUMS. This may change once the problems of urban transport become more pertinent.

Any farther reaching attempt to reconfigure the provision of urban mobility would need to involve a broad range of actors from technology supply and operation, from the public as well as from policy-making in an extensive process of learning. Although NUMS have clearly several promising features to offer, the concept is still far too immature to put them at the centre of a long-term strategy. In fact, in view of the uncertainties, it would be dangerous to initiate today a reconfiguration process that relies to a significant extent on NUMS.

Exploration and policy implications

Finally, the question needs to be raised how the current lock-in situation could be overcome. A purely technology-driven approach is not promising, because of the need to integrate user requirements and structural constraints into the process of technology shaping. Nor does top-down structural change (e.g. by liberalising transport markets in cities) seem appropriate, because it would not provide sufficient targeted incentive mechanisms to enable the learning and innovation process still needed for new urban mobility systems. A purely bottom-up approach of experimentation might lead to some small-scale niche development for new urban mobility systems, but if it is not accompanied by structural changes, it would remain insignificant. Consequently, adjusting carefully the structural conditions for NUMS and experimenting further with the concept in a co-evolutionary manner appears to be the most promising way forward for new urban mobility systems.

This points again to three potential socio-technical scenarios regarding the future of new urban mobility systems:

1. A top-down, liberalisation-driven scenario: the diffusion of new urban mobility systems would be likely only, if the liberalisation process is accompanied by additional co-ordination efforts between modes and systems, and by targeted technology support measures.
2. A bottom-up technology learning scenario: here the main structural settings would remain untouched, but new urban mobility systems would be improved through individual experiments in a bottom-up manner, possibly supported by targeted government policies. However, as long as no structural changes in mobility provision happen, it is quite likely that these new systems will remain small niche markets.
3. A co-evolutionary scenario: in this case structural changes would be combined with bottom-up learning and development efforts. This would probably be the most promising pathway for a widespread introduction of new urban mobility systems and for a substantial reconfiguration of the LSTS of urban transport.

CONCLUSIONS

The preceding two case analysis have highlighted key mechanisms and processes underlying the evolution of large socio-technical systems. They referred to issues of structural co-evolution, decision-making, system dynamics and policy. This exercise has shown how the conceptual framework of PET can be used constructively as a guidance tool for research. It helps describe the emergence of new sub-technologies within large socio-technical systems, and identify a range of potential intervention points for policy. More detailed policy implications are not formulated here, but could be developed as part of a more comprehensive system analysis and scenario building exercise along the lines of the PET framework (see Weber 1999 for the case of CHP in the UK and Germany).

This position is in line with the other research approaches in the social shaping of technology, arguing that very clear-cut *and* generalised policy recommendations are incompatible with the complexity, contingency and openness of socio-technical change. More specific results for technology and innovation policy can nevertheless be obtained by following a case-specific research methodology which is inspired and guided by the theoretical insights of social shaping research. The PET systems approach aims to devise such a methodology, based on a comprehensive multi-disciplinary theoretical foundation. Compared to other scenario building tools the PET framework has three major advantages. First of all, it regards the actor to whom advice should be given (here mainly policy actors) as an integral and endogenous part of the system under study. Secondly, it is highly explicit with regard to its underlying theoretical assumptions, and obliges the researcher who follows the framework and methodology to make his/her assumptions explicit to others. Finally, through the development of different future socio-technical scenarios in a highly explicit way, policy makers can be informed and learn in a forward-looking way without having to recur to the false promise of general policy 'principles' which apply across cases, countries and sectors.

Based on the two cases briefly discussed above, a number of preliminary conclusions can nevertheless be drawn with respect to the possibilities and limitations of reconfiguring and transforming large socio-technical systems by political means. First of all, the introduction of sub-technologies or sub-systems has far-reaching implications for the entire LSTS in question, as well as for the actors and stakeholders concerned. This means that the rigidities that need to be overcome to reconfigure a system are significant. However, the cases also show that by means of well-timed and multi-level policy strategies quite fast processes of change can be triggered, but the detailed results of such a comprehensive strategy can not be anticipated in detail due to the complexity of LSTS.

Another key insight concerns the interdependence and co-evolution of the different structural elements of an LSTS. Policy strategies need to address the technological, economic and informational as well as the political structures in order to overcome the in-built rigidities.

Policies for individual countries or technologies should consider all the different elements suggested by the PET system approach. A structural policy may be the key policy inroad in one context, while a bottom-up niche development process may be more promising in another. The key point is that both alternatives should be taken into consideration when exploring potential policy strategies to reconfigure an LSTS and promote the corresponding technologies. It is an empirical question to find out which of these will matter most.

The perceptions and expectations of individual actors are the key nodes that interconnect structural and behavioural elements of the system. Especially the bottom-up elements of an envisaged policy strategy require the involvement of actors and stakeholders. Learning processes and interactions are crucial to overcome established mental frameworks and perceptions, and create new promising visions and expectations.

From the case studies it can be concluded that combined policy strategies that address the structural, technological and behavioural levels simultaneously are most promising to induce system transformations. In order to increase the effectiveness of policy initiatives, nurturing self-reinforcing mechanisms and aiming for a good timing of support actions turned out to be critical.

NOTES

1 Earlier versions of this paper were presented at the EASST '98 conference in Lisbon, 30 September - 3 October 1998, and at a COST workshop in Amsterdam, 11/12 November 1999. The comments made by the participants at both events are gratefully acknowledged.
2 See for example the project on Simulating Self-Organising Innovation Networks (SEIN), financed by the European Commission (http://www.uni-bielefeld.de/iwt/sein/).
3 See in particular the current British Foresight programme, the German FUTUR exercise, and the EC-funded thematic network FOREN that aims to transfer foresight-experiences to the regional level.
4 'Technological' and 'technical' are often used with unclear meanings. Here, 'technical' is used when referring to the embodied elements (artefacts) of a technology, whereas 'technological' is the more encompassing concept which includes both embodied and disembodied elements of technology, i.e. artefacts as well as knowledge.
5 This has obvious implications for the notion of a technology's 'superiority' as used in earlier diffusion models because superiority can not be defined any more in absolute terms but only with respect to a specific actor/stakeholder and assessment dimension.
6 The PET-system includes several of the elements that are also described in national or sectoral innovation systems, but also explicitly considers political structures. The concept of 'technological regimes' also shares a number features with PET structures, but strongly emphasises the cognitive dimension (Kemp *et al.* 1994) and underestimates the constraining force of organisational, political and technical interdependencies at regime level.
7 Compare also the concept of 'development arenas' as used by Jørgensen and Sørensen (chapter 7 in this volume).
8 These three categories should not be interpreted too rigidly; depending on the specific technological example chosen, quite different actors can contribute to these types of decisions.
9 Callon (1992) uses the concept of 'intermediaries' in a similar way as 'means' is used here. He distinguishes four types of intermediaries, namely texts, technical artefacts, human beings and their skills, and money.
10 A richly detailed description of how information is translated into knowledge (i.e. how it is incorporated into a firm's effective competencies) is given in Clark & Juma (1992), Chapter 5.
11 According to Dosi (1988) and Kemp *et al.* (1991), a set of main external determinants of innovation and of adoption decisions can be distinguished. Innovation decisions predominantly depend on a) the further opportunities offered by a new technology, b) the perceived opportunities to appropriate the benefits of innovation, and c) the expected patterns of demand. Adoption decisions mainly depend on a) the price and quality of an innovation, b) the difficulties to transfer knowledge needed to make the technology work, and c) the economic risks and uncertainties involved in adopting the technology. With respect to political decisions, the main external determinants could be a) the role and

influence of political pressure groups and of public opinion, b) the expected costs and benefits of political measures, and c) the difficulties to implement them effectively. Examples of internal driving forces are the influence of powerful individuals, conflicts of interests among sub-units of firms, self-interestedness of management or – in the case of politicians – the desire of being re-elected.

12 From a different angle than the one adopted here, political measures can be regarded as being subjected to variation and selection processes in a quite similar way as technologies, i.e. as determined by the interests of political stakeholders, by financial constraints, or by the creativity of the political administration.

13 Selection by technology users ('adoption') represents the classical evolutionary element of economic analysis of technological change (c.f. Nelson & Winter 1982). It should be seen in close connection with the definition of selection conditions by government policies and the establishment of shared mental frameworks and evolving expectations of the different types of actors involved.

14 See also Summerton (1994) who suggests six underlying reasons for changes (or reconfiguration, as she calls it) in large (socio-)technical systems: congestion in the physical network of many systems, negative externalities, e.g. environmental impacts, customer pressures, political pressures to improve performance, broad political ideologies, political developments or contingencies in a broad sense (e.g. wars), changing competitive conditions

15 The empirical material is based on interviews with technology users and suppliers and with actors in the political realm, complemented by further statistical and secondary information (Weber 1999).

16 The empirical material in this section is drawn to a large extent from the EU-funded projects SNM-T (dealing with Strategic Niche Management) and FANTASIE (on forecasting and assessment of new transport technologies).

17 In-depth information about the example of Praxitele can be found in Simon (1998). The Praxitele experiment is based on earlier experiences made in La Rochelle with electric vehicle schemes, c.f. Simon and Hoogma (1998).

18 This is exactly the problem addressed by the approach of Strategic Niche Management, see Kemp *et al.* 1998 and Weber *et al.* 1999.

BIBLIOGRAPHY

Arthur, B. (1988), 'Competing technologies: An overview', in Dosi, G. et al. (eds), *Technical Change and Economic Theory*, London: Pinter, pp. 590-607.

Bijker, W., Hughes, T.P. and Pinch, T. (eds.)(1987), The Social Construction of Technological Systems. New Directions n the Sociology and History of Technology, Cambridge (MA): MIT Press.

Bijker, W.E. and Law, J. (eds) (1992), *Shaping Technology/Building Society. Studies in Sociotechnical Change*, Cambridge: MIT Press.

Braczyk, H.-J., Cooke, P. and Heidenreich, M. (eds.)(1998), *Regional Innovation Systems. The role of governance in a globalized world*, London: UCL Press.

Busshoff, H. (ed.) (1992), *Politische Steuerung. Steuerbarkeit und Steuerungsfähigkeit. Beiträge zur Steuerungsdiskussion*, Baden-Baden: Nomos.

Callon, M. (1992), 'The dynamics of techno-economic networks', in Coombs, R., Saviotti, P.P. and Walsh, V. (eds.), *Technological change and company strategy. Economic and sociological perspectives*, London: Academic Press.

Clark, N. and Juma, C. (1992), *Long-Run Economics. An Evolutionary Approach to Economic Growth*. 2nd edition, London: Pinter.

Dierkes, M. and Hoffmann, U. (eds) (1992), *New Technology at the Outset. Social Forces in the Shaping of Technological Innovations*, Frankfurt: Campus/Westview.

Dosi, G. (1982), 'Technological Paradigms and Technological Trajectories: A Suggested Interpretation of the Determinants and Directions of Technological Change', *Research Policy*, **11**, pp. 147-162.

Dosi, G. (1988), 'Sources, procedures and microeconomic effects of innovations', *Journal of Economic Literature*, 26, pp. 1120-1171.

Edquist, C. (1997), *Systems of Innovation. Technologies, Institutions and Organizations*, London: Pinter.

Geels, F.W. (2001), 'Sociotechnical scenarios as a tool for reflexive technology policies: Using evolutionary insights from technology studies', chapter 13 in this volume.

Geurts, J. and Mayer, I. (1996), Methods of Participatory Policy Analysis. Towards a Conceptual Model and Development, *WORC Report* 96.12.008/3, Tilburg University.

Gibbons, M., Limoges, C., Nowothy, H., Schwartzmann, S., Scott, P., Trow, M. (1994), *The New Production of Knowledge. The dynamics of science and research in contemporary societies*, London: Sage.

Godet, M. (1993), *From Anticipation to Action. A handbook of strategic prospective*, Paris: UNESCO.

Görlitz, A. and Druwe, U. (eds) (1990), *Politische Steuerung und Systemumwelt*, Pfaffenweiler: Centaurus.

Görlitz, A. (ed.)(1994), *Umweltpolitische Steuerung*, Baden-Baden: Nomos.

Görlitz, A. (1995), *Politische Steuerung. Ein Studienbuch*, Opladen: Leske & Budrich.

Harms, S. and Truffer, B. (1998), *The Emergence of a Nation-wide Carsharing Co-operative in Switzerland*. A case study for the project 'Strategic Niche Management as a tool for transition to a sustainable transportation system', Zürich: EAWAG.

Helpman, E. (1992), 'Endogenous macroeconomic growth theory', *European Economic Review*, Vol. 36, pp. 237-267.

Hughes, T.P. and Mayntz, R. (eds) (1988), *The Development of Large Technical Systems*, Frankfurt: Campus.

Hunecke, M. and Sibum, D. (1997), *Socioeconomic aspects of individual mobility*, Research Report EUR, Sevilla: JRC-IPTS.

Jasanoff, S., Markle, G.E., Petersen, J.C. and Pinch, T. (eds) (1995), *Handbook of Science and Technology Studies*, Thousand Oakes: Sage.

Jørgensen, U. and Sørensen, O. (2001), 'Arenas of Development: a Space Populated by Actor-worlds, Artefacts, and Surprises', chapter 7 in this volume.

Kemp, R., Olsthoorn, A., Oosterhuis, F.H. and Verbruggen, H. (1991), 'An Economic Analysis of Cleaner Technology: Theory and Evidence', in: *Proceedings of the 'Greening of Industry Conference'*, Noordwijk aan Zee, 17-19 November 1991.

Kemp, R., Miles, I. and Smith, K. (1994), *Technology and the Transition to Environmental Stability. Continuity and Change in Technological Systems.* Final Report from project 'Technological Paradigms and Transition Paths: The Case of Energy Technologies' (Research Report No. PL910282), Maastricht: MERIT.

Kemp, R., Schot, J. and Hoogma, R. (1998), 'Regime Shifts to Sustainability Through Processes of Niche Formation: The approach of Strategic Niche Management', *Technology Analysis & Strategic Management*, **10**, No. 2, pp. 175-195.

Knie, A. and Petersen, M. (1999): Intermodalität als wissenschaftsbasierte Dienstleistung: Das Unternehmen CHOICE, in: Buhr, R., Canzler, W., Knie, A. and Rammler, S. (eds): *Bewegende Moderne. Fahrzeugverkehr als soziale Praxis*, pp. 133-146.

Lundvall, B.-A. (1988), 'Innovation as an interactive process: from user-producer interaction to the national system of innovation', in Dosi, G., Nelson, R., Silverberg, G. and Soete, L. (eds), *Technical Change and Economic Theory*, London: Pinter.

Lundvall, B.-A. (ed.)(1992), *National Systems of Innovation. Towards a Theory of Innovation and Interactive Learning*, London: Pinter.

Maturana, H.R. and Varela, F.J. (1980), *Autopoiesis and Cognition. The Realization of the Living*, Dordrecht: Reidel.

Mayntz, R. and Scharpf, F. (eds) (1995), *Gesellschaftliche Selbstregelung und politische Steuerung*, Frankfurt: Campus.

Nelson, R. and Winter, S. (1982), *An Evolutionary Theory of Economic Change*, Cambridge (MA): Harvard University Press.

Nelson, R. (ed.)(1993), *National Innovation Systems: A Comparative Study*, Oxford: Oxford University Press.

Rip, A, Misa, T. and Schot, J. (eds) (1995), *Managing Technology in Society. The Approach of Constructive Technology Assessment*, London: Pinter.

Romer, P. (1990), 'Endogenous Technological Change', *Journal of Political Economy*, **98**, pp. 71-102.

Simon, B. (1998), *The Praxitèle Experiment of Self-service Rented Electric Vehicles*. A case study for the project 'Strategic Niche Management as a tool for transition to a sustainable transportation system', Maastricht/Paris: MERIT/CIRED.

Simon, B. and Hoogma, R. (1998), *The La Rochelle Experiment with Electric Vehicles*. A case study for the project 'Strategic Niche Management as a tool for transition to a sustainable transportation system', Maastricht/Enschede: MERIT/CIRED & University of Twente.

Summerton, J. (ed.)(1994), *Changing Large Technical Systems*, Westview: Boulder (CO), pp. 1.23.

Tushman, M.L. and Anderson, P. (1986), 'Technological Discontinuities and Organisational Environments', *Administrative Science Quarterly*, **31**, pp. 439-465.

Weber, K.M. (1998), 'Innovation, diffusion and political control of co-generation technology in the UK since privatization', in Coombs, R., Green, K., Richards, A. and Walsh, V. (eds), *Technological Change and Organisation*, Edward Elgar, pp. 210-238.

Weber, K.M. (1999), *Innovation Diffusion and Political Control of Energy Technologies. A comparison of combined heat and power generation in the UK and Germany*, Heidelberg: Springer/Physica.

Weber, K.M., Hoogma, R., Lane, B. and Schot J. (1999), *Experimenting with Sustainable Transport Innovations. A workbook for strategic niche management*, IPTS: Sevilla/Enschede.

13. Towards Sociotechnical Scenarios and Reflexive Anticipation: Using Patterns and Regularities in Technology Dynamics[1]

Frank W. Geels

1. INTRODUCTION

Anticipation is an anthropological concept addressing a basic element in decision making in the sense that human beings are oriented towards the future. Individual actors and collective actors, such as firms and government agencies, make decisions on the basis of assessments of the current situation and expectations about the future. A range of methods has been developed over the past decades to formalise and structure anticipation efforts, e.g. trend extrapolation and curve fitting, computer modelling, cross impact analysis, Delphi methods, scenarios and foresight exercises. The renewed efforts in strategic planning in firms made much use of scenarios (Amara, 1989; Dewulf, 1998). Foresight exercises became popular with government agencies in the 1990s (Martin, 1995). We see debates about the adequacy of anticipation tools. Anticipation of the future and articulation of strategic plans or policies were no longer seen as rational processes to be left to experts. Instead, it was now seen as a rational *and* interactive process which included uncertainty, communication, learning, vision building, creation of consensus and commitment (Wack, 1985a, 1985b; Martin, 1995, Street, 1997).

There is a close relationship between technological development and anticipation efforts, strategic plans and policies. For example, technology-related anticipations play important roles in corporate decisions about R&D portfolios, and in public policies regarding RTD-programmes or climate change.

Reflexive anticipations, in this context, are anticipations which take into account the real-world processes involved in technological change. Empirical technology studies has identified a whole range of such processes, e.g. building networks of actors, negotiations within networks about design and functional specifications of technologies, learning processes (learning by doing, learning by using) [See Russell and Williams, Chapter 3 in this volume]. Users are not passive consumers of technology, but they can change and modify artifacts. The use of technology in real-life settings may require 'domestication' and adaptation of the setting (See Sørensen, chapter 2 in this volume). In short, technology and society co-evolve. Reflexive anticipation takes such insights into account.

When technology-related anticipation is based on too simple assumptions about technological development this can create several problems. First is the failure of strategies or policies which are based on these anticipations. We have identified seven pitfalls for technology-related anticipations, all related to the neglect of co-evolution of technology and society (Geels & Smit, 2000). The second problem is that such anticipations, because of their implicit assumptions about technology dynamics, do not allow for discussion of certain types of strategy or policy. Many technology-related anticipations are based implicitly on a 'linear model' of technological development. This stimulates discussions of only two kinds of technology policy: supply-push policies (e.g. R&D subsidies) or market-pull policies (e.g. purchase subsidies, regulations). Other kinds of policies run the risk of being neglected. On the basis of such simplistic technology anticipations it is difficult to address aspects of the new emerging perspectives on technology policy as identified by Russell and Williams (chapter 3 in this volume) a broader view of possible means and points of intervention for steering of technology, a wider notion of how technology steering is achieved and by whom, a different conception of the objectives and modes of intervention.

Simplistic anticipations are most likely to fail when products and services are novel and diverge from existing social and technical templates. Garud (1994) makes the distinction between 'fluid' and 'specific' structures, and Callon (1998) talks of 'hot' and 'cold' situations. While anticipations based on linear models may work satisfactorily in 'cold' and 'specific' situations, they often fail in 'fluid' and 'hot' situations. In these latter situations, reflexive technology policies are called for, and reflexive anticipations can be one way of facilitating such policies.

This chapter aims to contribute to solving the above problem. It claims that an important step towards reflexive technology anticipation is possible by using insights from technology studies, in particular, regarding patterns and regularities in technology dynamics. To this end, a list of patterns and regularities will be articulated. Some of them will be empirically illustrated. Furthermore, this chapter will situate particular patterns and regularities within a broader conceptual framework, in order to give them more general plausibility.

The list of patterns and the conceptual frameworks can be used to develop a new anticipation tool, namely sociotechnical scenarios (STSc), based on a more sophisticated understanding of technology dynamics.

These sociotechnical scenarios can be particularly useful in 'fluid' and 'hot' situations, i.e. when the dominance of existing technologies is challenged by newly emerging technologies. Technology studies have shown that such situations are characterised by indeterminacy, and uncertainty about the eventual outcome. There is much variety; different technical options are tried in different local settings. The complexity and contingency of interactions between local actors imply that outcomes are uncertain. Different paths may be taken. Precise predictions are therefore impossible. Nevertheless, I think it is possible to go beyond merely recognising the complexity and contingency in radical change. While recognising this, I further try to distil *patterns and regularities* and develop insights in such patterns that would make it possible to speculate somewhat more plausibly about transition paths and structural change.

Yet, even such grounded anticipations cannot claim to offer precise predictions. Sociotechnical scenarios are not a 'truth machine' which predicts the future. STSc cannot be used to deliver traditional 'advice' or consultancy, e.g. in terms of 'how to do it' recipes. Sociotechnical scenarios do not reinforce a technocratic policy style, because they keep complexity visible, and stimulate wider anticipation. STSc can be seen as a 'barbed' tool, which helps the policy maker to some extent but also stimulates reflexivity. Like a Trojan horse, STSc aim to offer help to policy makers but at the same time introduce STSc insights in the policy community. STSc helps structure the thinking of actors but does not replace it.

Scenarios usually concern longer time periods (e.g. 10-20 years) and often cover wide areas, e.g. sectors or technology domains. Thus, my search for useful insights about patterns and regularities was guided by the following requirements: i) provide insight to processes of radical and structural change, ii) cover long-term processes, iii) cover sector-wide developments.

Guided by these requirements the aim of this chapter is, first, to take stock of findings and insights from technology studies with regard to patterns and regularities. The resulting summary of patterns is a new contribution to the field of technology studies. Second, this chapter aims to give an empirical illustration of some of the patterns and mechanisms, and show how they contribute to dynamics in technological development.

The taking stock of findings is done on the basis of different case-studies and conceptual works. This leads to a problem of generalisation. Because the patterns and regularities are taken from different (local) backgrounds, they have limited general validity. In order to increase their general validity and applicability, I follow a strategy which has been identified by sociology of knowledge, namely to relate local findings to a theoretical framework (Rip, 2000). 'Theory' is not about reality *out there*. Instead, 'theory' offers a framework by which local knowledge can be made more robust. Accordingly,

I aim to make the patterns and regularities more robust by placing them within a multi-level theoretical framework. This multi-level framework has already been developed and described in the literature. Because it describes regime-shifts at a sectoral level, it does meet all three requirements formulated above.

2. PROBLEMS IN FORMALISED ANTICIPATIONS

The formalisation of technology-related anticipations traditionally took the form of technological forecasting (e.g. trend extrapolation, modelling). This worked well in the case of incremental technological change, when situations are stable and crystallised (Ayres, 1989; Huss, 1988; Coates *et al.*, 1994; Sapio, 1995). In 'fluid' and 'hot' situations, however, with the possibility of structural change and technological transitions, traditional technological forecasting techniques run into problems. This is experienced as a problem in the futures community. Theorists from this community feel that one cause of this problem is that futurists have limited knowledge of the underlying processes which are involved in technological development and radical change.

> Discontinuities are so difficult to forecast. (...) We don't know enough. about the underlying structure or the key driving forces *(Amara, 1988: 395-396; see also Ayres, 1989: 50*)

In fact, there is an articulated need in the futures community for new anticipation methods, based on insights in underlying processes of technological development.

> Technology and society are interrelated, and both are changing rapidly. (...) We must understand these interrelationships to forecast and manage technology effectively. Unfortunately, a macro-theory of sociotechnical change has not yet been devised. (...) There is no theory that is broad enough to embrace all sociotechnical change. Yet, this leaves the very real problem of establishing a conceptual framework that is robust enough to deal with the bewildering variety of complex situations that characterize sociotechnical change (Porter et al., 1991: 17-19).

I aim to address this need by developing sociotechnical scenarios based on a thorough conceptualisation of technological development. Yet, existing scenarios too suffer from problems. I diagnose two main problems:[2]

1 an implicit linear model of technological development;

2 undue emphasis on macro-logic and neglect of meso-logic.

The first problem is that scenarios often embody simplistic assumptions about the dynamics and drivers of technological development. The simplest assumption is that technological development comes from outside society and economics; something which simply happens, like manna from heaven. In computer models, technical development is often conceptualised as an independent exogenous parameter. This is close to technological determinism. In the recent Shell scenarios (Shell, 1996), for instance, information and communication technology is seen as TINA ('there is no alternative'), something which simply happens.

A somewhat improved conceptualisation is that technical development depends on investments in tangible and intangible assests, e.g. R&D, education, software (Minne, 1993). These investments, in turn, can be made dependent on the willingness of firms to take risk, the time (horizon) of managers, etc. In this conceptualisation, technical development is thought to depend on supply-side factors. For example in the economic theory of endogenous growth (e.g. Grossman & Helpman, 1991), economic growth and technical development are thought to depend on knowledge production. This is a linear 'pipeline' model: if you push and invest in research and knowledge creation on one side, technical development flows out on the other side.

Other scenarios (e.g. Dutch Central Planning Bureau, 1997) also make technical development dependent on demand-side factors, e.g. economic growth and consumers' purchasing power, cultural preferences (e.g. for sustainable products) or purchase subsidies from governments. These factors influence the direction of technical development and the speed of diffusion of products. This, too, is a 'pipeline' model,: if society pulls on one side, technology is developed on the other side.

If (government) policies are discussed in the scenarios, these fall basically in two categories: supply-push policies (e.g. R&D subsidies) or market-pull policies (e.g. purchase subsidies, regulations). Process-related policies (e.g. network building, experimentation and learning) fall outside the scope, due to the implicit linear/pipeline model.

The second problem in scenarios is the so-called macro-bias, implying too much emphasis on 'macro-logic'. Scenario logic refers to the basic differences between the number of scenarios which are constructed. This difference is constructed by varying two factors that are very uncertain and have high impact. By 'scoring' these two factors differently, four different scenarios are constructed as a '2 by 2' matrix based on different 'logics'. Other uncertain factors are 'scored' differently in each scenario, according to the basic logic. Wilson (1997: 37) describes this as 'putting the flesh on the skeleton'. This metaphor is not neutral but implies some causative model. The term macro-bias indicates that the construction of scenario logics is usually derived from macro-elements. As a result, the dynamics and outcomes of the scenarios depend too much on such macro-aspects (e.g. econonomic growth, environmental awareness, oil price). The 'logic' of the scenarios is top-down in the sense that processes and actions at the meso- and micro- level are determined by macro-elements. The related problem is that the dynamics and outcomes are unsurprising and somewhat tautological. It is no surprise that environmentally friendly technologies become dominant in scenarios with high environmental awareness and high economic growth.

An example is the foresight and scenario exercise which was done for the Dutch Department of Environment about future trends for sustainable technologies (Weterings *et al.*, 1997). In the first step, the state of the art of a whole range of sustainable technologies was described for sectors like energy supply, transportation, food production, housing. In the second step, three macro-scenarios about the future of the Netherlands were given (European renaissance, Balanced growth, Global shift). In the third step technology scenarios were made by varying the development and diffusion of the sustainable technologies, according to the macro-scenarios. Not surprisingly, sustainable technologies diffused easily in Balanced growth, which was characterised by a 'mentality shift', strong economic growth and much R&D. Sustainable technologies were hardly developed and used in Global shift, which was characterised by strong economic competition, liberalism and little R&D.

Technology scenarios of this sort have been criticised as not being very useful for sectoral policies (Schoonenboom & Van Latensteijn, 1997). Such scenario approaches give some insight into policy options. However scenario outcomes are externally determined, by national and international forces, even when they concern sectoral issues. Such scenarios are straightforward to make, but do not show interesting dynamics, and the outcomes may become tautological.

To address this problem, scenarios should be constructed which also have a 'meso- and micro logic'. Their dynamics and outcomes should also depend on sectoral dynamics, where different players (universities, producers, users, public authorities) are involved in learning processes and strategic games. Only when such 'meso- and micro logics' are included can we hope to stimulate reflexive anticipation. It is here that the consideration of patterns and regularities can make a substantial contribution (see section 5).

The articulation of both problems in technological forecasting and technology scenarios, can be translated into requirements or directions for improvement. These are helpful in developing sociotechnical scenarios. First, we need a macro-theory of sociotechnical change, which is robust and broad enough to deal with a wide variety of situations. This conceptual framework needs to go beyond the linear model of innovation and meet the requirements articulated above (providing insight to processes of radical and structural change, covering long-term processes, and sector-wide developments). Second, the sociotechnical scenarios need to have 'meso-logics' as well as 'macro-logics'.

Technology studies can contribute concretely to both requirements. The analysis of present and ongoing trends can be done with a multi-level conceptual framework of sociotechnical development (section 3). Meso- and micro-logics can be created by paying attention to processes and dynamics in niches and regimes as well as patterns and regularities in technology dynamics. Because these regularities provide insights about the interaction between processes at several levels, they form the 'glue' with which dynamic multi-level scenarios can be made. Scenarios with different 'internal logics' can be created by varying the importance of the regularities and mechanisms. In this way the internal dynamics of the scenarios can be addressed.

3. A MULTI-LEVEL FRAMEWORK OF SOCIO-TECHNICAL CHANGE

A theoretical perspective that meets the formulated requirements is a multi-level framework that consists of three levels:

- a 'micro'-level of *technological niches*, which are 'protected' spaces in which actors learn in various ways about new technologies and their use,
- a 'meso'-level of *technological regimes*, which are rule-sets that are built up around a dominant technology and grant it stability,
- a 'macro'-level of *sociotechnical landscapes*, which consist of a range of contextual factors that influence technological development but that cannot be changed by technology actors.

This framework has been developed at Twente University in The Netherlands (Rip & Kemp, 1998; Kemp, Schot & Hoogma, 1998; Schot, Hoogma & Elzen, 1994; Van de Poel, 1998).

The meso-level of technological regimes is crucial in the multi-level framework. Technological regimes refer to the cognitive, normative and formal rules which are shared by the dominant actors in a sector or technology domain (e.g. producers, users, government). Technological regimes are an explanation for the occurence of *technological trajectories*, i.e. the structured and ordered development of technologies. The concept of technological regimes is derived from evolutionary economics (Nelson & Winter, 1977, 1982). Nelson and Winter (1977) observed that engineers focused on particular problems and had certain cognitive notions of how to deal with them.

The regime concept from evolutionary economics needs to be changed and made more sociological in order to bring it closer to empirical studies. The emphasis on the sociological concept of 'rules' aims to widen the concept of 'search heuristics' from evolutionary economics. Rip and Kemp (1998: 338) suggested a broader definition, which takes on board the technical, economic, organisational and social aspects. They define a technological regime as: the rule-set or grammar embedded in a complex of engineering practices, production process technologies, product characteristics, skills and procedures, ways of handling relevant artifacts and persons, ways of defining problems, all of them embedded in institutions and infrastructures.

While the cognitive rules of Nelson and Winter are embedded in the minds of engineers, the rules of this broader definition are embedded more widely: in the knowledge base, in engineering practices and beliefs, in management systems and corporate governance structures, in manufacturing processes, in the characteristics or products, institutions and infrastructures that make up a technological regime.

Examples of such rules are: the search heuristics of the engineers, technical standards, the rules of the market in which firms operate, the user requirements to be accommodated at any give time, and the rules laid down by governments, investors and insurance companies. These rules guide (but do not fix) the kind of research activities that companies are likely to undertake, the solutions that will be chosen and the strategies of actors (suppliers, government and users).

The concept of technological regimes helps understand inertia in sociotechnical change (e.g. car manufacturers stick to gasoline engines), due largely to the embeddedness of existing technologies in broader technical systems, production and consumption patterns, belief systems and values, which creates economic, technological, cognitive and social barriers for new technologies.

Because of the inertia of incumbent technologies, many new technologies remain on the shelf. For instance, the new product may require different knowledge and capabilities, new production techniques and skills that may not be available. Their use and development may require complementary inventions and changes in organisation (in production routines, in plant and factory lay out) plus changes in the institutional context (in regulation, fiscal policies and social norms and values). These are known to come about slowly.

Radically new technologies may also have a hard time to penetrate the market because of strategic opposition from firms vested in the old technology, as such technologies may erode the value of past investments and cause anxiety in the organisation (see section 4 for examples from the automobile regime).

This raises questions about how regime barriers may be overcome: how may the technology come into its own, develop from an idea or prototype into a successful product?

A niche can be seen as a specific domain for application in which producers and users – sometimes together with third parties such as governments – form an alliance to protect new technologies against too harsh market selection.

Niche developments happen in two (partly overlapping) forms: technological niches, and market niches. Niche-development starts in protected spaces, where regular market conditions do not prevail because of special conditions created through subsidies and an alignment between various actors. These *technological niches* are often played out in the form of experiments like those with electric vehicles in various European countries and cities (Rochelle, Rugen, Gothenborg etc.). Technological niches can develop into *market niches* – application in specific markets in which regular market transactions prevail.

These niches are important for the development of a new technology, underpinning the take-off of a new regime and the further development of a new technology. They provide space for a such key processes as: coupling of promises expectations, learning and articulation processes, and network formation (based on Schot *et al.* 1996; see also Elzen *et al.* 1996).

3.1 Promises and expectations

In the early stages of development, the advantages of a new technology are often not evident. Its value still has to be proven and there are many resisting forces. In order to get the new technology on the agenda, actors make promises and raise expectations about new technologies. Promises are especially powerful if they are shared, credible (supported by facts and tests), specific (with respect to technological, economic and social aspects), and coupled to certain societal problems which the existing technology is generally not expected to be able to solve. Once certain promises have been accepted and placed on the agenda, activities need to be developed to substantiate the expectations, for example by conducting research or doing experiments (Van Lente & Rip, 1998).

3.2 Learning and articulation processes

In niches one can learn about regime barriers and how they may be overcome. There is a wide range of dimensions for learning:

- Articulation of technical aspects and design specifications: what adjustments to the technology are required? What is the scope for learning economies, and for overcoming initial limitations?
- Articulation of government policy: what changes in fiscal policies and legislation are necessary to make an application of the technology possible or stimulate its use? Should the government assume a different role?
- Articulation of cultural and psychological meaning: what symbolic meaning can be given to the new technology? For example, can it be labelled and promoted as a safe and environmentally benign technology, as a 'feminine' technology and/or as a technology that accords with a modern lifestyle?
- Articulation of the market: for whom is the new technology produced? What are the consumers' needs and requirements? How can the technology be effectively marketed?
- Articulation of the production network: who should produce and distribute the new technology and complementary products (fuel in the case of transport)?
- Articulation of the infrastructure and maintenance network: what complementary technologies, capabilities and infrastructure must be developed, by who? Who looks after the maintenance of the new technology? Who is responsible for recycling or waste?
- Articulation of societal and environmental effects: what effects does the new technology have on society and the environment?

3.3 Network formation

The development of a niche often requires the formation of a new actor network. Actors with vested interests in other technologies will generally not be interested in stimulating a new, competing technology. They may participate in the developments for defensive reasons but will show no real initiative. There are many examples of actors trying to slow down or even stop the niche-development (see section 4.2.). In order to expand the niche, specific new actors must be involved. New network relations should be developed in which the new technology can function as desired.

The third level of socio-technical landscape refers to a set of deeper structural trends and changes (see Rip & Kemp 1998). The function of the concept 'socio-technical landscape' in the conceptual framework is that it accounts for technology-external factors that influence its development. The socio-technical landscape contains a heterogeneous set of factors, ranging from macro-economic factors such as oil prices and economic growth, to political coalitions, from broad cultural and normative values to large environmental problems. The metaphor 'landscape' is chosen because of the literal connotation of something around us that we can travel through. The landscape provides a 'gradient' for technological developments at the regime and niche-level. One element of the landscape is the large-scale material context of society, e.g. the material and spatial arrangements of cities, factories, highways, land-use, gas and electricity infrastructures. Examples of other elements are political culture, macro-economic trends, oil price, war, values. In the description of 'socio-technical landscape' the exact analytical content has been left open. It is not so much the exact content that I want to emphasise, but the function this concept fulfils in the multi-level framework. Both regime and landscape are structures or contexts for interactions of actors, but in a different way (see Figure 13.1).

Figure 13.1 Static multi-level framework

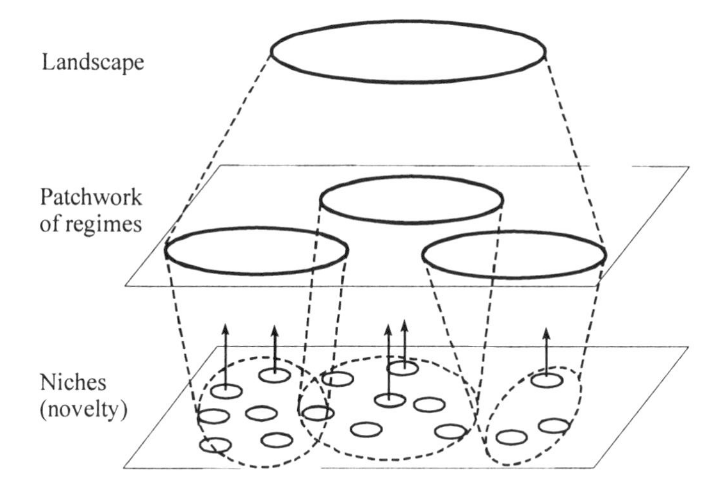

Technical regimes are bounded networks of actors that share and use a set of semi-coherent rules. The rules have a direct effect on the technical search process. They are internalised in the behaviour of actors that produce, use and regulate technology. The socio-technical landscape has an indirect effect on technology, because it forms the wider context in which multiple actors make their assessments and decisions. Hence, we have called the socio-technical landscape a technology-external context.

Regime-shifts occur as follows. A novelty emerges in a local practice and becomes part of a technological niche when system builders are able to form a network of actors that share certain expectations about the future success of the novelty, and are willing to fund further development and learning processes. The technological niche is formed against the background of the existing regime and landscape. The further success of niche formation is on the one hand linked to processes within the niche (micro-level) and on the other hand to developments at the level of the existing regime (meso-level) and the sociotechnical landscape (macro-level). In the niche the technology is improved. Under certain circumstances it may find a place in certain market niches. When the market niches expand, the technology may transform or substitute the existing regime. In a later stage, the new regime may even trigger changes at the landscape level.

So it is the multi-level coincidence and coupling over time of several developments (successful processes within the niche, reinforced by changes at regime level and at the level of the sociotechnical landscape) which determines whether a regime-shift will occur. The overall multi-level dynamic is schematically represented in Figure 13.2.

Figure 13.2 A dynamic multi-level view of regime shifts (Rip & Kemp, 1996)

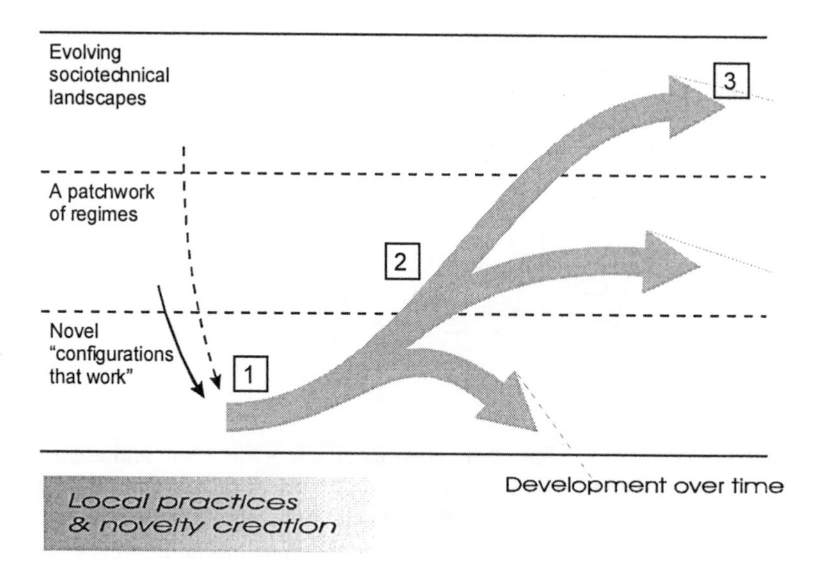

[1] Novelty, shaped by existing regime
[2] Evolves, is taken up, may modify regime
[3] Landscape is transformed

4. EMPIRICAL ILLUSTRATIONS FROM THE TRANSPORTATION REGIME

This section gives an empirical illustration of the usefullness of the multi-level framework for describing and analysing dynamics in technological development. It examines the transportation sector from the late 1980s to the late 1990s, and focuses in particular on the interaction between the incumbent gasoline vehicles regime and the battery-electric vehicle (BEV) niche.[3] All major automobile manufacturers have invested in BEV-development programmes since the mid-1970s. Prior to the 1990s this did not result in anything more than experimental prototypes. In the 1990s, however, the BEV-niche greatly expanded, because of developments in the state of California and in some European countries (France, Switzerland).

The empirical description consists of two parts. First, the expansion of the BEV-niche in interaction with landscape developments is described, thus illustrating the multi-level framework. Subsequently, I will analyse in more detail the dynamics in the American niche, and illustrate some patterns with regard to the interaction between the niches and the incumbent regime. This latter empirical illustration serves as a stepping stone for section 5 which aims at more general patterns and regularities.

4.1 The expansion of the BEV-niche as a result of multi-level linkages

The most important stimulus for the expansion of the BEV-niche was the very strict mandate that the California Air Resources Board (CARB) proclaimed in 1990 for large car manufacturers. Four categories of pollution norms were introduced: 'transitional low-emission vehicle' (TLEV), 'low emission vehicle' (LEV), 'ultra-low-emission vehicle' (ULEV) and 'zero-emission vehicle' (ZEV). These categories were used to make a future schedule in which increasingly strict norms were to be met. The mandate required that the sales of large manufacturers would consist of 2 % zero-emission vehicles in 1998, 5% in 2001 and 10% in 2003. Since the ZEV requirements in the mandate could only be met with electric vehicles, this guaranteed a market of tens of thousands electric cars a year. In 1996 CARB revised its earlier mandate, letting go of the ZEV-requirements for 1998 but maintaining them for 2003.

The *California* BEV-niche was the result of the cumulation of several processes and factors related to the particular political landscape and culture. First, in the Unites States legislative bodies (federal and state of California) take a strong stance towards industry, with a tradition of forcing innovations by strict laws and emission norms. As a result, the relationship between the

US car industry and government used to be antagonistic. Furthermore, California received special institutional and legal powers to deal with air pollution, because of specific geographic characteristics, which mean that air pollution only disperses very slowly. From the 1970s they have required cars to meet strict emission norms. The industry often felt constrained by these norms. Second, the state of California was affected by the end of the Cold War, which resulted in the 'downsizing' of the military and created unemployment problems in former high-tech military industries. This reinforced efforts by politicians to stimulate newly emerging industries to reduce unemployment and make good use of the pool of high-tech competencies. Because battery-electric vehicles were seen as a promising new technology, Californian politicians were willing to support them. Thus, the strict CARB regulations were not only inspired by environmental considerations, but also by industrial and political ones. A third process, occurring in networks at the niche level, was the *Impact* BEV-project at General Motors. Started as an experimental solar-powered vehicle project, the Chairman of GM, Smith, was so enthusiastic about the design that he had it transformed into a BEV. The *Impact* was exhibited at the Los Angeles Motor Show, early in 1990. It triggered excitement and interest from the public and also CARB officials. Although it was difficult for CARB officials to appraise the technical and commercial aspects of the car, the *Impact* convinced them that BEVs were feasible (Wallace, 1995). Within the distrustful, adversarial relationship between car industry and regulators, CARB officials thought that the main problem was the reluctance of car manufacturers to bring electric vehicles onto the market. From this perception it seemed logical that car manufacturers needed to be pushed: hence the CARB mandate. In sum, the Californian BEV-niche emerged as the result of the linkage between processes at different levels.

In *Europe*, too, BEV-niches emerged in the 1990s. In the EU, national governments try to protect their car industries from too strict regulations (especially Germany, France and Italy) which makes it difficult to design strict regulations. Moreover there is a consensual tradition of only applying strict industrial regulations when the industry indicates they are ready. At the municipal level, however, there is more activity in Europe than in America, partly since many transport-related problems are located in cities (especially old European cities with narrow streets, where congestion and local air quality are increasingly problematic). These problems are increasing because of cultural and normative changes at the landscape level. For example, environmental concerns have gained increasing prominence on policy agendas (e.g. air quality and congestion in cities, noise, safety). As a result cities, especially European cities, are increasingly taking an active role in dealing with transport-related problems. In the last couple of years, many European cities have started experimenting with new transportation systems (eg. La Rochelle, Paris, Strasbourg, Stockholm, Göteborg, Malmö, Mendrisio), including BEV experiments. In many cases they cooperate with car manufacturers, national governments, utilities etc. This has created technological niches of several kinds, where learning processes are taking place with regard to new technologies, infrastructures, users and conditions of implementation. In France and Switzerland these local experiments linked up with and were stimulated by national governments and industrial networks: these multi-level linkages resulted in the expansion of BEV-niches.

Particularly in *France*, the government has strongly stimulated the development of electric vehicles (see Callon, 1980, for an analysis of early French attempts). Together with the French car industry and the national electric utility, EDF, a national programme has been implemented, involving not only R&D, but also concrete societal experiments in cities, designed to learn about the practical performance of electric vehicles and user experiences. Because of specific and robust visions and expectations of both industry, government and the electric utility, BEV development is beginning to take off in France. Users are mostly fleet owners at the moment, but consumers have also shown interest.

In *Switzerland* a particular technological niche – of small and light-weight cars – is being developed. Some of these will be electric vehicles. In the municipality of Mendrisio a large-scale societal experiment with light-weight electric vehicles started in 1995. In 6 years, 350 vehicles will be sold to consumers (with a 50% subsidy). The main goal is to find out about vehicle performance and user experiences. The experiment is supported by the Swiss government, as part of a major CO_2-reduction and energy saving programme ('Energy 2000') by the Department of Energy.

4.2 Specific patterns in the American BEV-niche

The key actors in the BEV-niche and their strategies are described below.
Small car producers. They have made the bulk of functioning BEV. Many small BEV-companies were started on the basis of expectations about an emerging BEV market. Sales figures of hundreds or thousands of vehicles presented a large enough market for them. For example the CALSTART consortium was started in 1992 by 40 members. In 1996 this number had risen to 200. It well illustrates the heterogeneous network of actors interested in alternative vehicles (BEV, HEV, NGV), including utilities (gas and electric), public transportation companies, R&D firms (from the battery sector), small BEV producers, BEV importers. The participating organisations are already 'jockeying for position', hoping to become important suppliers, when the BEV market takes off.

Battery manufacturers. Batteries are perceived by many actors as the main bottleneck for BEV's. Battery producers are dedicating great efforts to overcoming these bottlenecks, especially since the 1990 CARB mandate. Optimistic forecasts estimated a possible future market of millions of BEV, representing a multi-billion dollar battery market. Three generations of batteries are distinguished, each with its specific pros and cons. Current lead and nickel-cadmium batteries are relatively cheap, but have a low energy content, which limits the driving range. Second generation nickel-metal hydrides (NiMH) have an energy density twice that of lead batteries. The third generation Lithium-batteries are expected to have energy storage capabilities five times higher than current lead batteries. Many BEV-actors expect these to be the best in 5-10 years time.

Electric utilities. In many countries electric utilities play an active role in stimulating a market for electric vehicles. They are interested in BEV for different reasons: i) stimulating growth of the electricity market, ii) possibilities for load levelling (when BEVs are charged during the night), iii) public relations and environmental image. Electric utilities are mainly involved in R&D on the charging infrastructure for BEV's, as well as engaging in real-life experiments and giving subsidies to users to develop the BEV market. They also tend to take the role of network manager in bringing different actors together, e.g. the CALSTART consortium. In the US, almost all large utilities have BEV programs. Especially when large car manufacturers were opposing the ZEV-requirements from the CARB mandate, the utilities took it upon themselves to demonstrate that electric vehicles were a viable alternative. The Californian utilities have been very active. However, many of the programmes came under pressure when the electricity market was liberalised.

Cities. Persistently confronted with congestion and air quality problems, many European city councils provide funds for experiments with BEV in their cities. American cities are not very active however.

National governments. Especially US governments (California!) have formulated strict requirements for lower car emissions. Many national governments stimulate R&D efforts to develop BEV.

Large car manufacturers. Although large car manufacturers are, of course, important actors in the incumbent gasoline car regime, they also play a role in the BEV-niche. As part of diversification and hedging strategies they invest some of their R&D funds to explore the BEV-option. Yet, they are unwilling to mass-produce BEV. Assuming the indispensability of the gasoline car, their main strategy for dealing with environmental problems is an *efficiency revolution*. The fuel efficiency of internal combustion engines may be improved by 50% and the use of lighter materials and new body designs has large potential. This illustrates the pattern of 'slow regime change': the development activities of large car manufacturers are largely directed towards incremental improvements of existing cars. This way they are protecting their sunk costs (in production lines, competencies, factories, etc). This conservative attitude is reinforced by marketing arguments: if new cars do not sell, or perform less well than expected, their introduction may damage the image and brand name of a firm. This doubt about customer behaviour leads to a 'cartel of fear'. Despite their overall incremental strategy, large car manufacturers are also involved in exploring alternative technologies. Most attention, at this point, is given to alternative fuels (natural gas, methanol and ethanol). For radically different power systems (e.g. BEV), the car industry expects only small market niches in 10-15 years time. They expect the performance of BEV to be too low for consumers to accept them (e.g. limited driving range, high costs, slow acceleration, limited battery lifetime).

Having described the current and historical perceptions, expectations and strategies of actors I will now consider dynamics and describe how the attitude of the car industry towards BEV has changed substantially over the last couple of years, focusing on the United States. In this story, I distinguish three phases: before 1990, 1990-1995, and the period after 1995.

Until 1990, the attention to BEV was generally low. The relationship between the US car industry and government can be characterised as antagonistic.

In 1990, the California Air Resources Board (CARB) proclaimed a very strict mandate for large manufacturers. This mandate guaranteed a market for tens of thousands of electric vehicles a year. Car manufacturers were very sceptical about the feasibility of the mandate. Nevertheless, they increased their R&D efforts and made prototypes. These prototypes, however, were offered at very high prices and were used in struggles against the CARB. This illustrates a pattern of 'slowing down through participation'. Chrysler, for instance, built an electric van that was offered for sale for $100,000. Ford built a small electric van, in which the battery alone cost $50,000. Yet, despite their struggles against CARB, US car manufacturers *did* develop 'electric activities'. Besides their separate activities, the Big Three (General Motors, Ford and Chrysler) all participated in the Advanced Battery Consortium (see below). While the large manufacturers were opposing the CARB mandate, the mandate created a window of opportunity for the emergence of many small BEV manufacturers. These regime outsiders played an important role in stimulating the dynamics in the BEV-niche. The most important effect was that they were producing and selling *working* electric vehicles while the large manufacturers were opposing CARB and presenting 'bad' prototypes. Via this 'demonstration effect' they proved the feasibility of BEVs, thus undermining the position of the large manufacturers. This was technically possible because of substantial improvements in battery technology between 1990 and 1995, which, in turn, stimulated the expansion of social networks supporting battery technology.

An influential demonstration effect was created by the *Sunrise*, produced in 1996 by the small BEV car manufacturer Solectria. The *Sunrise* was a 4-person sedan vehicle, designed from scratch with low weight and good aerodynamics. In 1996 this BEV broke the world record for electric vehicle driving range, covering 600 km. The *Sunrise* strengthened expectations around NiMH batteries and BEV more generally. In 1997, most actors in the BEV-niche expected this battery type to be used in the near future. Many R&D activities were developed in this direction, and new strategic networks were emerging. An example is the 'Advanced Battery Consortium' (ABC) in which the big American car manufacturers participated, which targeted NiMH-batteries. New alliances and networks are also emerging between battery producers for consumer electronics and the car industry. The American Ovonic Battery Company (OBC) started a joint venture with General Motors in 1994, called GM-Ovonics. In 1996, Honda also started cooperating with OBC. Thus, in recent years, large car manufacturers have begun building networks with strategic partners in the battery industry. This move was stimulated by the working prototypes produced by small manufacturers, and indicates the importance of regime outsiders.

The changes in social networks and improvements in battery technology occurred in tandem with changes in the relationship between CARB and the car industry, as well as with changes in the BEV-strategies of car manufacturers. Confronted with continuing opposition from large manufacturers, CARB appointed an independent committee, which concluded in 1995 that the target of 2% electric vehicles in 1998 was too high. CARB and the large manufacturers therefore initiated discussions which resulted in a revised CARB mandate (1996) and a Memorandum of Agreement, stating that i) the end goal of 10% ZEV in 2003 was maintained, ii) manufacturers would not agitate against BEV in public, iii) manufacturers would produce more low-emission vehicles (LEV) in order to compensate for the deferred ZEV-reductions. The agreement was also important in the sense that it increased mutual understanding between industry and government: 'together we can solve problems'. The new mandate *and* the working BEV's produced by small manufacturers, also led to a change in the expectations of industry. General Motors announced in 1995 that it would start producing and marketing the *EV1* (based on the *Impact*). Toyota and Honda announced that they would bring hybrid electric vehicles to market in 1997. In 1999 a couple of thousand hybrids had been sold. The scepticism which characterised the industry in 1995, has turned into moderate expectations of the prospects of electric vehicles. General Motors has even started profiling itself as the 'leader in the newly developing EV-business'.

5. SOCIO-TECHNICAL REGULARITIES AND PATTERNS

The preceding section provided an empirical illustration of the multi-level framework for conceptualising technological change and regime shifts presented in section 3, and showed that this general framework can be filled in with more specific patterns and regularities. These patterns add extra dynamics with regard to processes at different levels and the interaction between the levels. It has been suggested that such patterns and regularities can also be used in the construction of sociotechnical scenarios (Deuten, Rip & Smit 1998). By varying the importance of the patterns, it is possible to create scenarios with different 'meso logics'.

This section aims to contribute to the creation of such a list of patterns and regularies. I will first extract some patterns from the empirical illustration in section 4, and then add some further patterns from technology dynamics literature. Some of the patterns are given in the form of theoretical notions and others in the form of examples.

Niches created by landscape developments

Section 4 described how BEV-niches were created in California, France and Switzerland in interaction with landscape developments. Another example is that the trend in some European countries towards compact cities requires a new transport system because of current congestion and air quality problems. As the present transport regime cannot meet this need easily, it opens up a niche for alternative vehicles, e.g. small cars, electric cars.

Patterns and regularities from interactions between the niche- and regime-level

Slow regime change. In general, regimes change slowly and gradually. Actors within the regime have a strong tendency to look for incremental solutions by improving the dominant technology, even when confronted by persistent problems in their domain. It makes economic sense for them to stick with the dominant technology, as they have so many 'sunk costs' and vested interests, especially in capital intensive industries. Section 4.2. described the strong preference of big car manufacturers for incremental innovations to gasoline cars. Because of their market power and political alliances, regime changes in such industries tend take a lot of time (Grübler *et al.* 1993).

Speeding up niche developments by domino effects in strategic games. Actors in the regime usually take an initial 'wait and see' attitude towards new technologies. They may invest in new technologies as a hedging strategy and to gain some experience. The 'jockeying for position' is part of strategic games; it usually results in slow development, but can also cause acceleration. When one actor decides to take a bold step forward, he may drag others along, for fear of being left behind. An example is that all big car manufacturers engaged in alternative technologies in the last couple of years (electric vehicles, hybrids, fuel cells). While car manufacturers were initially watching and waiting for each other, the actions of Toyota in 1997 sped things up. Toyota started marketing hybrid cars in 1997, and this stimulated other car manufacturers to speed up their own developments. Their R&D budgets suddenly soared, speeding up niche developments.

Slowing down niche developments through participation. Actors in the regime can participate strategically in developments in technological niches in order to slow these developments down. For years, car manufacturers have followed this strategy to slow down the niche of electric vehicles. There are many tactics for slowing down. One tactic which established manufacturers used was to present prototypes of electric vehicles which had low performance, leading to disappointment (see section 4). Such 'disappointment effects' are even larger when expectations have been whipped up beforehand. Low performance prototypes are also used in negotiations with policy makers, arguing that alternative technologies simply are 'no good'.

The importance of regime outsiders in creating niches. The network of a new technological niche is usually built by a 'product champion' who are often outsiders with less vested interests. Actors from the dominant regime can participate in niches, but seldom take a strong lead, as they have many vested interests in the incumbent technology. Section 4.2. showed how small car manufacturers were crucially important as outsiders in stimulating the BEV-niche.

Further examples of how landscape developments can influence regimes and niches

Impacts of pervasive technologies. The rise of a new pervasive technology influences competition between technologies at the regime and niche level. For example, through the application of ICT the energy efficiencies of gasoline cars can be improved and new possibilities become visible to create a better flow of traffic (intelligent vehicle/highway systems), which, paradoxically, made the existing regime appear stronger against emerging technologies such as electric vehicles.

Impact of landscape on expectations. For example, changing oil prices can alter expectations about the future of a regime. While the future of the gasoline car has been reinforced by low prices, rising gasoline prices (e.g. as a result of eco-tax) could reverse this.

Further patterns in interactions between the niche- and regime-level

Sailing ship effect. Actors in an established regime spend much effort at improving the performance of the dominant technology, particularly in the face of challenges from newly emerging niche technologies. This is called the 'sailing ship effect', after the large improvements in sail ships (in size, hull, sail surface and speed) when challenged by steam ships (Mokyr, 1990). As a result, old and new technologies may co-exist for along time. Similarly telephone companies made large investments in improving copper cables, when challenged by glass fibre options (which promised larger capacities). As a result, developments in glass fibre were (temporarily?) out competed.

Missing the wave of new technologies. Dominant players in the regime are so committed to the dominant technology that they tend to be short sighted with respect to emerging technologies. Furthermore, the existing social network may constantly reproduce existing perceptions concerning problems and solutions ('patting each other on the back'), at the expense of new and fresh ideas. There are plenty of examples where large firms overlooked important new technologies (e.g. IBM missed the personal computer, Microsoft initially failed to appreciate the importance of the Internet) (Bower & Christensen 1995).

Buying up the winners. Big firms within the regime can refuse to take part in niche developments but follow them from a distance. Once developments start taking off they can buy up the key firms. This tactic can make good economic sense. Big firms let others bare the risks and costs of experimentation, and buy the needed competencies when the experiments turn in positive directions. [4]

6. TOWARDS SOCIOTECHNICAL SCENARIOS: USING THE MULTI-LEVEL FRAMEWORK AND LIST OF REGULARITIES AS BUILDING BLOCKS

This chapter illustrates how insights from technology studies can help solve problems in other fields (e.g. futures studies). Many existing projections and scenarios are based on unduly simplistic ideas of technological development, and, as a result, do not provide a basis for discussing particular kinds of technology policies. Examples of such neglected policies are those which revolve around network building, learning processes, vision building. For such reflexive technology policies, which are based on real-life and ongoing technology dynamics, tools are needed which enable reflexive anticipation. Socio-technical scenarios can be such a tool. This chapter did not develop sociotechnical scenarios, but articulated two important ingredients or building blocks for making them. Firstly, a conceptual framework has been offered which fulfilled the following requirements: i) provide insight to processes of radical and structural change, ii) cover long-term processes, iii) cover sector-wide developments. Secondly, a list of patterns and regularities in technical development has been generated. Some elements of this list are based on an empirical example. Theoretical explanations have been added to give the patterns more general validity. Other elements of the list have been collected from technology studies literature.

The conceptual framework and list of patterns can be used to solve an important problem in existing scenarios, the lack of dynamics at meso- and micro-levels. With the conceptual framework and list of patterns and regularities interesting multi-level scenarios about sociotechnical development can be made. In that sense, this chapter has fulfilled a design role to develop the new tool of sociotechnical scenarios. Some preliminary experience has been gained with this tool (Deuten *et al.*, 1998; Elzen *et al.*, 1998). The next step is to build on this chapter and further develop sociotechnical scenarios.

NOTES

1 I would like to thank Knut Sørensen, Arie Rip, and Robin Williams for their useful comments on earlier versions of this chapter.

2 The examination of scenarios is based on: i) futures journals (mainly: Technological Forecasting and Social Change, Futures and, Technology Analysis and Strategic Management), ii) actual scenarios, mainly from the Dutch context (in particular, scenarios from the Dutch Central Planning Bureau (CPB), Ministry of Traffic and Transportation (V&W), Ministry of Housing, Public Planning and Environment (VROM)), iii) interviews with scenario builders.

3 The empirical illustrations are based on our report for the Dutch electric utilities, in which we made sociotechnical scenarios for the transportation system in the 21st century (Elzen, Geels, Hoogma, Schot & Te Velde, 1998). The empirical material in this report is based on

200 interviews from the period 1993-1995, and 50 additional interviews in 1996. On this basis, five country studies have been produced (Hoogma, 1995; Hoogma *et al.*, 1995; Hoogma & Schot, 1995; Elzen, 1995a and 1995b).

4 A failed example from the transportation sector is the Canadian firm Ballard. Since 1993 Ballard has been working on fuel cells (for buses and cars). When they were becoming successful a couple of years ago, Daimler-Benz tried to buy them. The attempt failed because many shares were held by Ballard employees who didn't want to sell. Daimler then entered into a partnership with Ballard, which can also be seen as part of 'jockeying for position'.

BIBLIOGRAPHY

Amara, R. (1988), 'What have we learned about forecasting and planning', *Futures*, August, pp. 385-401.

Amara, R. (1989), 'A note on what we have learned about the methods of futures planning', *Technological Forecasting and Social Change*, **36**, pp. 43-47.

Ayres, R. U. (1989), 'The future of technological forecasting', *Technological Forecasting and Social Change*, Vol. 36, no-1-2, pp. 49-60.

Bower, J.L., and C.M. Christensen (1995), 'Disruptive Technologies: Catching the Wave,' *Harvard Business Review*, Jan-Feb pp. 43-53.

Callon, M, (1980), `The State and Technical Innovation: a case-study of the electric vehicle in France', *Research Policy*, **9**, pp. 358-76.

Callon, M. (ed.), (1998), *The laws of the market*, Oxford: Blackwell.

Coates, J.F., Mahaffie, J.B., and Hines, A., (1994), 'Technological Forecasting 1970-1993', *Technological Forecasting and Social Change*, **47**, pp. 23-33.

Deuten, J.J., A. Rip en W.A. Smit, (1998), *Nieuwe werelden... Met micro-optica: Scenario's en scripts, en wat deze impliceren voor micro-optica onderzoek aan de Universiteit Twente* (Enschede: Universiteit Twente).

Dewulf, G., (1998), 'De kunst van het ontwerpen van strategische scenario's', in Hermkens, P., Van Rijsselt, R., Sanders, K., (eds), *Differentiatie en samenleving: Opstellen voor Henk Becker*, pp. 201-215 Thela: Thesis.

Dutch Central Planning Bureau (1997), *Economie en fysieke omgeving: Beleidsopgaven en oplossingsrichtingen, 1995-2020*, The Hague: Central Planning Bureau.

Elzen, B., (1995a), *Towards cleaner cars and transport: Country study Japan*, Working document, Enschede: University of Twente.

Elzen, B., (1995b), *Towards cleaner cars and transport: Country study USA*, Working document, Enschede: University of Twente.

Elzen, Boelie, Remco Hoogma en Johan Schot (1996), *Mobiliteit met Toekomst: Naar een vraaggericht technologiebeleid* (Mobility with a Future. Towards a demand-oriented technology policy), report to the Ministry of Traffic and Transport (in Dutch).

Elzen, B., Geels, F.W., Hoogma, R., Schot, J.W. and Te Velde, R., (1998), *Strategieën voor innovatie: Experimenten met elektrische voertuigen als opstap naar marktontwikkeling*, Report for the Dutch Electricity Utilities (Sep), 208 pp.

Garud, R., (1994), 'Cooperative and competitive behaviors during the process of creative destruction', *Research Policy*, Vol. 23, pp. 385-394.

Geels, F.W. and Smit, W.A., (2000), 'Failed technology futures: Pitfalls and lessons from a historical survey', *Futures*, Vol. 32, pp. 867-885.

Grossman, G.M. and Helpman, E., (1991), *Innovation and growth in the global economy*, Cambridge MA: MIT Press.

Grübler, A., Nakicenovic, and Schäfer, A., (1993), Dynamics of transport and energy systems: History of development and a scenario for the future, *Research Report*-93-19, Vienna: International Institute for Applied Systems Analysis, 23 pp.

Hoogma, R., (1995), *Towards cleaner cars and transport: Country study France*, Working document, Enschede: University of Twente.

Hoogma, R. and Schot, J., (1995), *Towards cleaner cars and transport: Country study Sweden*, Working document, Enschede: University of Twente.

Hoogma, R., De la Bruhèze, A. and Schot, J., (1995), *Towards cleaner cars and transport: Country study* Germany, Working document, Enschede: University of Twente.

Huss, W.R., (1988), 'A move toward scenario analysis', *International Journal of Forecasting*, **4**, pp. 377-388.

Kemp, R, J. Schot, and R. Hoogma (1998), 'Regime shifts through processes of niche formation: the approach of strategic niche management,' *Technology Analysis and Strategic Management*, **10**, pp. 175-196.

Martin, B., (1995), 'Foresight in science and technology,' *Technology Analysis & Strategic Management*, **7**, pp. 139-168.

Minne, B., (1993), *Science and technology in scenarios*, Research Memorandum no. 110, Dutch Central Planning Bureau, The Hague.

Mintzberg, H. (1994), *The rise and fall of strategic planning: reconceiving roles for planning, plans, planners*, New York: Free Press; Toronto: Maxwell Macmillan Canada.

Mokyr, J. (1990), *The Lever of Riches*, New York: Oxford University Press.

Nelson, Richard R., and Sidney G. Winter (1977), `In Search of a Useful Theory of Innovation', *Research Policy* **6**, 36-76.

Nelson, Richard R., and Sidney G. Winter (1982), *An Evolutionary Theory of Economic Change*, Cambridge (Mass.): Bellknap Press.

Ogilvy, J. and Schwarz, P., (1997), 'Rehearsing the future through scenario planning', in: IPTS, Scenario building: Convergences and differences, *Proceedings of Profutures workshop*, pp. 62-68.

Porter, Alan, L., A. Thomas Roper, Thomas W. Mason, Frederick A. Rossini, and Jerry Banks, (1991), *Forecasting and management of technology*, New York.: John Wiley & Sons.

Rip, A., and R. Kemp (1996): *Towards a Theory of Socio-Technical Change*, mimeo UT, report prepared for Batelle Pacific Northwest Laboratories, Washington, D.C. An edited version has been published as book chapter, 'Technological Change', in: Rayner, S., and E.L. Malone (1998): Human Choice and Climate Change. An International Assessment, **2**, Washington D.C: Batelle Press, pp. 327-400.

Rip, A., and R. Kemp (1998): "Technological Change", in: Rayner, S., and E.L. Malone (eds): *Human Choice and Climate Change*, Columbus, Ohio: Battelle Press, Volume 2, Chapter 6, pp. 327-399.

Rip, A., and J. Schot (1999): 'Anticipation on Contextualization: Loci for Influencing the Dynamics of Technological Development,' Sauer, D., and C. Lang (Hrsg.): Paradoxien der Innovation. Perspektiven sozialwissenschaftlichter Innovationsforschung, Frankfurt/New York: Campus Verlag, 1999. Revised edition as Chapter 4 in this volume.

Rip, A., (2000), 'Indigenous knowledge and western science, in practice', Paper presented at Cardiff conference, *Demarcation socialised: or, can we recognise science when we see it?* August 25-27.

Russell, S and Williams, R (2001) 'Concepts, spaces and tools for action? exploring the policy potential of the social shaping perspective', Chapter 4 in this volume.

Russell, S and Williams, R (2001) 'Social Shaping of Technology: Frameworks, Findings and Implications for Policy', Chapter 3 in this volume.

Sapio, B., (1995), 'SEARCH (Scenario Evaluation and Analysis through Repeated Cross impact Handling): A new method for scenario analysis with an application to the Videotel service in Italy', *International Journal of Forecasting*, **11**, pp. 113-131.

Sahal, D, (1985), 'Technological guideposts and innovation avenues', *Research Policy*, **14**, pp. 61-82.

Schoonenboom, I.J. and Van Latensteijn, H.C., (1997), 'Toekomstonderzoek en beleid: Goede intenties en kwade kansen', in: *Toekomstonderzoek en strategische beleidsvorming: Probleemverkenningen en praktijktoepassingen*, Report from NRLO, no. 3, Nationale Raad voor Landbouwkundig Onderzoek, The Hague, pp. 9-24.

Schot, J., R. Hoogma, and B. Elzen, (1994), 'Strategies for Shifting Technological Systems. The case of the automobile system', *Futures*, **26**, pp. 1060-1076.

Schot, J., (1996), 'De inzet van constructief technology assessment', in: Kennis en Methode, nr. 3, pp. 265-293.

Schot, J., and A. Rip, (1997), 'The Past and Future of Constructive Technology Assessment', *Technological Forecasting and Social Change*, Vol. 54, pp. 251-268.

Schot J.W., 'The Usefulness of Evolutionary Models for Explaining Innovation. The Case of the Netherlands in the Nineteenth Century', *History of Technology*, (1998) **14**, 173-200.

Schwarz, P., (1993), *De kunst van het lange termijn denken*, (translation of 'The art of the long view'), Amsterdam.

Shell, (1996), *Global Scenarios, 1995-2020*, Shell, The Hague.

Sørensen, K H (1994): "Technology in use. Two essays on the domestication of artifacts", *STS Working paper* 2/94, Trondheim: Centre for technology and society.

Sørensen, K. H. (2001), 'Social Shaping on the Move? On the Policy Relevance of The Social Shaping of Technology Perspective', chapter 2 in this volume.

Street, P., (1997), 'Scenario workshops: A participatory approach to sustainable urban living?', *Futures*, **29**, pp. 139-158.

Van de Poel, I. (1998): *Changing Technologies. A comparative study of eight processes of transformation of technological regimes*, Twente University Press, Enschede, PhD thesis.

Van der Meulen, B., (1999), 'The impact of foresight on environmental science and technology policy in the Netherlands', *Futures*, **31**, pp. 7-23.

Van Gurchom, Manfred, Henri ter Hofte, and Olaf Tettero, (1994), TRC (Telematica Research Centrum), Methodische aanpak scenario ontwikkeling TGRP, TRC Series IRS/94007.

Van Lente, H. and Rip, A., (1998), 'Expectations in technological developments: An example of prospective structures to be filled in by agency', in: Disco, C. and Van der Meulen, B. (eds), *Getting new technologies together*, Berlin: Walter de Gruyter, pp. 203-229.

Von Reibniz, U., (1988), *Scenario techniques*, Hamburg: Mc-Graw-Hill Book Company GmbH.

Wack, P., (1985a), 'Scenarios: Uncharted waters ahead', *Harvard Business Review*, September-October, pp. 73-89.

Wack, P., (1985b), 'Scenarios: Shooting the rapids', *Harvard Business Review*, November-December, pp. 139-150.

Wallace, D., (1995), *Environmental Policy and Industrial Innovation: Strategies in Europe, the USA and Japan*, London, Earthscan Publications Ltd.

Weber, M., R. Hoogma, B. Lane, and J. Schot, (1999), *Experimenting with Sustainable Transport Innovations. A workbook for Strategic Niche Management,* Seville/Enschede: IPTS.
Weterings, R., Kuijper, J., Smeets, E., Annokkée, G.J., and Minne, B., (1997), *81 mogelijkheden voor duurzame ontwikkeling*, Eindrapport van de milieugerichte technologieverkenning, in opdracht van Ministerie van Volkshuisvesting, Ruimtelijke Ordening and Milieubeheer.
Wilson, I., (1997), 'Linking intuition and structure: An integrated approach to scenario development', in: IPTS, *Scenario building: Convergences and differences, Proceedings of Profutures* workshop, pp. 31-41.

Index